BEFORE COPERNICUS

McGILL-QUEEN'S STUDIES IN THE HISTORY OF IDEAS
Series Editor: Philip J. Cercone

1 Problems of Cartesianism
Edited by Thomas M. Lennon, John M. Nicholas, and John W. Davis

2 The Development of the Idea of History in Antiquity
Gerald A. Press

3 Claude Buffier and Thomas Reid: Two Common-Sense Philosophers
Louise Marcil-Lacoste

4 Schiller, Hegel, and Marx: State, Society, and the Aesthetic Ideal of Ancient Greece
Philip J. Kain

5 John Case and Aristotelianism in Renaissance England
Charles B. Schmitt

6 Beyond Liberty and Property: The Process of Self-Recognition in Eighteenth-Century Political Thought
J.A.W. Gunn

7 John Toland: His Methods, Manners, and Mind
Stephen H. Daniel

8 Coleridge and the Inspired Word
Anthony John Harding

9 The Jena System, 1804–5: Logic and Metaphysics
G.W.F. Hegel
Translation edited by John W. Burbidge and George di Giovanni
Introduction and notes by H.S. Harris

10 Consent, Coercion, and Limit: The Medieval Origins of Parliamentary Democracy
Arthur P. Monahan

11 Scottish Common Sense in Germany, 1768–1800: A Contribution to the History of Critical Philosophy
Manfred Kuehn

12 Paine and Cobbett: The Transatlantic Connection
David A. Wilson

13 Descartes and the Enlightenment
Peter A. Schouls

14 Greek Scepticism: Anti-Realist Trends in Ancient Thought
Leo Groarke

15 The Irony of Theology and the Nature of Religious Thought
Donald Wiebe

16 Form and Transformation: A Study in the Philosophy of Plotinus
Frederic M. Schroeder

17 From Personal Duties towards Personal Rights: Late Medieval and Early Modern Political Thought, c. 1300–c. 1650
Arthur P. Monahan

18 The Main Philosophical Writings and the Novel *Allwill*
Friedrich Heinrich Jacobi
Translated and edited by George di Giovanni

19 Kierkegaard as Humanist: Discovering My Self
Arnold B. Come

20 Durkheim, Morals, and Modernity
W. Watts Miller

21 The Career of Toleration: John Locke, Jonas Proast, and After
Richard Vernon

22 Dialectic of Love: Platonism in Schiller's Aesthetics
David Pugh

23 History and Memory in Ancient Greece
Gordon Shrimpton

24 Kierkegaard as Theologian: Recovering My Self
Arnold B. Come

25 Enlightenment and Conservatism in Victorian Scotland: The Career of Sir Archibald Alison
Michael Michie

26 The Road to Egdon Heath: The Aesthetics of the Great in Nature
Richard Bevis

27 Jena Romanticism and Its Appropriation of Jakob Böhme: Theosophy – Hagiography – Literature
Paolo Mayer

28 Enlightenment and Community: Lessing, Abbt, Herder, and the Quest for a German Public
Benjamin W. Redekop

29 Jacob Burckhardt and the Crisis of Modernity
John R. Hinde

30 The Distant Relation: Time and Identity in Spanish-American Fiction
Eoin S. Thomson

31 Mr Simson's Knotty Case: Divinity, Politics, and Due Process in Early Eighteenth-Century Scotland
Anne Skoczylas

32 Orthodoxy and Enlightenment: George Campbell in the Eighteenth Century
Jeffrey M. Suderman

33 Contemplation and Incarnation: The Theology of Marie-Dominique Chenu
Christophe F. Potworowski

34 Democratic Legitimacy: Plural Values and Political Power
F.M. Barnard

35 Herder on Nationality, Humanity, and History
F.M. Barnard

36 Labeling People: French Scholars on Society, Race, and Empire, 1815–1849
Martin S. Staum

37 The Subaltern Appeal to Experience: Self-Identity, Late Modernity, and the Politics of Immediacy
Craig Ireland

38 The Invention of Journalism Ethics: The Path to Objectivity and Beyond, Second Edition
Stephen J.A. Ward

39 The Recovery of Wonder: The New Freedom and the Asceticism of Power
Kenneth L. Schmitz

40 Reason and Self-Enactment in History and Politics: Themes and Voices of Modernity
F.M. Barnard

41 The More Moderate Side of Joseph de Maistre: Views on Political Liberty and Political Economy
Cara Camcastle

42 Democratic Society and Human Needs
Jeff Noonan

43 The Circle of Rights Expands: Modern Political Thought after the Reformation, 1521 (Luther) to 1762 (Rousseau)
Arthur P. Monahan

44 The Canadian Founding: John Locke and Parliament
Janet Ajzenstat

45 Finding Freedom: Hegel's Philosophy and the Emancipation of Women
Sara MacDonald

46 When the French Tried to Be British: Party, Opposition, and the Quest for Civil Disagreement, 1814–1848
J.A.W. Gunn

47 Under Conrad's Eyes: The Novel as Criticism
Michael John DiSanto

48 Media, Memory, and the First World War
David Williams

49 An Aristotelian Account of Induction: Creating Something from Nothing
Louis Groarke

50 Social and Political Bonds: A Mosaic of Contrast and Convergence
F.M. Barnard

51 Archives and the Event of God: The Impact of Michel Foucault on Philosophical Theology
David Galston

52 Between the Queen and the Cabby: Olympe de Gouges's Rights of Women
John R. Cole

53 Nature and Nurture in French Social Sciences, 1859–1914 and Beyond
Martin S. Staum

54 Public Passion: Rethinking the Grounds for Political Justice
Rebecca Kingston

55 Rethinking the Political:
The Sacred, Aesthetic Politics,
and the Collège de Sociologie
Simonetta Falasca-Zamponi

56 Materialist Ethics and
Life-Value
Jeff Noonan

57 Hegel's *Phenomenology*:
The Dialectical Justification of
Philosophy's First Principles
Ardis B. Collins

58 The Social History of Ideas
in Quebec, 1760–1896
Yvan Lamonde
Translated by Phyllis Aronoff and Howard Scott

59 Ideas, Concepts, and Reality
John W. Burbidge

60 The Enigma of Perception
D.L.C. Maclachlan

61 Nietzsche's Justice:
Naturalism in Search of
an Ethics
Peter R. Sedgwick

62 The Idea of Liberty in Canada
during the Age of Atlantic
Revolutions, 1776–1838
Michel Ducharme
Translated by Peter Feldstein

63 From White to Yellow:
The Japanese in European
Racial Thought, 1300–1735
Rotem Kowner

64 The Crisis of Modernity
Augusto Del Noce
Edited and translated by
Carlo Lancellotti

65 Imprinting Britain:
Newspapers, Sociability,
and the Shaping of British
North America
Michael Eamon

66 The Form of Politics:
Aristotle and Plato
on Friendship
John von Heyking

67 War as Paradox:
Clausewitz and Hegel on
Fighting Doctrines and Ethics
Youri Cormier

68 Network Democracy:
Conservative Politics and the
Violence of the Liberal Age
Jared Giesbrecht

69 A Singular Case:
Debating China's Political
Economy in the European
Enlightenment
Ashley Eva Millar

70 Not Even a God Can Save
Us Now:
Reading Machiavelli after
Heidegger
Brian Harding

71 Before Copernicus:
The Cultures and Contexts
of Scientific Learning
in the Fifteenth Century
Edited by Rivka Feldhay
and F. Jamil Ragep

BEFORE COPERNICUS

The Cultures and Contexts of Scientific Learning in the Fifteenth Century

Edited by

Rivka Feldhay and F. Jamil Ragep

McGill-Queen's University Press
Montreal & Kingston • London • Chicago

© McGill-Queen's University Press 2017

ISBN 978-0-7735-5009-4 (cloth)
ISBN 978-0-7735-5010-0 (paper)
ISBN 978-0-7735-5011-7 (ePDF)
ISBN 978-0-7735-5012-4 (ePUB)

Legal deposit second quarter 2017
Bibliothèque nationale du Québec

Printed in Canada on acid-free paper that is 100% ancient forest free (100% post-consumer recycled), processed chlorine free

McGill-Queen's University Press acknowledges the support of the Canada Council for the Arts for our publishing program. We also acknowledge the financial support of the Government of Canada through the Canada Book Fund for our publishing activities.

Library and Archives Canada Cataloguing in Publication

Before Copernicus: the cultures and contexts of scientific learning in the fifteenth century / edited by Rivka Feldhay and F. Jamil Ragep.

(McGill-Queen's studies in the history of ideas; 71)
Includes bibliographical references and index.
Issued in print and electronic formats.
ISBN 978-0-7735-5009-4 (cloth). – ISBN 978-0-7735-5010-0 (paper). – ISBN 978-0-7735-5011-7 (ePDF). – ISBN 978-0-7735-5012-4 (ePUB)

1. Astronomy – Europe – History – 15th century. 2. Science – Europe – History – 15th century. 3. Learning – Europe – History – 15th century. 4. Europe – Intellectual life – 15th century. I. Ragep, F. J., author, editor II. Feldhay, Rivka, author, editor III. Series: McGill-Queen's studies in the history of ideas; 71

QB29.B44 2017   520.9409'024   C2017-900389-5
                                         C2017-900390-9

This book was typeset by Marquis Interscript in 10/12 New Baskerville.

In memory of Yehuda Elkana (1934–2012)

Contents

Note on Conventions

Throughout this volume, all transcriptions of Arabic names follow the convention of the *Encyclopaedia of Islam*, 3rd edition, edited by Kate Fleet, Gudrun Krämer, Denis Matringe, John Nawas, and Everett Rowson (Leiden: Brill, 2007–).

The first of two dates separated by a backslash is the *hijra* date; the second is the corresponding Common Era (CE) date. When only one date is given, it is the Common Era date unless otherwise indicated.

For purposes of alphabetizing in the bibliography, the Arabic article "al-" is ignored.

Tables and Figures

TABLES

FIGURES

Acknowledgments

The editors wish to extend their sincere thanks to the many people and organizations that have contributed to realizing this collection of essays.

First and foremost, we thank the Max Planck Institute for the History of Science (MPIWG, Berlin), which provided the venue for our project, which began as an exploration of "Knowledge and Belief" more than fifteen years ago. The director of the MPIWG's Division 2, Dr Lorraine Daston, instigated and supported that project, as well as its metamorphosis into *Before Copernicus*, from its very beginning. Our boundless admiration for her is matched only by our gratitude for the financial, intellectual, and emotional support she has provided over those many years. Three of the four workshops underpinning the project were held in Berlin in 2005, 2006, and 2007; we are grateful to the staff of the MPIWG for the logistical support that helped to make those meetings such a success.

The fourth workshop was held in Montreal in 2009 under the auspices of the MPIWG and the Institute of Islamic Studies, McGill University. Other financial support for the publication of the volume has come from the Cohn Institute for the History and Philosophy of Science and Ideas, Tel Aviv University, and from the Canada Research Chair in the History of Science in Islamic Societies, McGill University.

This has been a book long in the making, so we must thank our remarkable group of contributors for sticking with us for a decade. It was an inspiration to be with them as we navigated the Scylla of disciplinary boundaries and the Charybdis of cross-cultural barriers, listened to tales of unfamiliar names, confronted unaccustomed methodologies, and delighted in one another's enthusiasms. We certainly learned a lot, and we hope our colleagues feel likewise. We also thank Maria Mavroudi and İhsan Fazlıoğlu for participating in the project at an earlier stage; both of them contributed much to our workshops and deliberations.

The staff at McGill-Queen's University Press have been outstanding with their customary professionalism and creativity. We are deeply indebted to our editor, Kyla Madden, who somehow manages in these times to combine extraordinary kindness with remarkable efficiency. Ryan Van Huijstee, our production editor, made the often laborious steps to final product as painless as possible, and our copy editor, Robert Lewis, smoothed over the embarrassingly large number of rough edges that somehow remained in our "final" versions.

We thank Sally Ragep for all her help throughout the process in proofreading, editing, and improving countless versions of the essays that were sent her way. We also thank Miki Elazar for his assistance with the index. And our heartfelt gratitude to the two anonymous referees whose selfless devotion to helping us improve our book left us in awestruck admiration.

Finally, we are most grateful to the libraries and individuals who have granted us permission to reproduce or adapt images: the History of Science Collections, University of Oklahoma Libraries; Houghton Library, Harvard University; the Regiomontanus Defensio Theonis Project, Dartmouth College; the Turkish Institution for Manuscripts and the Süleymaniye Manuscripts Library; Robert Hatch; and Noel Swerdlow.

Rivka Feldhay and F. Jamil Ragep
November 2016

BEFORE COPERNICUS

Introduction

F. Jamil Ragep and Rivka Feldhay

NICHOLAS COPERNICUS HAS BECOME such an iconic figure, and the subject of a not inconsiderable amount of historical attention, that it is surprising to discover how little is understood of his intellectual and social background.[1] This deficiency perhaps becomes less surprising when we note that Copernicus has often been portrayed as the lone genius who, as a "sleepwalker," gave rise to the scientific revolution – and, in some accounts, modernity itself – without himself actually being a revolutionary or modern.[2] He thus becomes a figure without history, and hardly in need of context, whose flash of inexplicable genius is all that is necessary to spark the true genius of the Keplers, Galileos, and Newtons. Hence many accounts of the Copernican revolution have begun on his deathbed, where his masterpiece *De revolutionibus orbium coelestium* was supposedly delivered to him in 1543.[3]

This view of Copernicus began to change over a half-century ago with the publication of Thomas Kuhn's *The Copernican Revolution: Planetary Astronomy in the Development of Western Thought.* Although subsequent scholarship has questioned a number of its claims, the book did have the virtue of providing a background and context against which to view Copernicus's innovation.[4] The now familiar (if overused) terminology of "paradigm," "anomaly," "crisis," and "revolution" seemed an appropriate way to view one of the most earth-shaking events in the history of science.[5] The exact nature of the crisis, however, remained elusive, in large part because the fifteenth-century background to Copernicus was and remains to a large extent *terra incognita.* The Copernican scholar Edward Rosen provided innumerable details regarding Copernicus and his work but, like most post-Enlightenment devotees, was mostly uninterested in this immediate foreground or in the more arcane technical aspects of Copernicus's work. Otto Neugebauer, along with his associates and students, was to provide much more of this context, especially the mathematical details that connected Copernicus's work with both the "Western" tradition and crucially the "Eastern," Islamic one, which he insisted was a critical part of the background: "There is no better way to convince oneself of the inner coherence of ancient and mediaeval astronomy than to place side by side the *Almagest*, al-Battānī's *Opus astronomicum* and Copernicus's *De revolutionibus*. Chapter by chapter, theorem by theorem, table by table, these works run parallel."[6]

The most incisive and exciting breakthrough in Copernican studies was to come from Noel Swerdlow, then a younger member of the "Neugebauer school," who gave a remarkable analysis of the so-called *Commentariolus*, which we can date to around 1510. Swerdlow contended that Copernicus had come to his Sun-centred cosmology by a technical route. Copernicus, in providing alternatives to Claudius Ptolemy's models that violated the accepted celestial physics requiring uniformity and circularity of motion, and influenced by Johannes Regiomontanus's propositions allowing Mercury's and Venus's epicycles to be transformed into eccentrics, was led to a "Tychonic" cosmography, with the Sun moving about the Earth while being at the centre (more or less) of the orbs of the retrograding planets. Swerdlow speculated that at this point Copernicus, unwilling to accept an intersection of the solid orbs of the Sun and Mars that this cosmography entailed, opted for one with a (again, more or less) static Sun and a moving Earth in which the orbs were discretely nested.[7]

There were a number of implications of Swerdlow's reconstruction that would not meet with universal approval. One was the issue of Copernicus's adherence to solid orbs, upon which the reconstruction depended. This led to rather vituperative exchanges between Swerdlow and Rosen.[8] In the ensuing years, there were quite a few articles dealing with solid-sphere astronomy and questioning whether Copernicus believed in solid orbs.[9] Another issue was the influence of late-medieval Islamic astronomy on Copernicus. This influence was a lynchpin in Swerdlow's reconstruction since so many of Copernicus's models were similar or identical to those of these Islamic predecessors. But more importantly, those models represented a response to the very issue Copernicus cited as a motivating factor in the *Commentariolus*, namely the irregular motion brought on by Ptolemy's equants.[10]

Critics of Swerdlow's reconstruction have been uneasy with its "internalist" character. There must, they contend, be more to this monumental cosmological shift than a strictly mathematical/astronomical explanation.[11] And one might argue that there were certainly other ways to deal with the problem of the equant and other Ptolemaic violations; after all, ʿAlāʾ al-Dīn ibn al-Shāṭir, from whom Copernicus seems to have borrowed so much in the *Commentariolus*, dealt with the Ptolemaic "difficulties" while retaining a geocentric cosmography.

In recent years, there have been several attempts to offer narratives that would provide the missing cause or motivation for Copernicus. Mario Di Bono has renewed attention to the Paduan Aristotelians, a group brought to prominence by John H. Randall Jr.[12] André Goddu has highlighted the Krakovian Aristotelians.[13] And Robert Westman, in his lifework, sought to account for Copernicus's innovation by associating it with an astrological "crisis" brought on by questions of the ordering of the planets.[14]

These contributions, and others, have certainly enriched the conversation and resulted in any number of lively debates, book reviews, and exchanges.[15] But we have been uncomfortable with their predominant attempt to reduce the "Copernican question" to one of finding *the* univocal explanation that somehow supersedes all others. This approach has had the unfortunate corollary of many recent works ignoring vast areas of relevant research that somehow do not fit into one's narrative. It is perhaps not coincidental that the most recent discussions of Copernicus have almost all taken a Eurocentric turn, with the question of cross-cultural influence mostly set aside.[16]

It was these considerations that led us almost a decade ago to seek another way to deal with the Copernican transformation, namely by assembling a group of scholars who could discuss the background to Copernicus in a multicultural and multidisciplinary way. In contrast to most discussions of the Copernican revolution, our endpoint was the earliest manifestation of the Copernican cosmography, the *Commentariolus*. Thus we were mainly concerned with the fifteenth-century background and context. Here are a number of observations that guided our discussions and research:

1 Copernicus's stated purpose in the *Commentariolus* is to find "a more reasonable model composed of circles ... from which every apparent irregularity would follow while everything in itself moved uniformly, just as the principle of perfect motion requires."[17]
2 Copernicus does not refer in the *Commentariolus* to the "marvelous symmetry" brought on by his new ordering of the planets, as he does in *De revolutionibus*. Although one must be cautious when speaking of motivation, it is curious that Copernicus does not explicitly put forth in the *Commentariolus* what is arguably his most compelling argument.
3 Copernicus's models (taking into account both the *Commentariolus* and *De revolutionibus*) contain both eccentrics and epicycles.

4 There is strong evidence that Copernicus adheres to solid-sphere astronomy.
5 There is no indication that Copernicus ever resorted to a strictly Aristotelian, Averroist, Biṭrūjian, or Paduan "homocentric" astronomy.[18] Copernicus does insist on a single centre for his main orbs and otherwise uses only epicycles in the *Commentariolus*, whereas he uses eccentrics with their multiple centres in *De revolutionibus*.
6 The number of similarities between the planetary models in the *Commentariolus* and those of Ibn al-Shāṭir (fourteenth-century Damascus) is significant.[19]
7 Discussions of the possibility of the Earth's motion can be found in both Islam and Christendom prior to Copernicus.

There are, of course, any number of conclusions that one might reach from these observations. Here are ours:

- From 1 and 2, we take seriously Copernicus's stated motivation in his earliest work and tend to see his justifications in *De revolutionibus* as post hoc. Of course, these justifications may have been what convinced him to *accept* the heliocentric theory, but they were not the initial motivations.[20] For this reason, we are not convinced that the ordering of the planets plays the motivating role advocated by Bernard Goldstein and Robert Westman.[21]
- From 3, 4, and 5, we are led to conclude that Copernicus is firmly within the tradition of Ptolemy's *Almagest* and *Planetary Hypotheses*, the *hay'a* tradition of Islamic astronomy,[22] and the fifteenth-century revival of both Ptolemaic astronomy and cosmology represented by Peurbach's *Theoricae novae planetarum* and Regiomontanus's *Epitome of the Almagest.*
- Further to 5, Copernicus's determination in the *Commentariolus* to have a single centre, namely the mean Sun, for each planetary system, despite the problems this approach entailed, indicates a certain type of quasi-homocentrism at the earliest stages of his career. This idea gave way to a more traditional Ptolemaic approach that accepted eccentrics and multiple centres when he was writing *De revolutionibus.*[23]
- From 6 (and other evidence), we are led to accept a significant influence of post-1200 Islamic astronomy on Copernicus. However, the existence of a longstanding tradition of Islamic criticism of Ptolemy and the proposal of alternative models that mostly satisfied those criticisms within a geocentric cosmography highlight that it was not necessary for Copernicus to make his momentous transformation in

order to satisfy his stated goal of a cosmography with uniform circular orbs. It thus seems that there were aspects of Copernicus's intellectual and cultural context that led him to his decision to put the Earth in motion.

- From 7, we have been led to explore discussions of the Earth's motion in multicultural contexts and the possibility that Copernicus may have been aware of, and influenced by, one or more of them.[24]

Our point of departure was the realization that Swerdlow's reconstruction, however satisfying it might be on a technical level, does not explain why Copernicus opted for a heliocentric alternative when he might have fulfilled his stated goal of a reformed astronomy with uniform, circular motions within a geocentric framework. This latter approach was, after all, the one that a number of Islamic astronomers had already employed to a large extent. Not incidentally, a reformed geocentric system would have secured his fame, earned him the gratitude and admiration of his contemporaries and successors, and spared him and those successors a considerable amount of grief.

But that is the point; he chose another alternative. One might say this was his individual choice, that his exceptionalism and expertise (a "community of one," to paraphrase Swerdlow)[25] provide the only explanation necessary.

But we think otherwise, mainly because our concept of what constitutes knowledge and how knowledge is related to the particular identity of a scholar differs from that implied by Swerdlow's writings from the 1970s. Knowledge, for us, is not an inert structure, stamped with an "inner logic" that is imprinted in the intellect of an idiosyncratic genius. As stated in point 1 above, Copernicus's problem was to mediate between a series of visual appearances (i.e., irregular motions of the stars) and a fundamental principle of astronomy that contradicted the appearances (i.e., uniform, circular motion). The knowledge accumulated around this fundamental problem consisted of a complex series of practices whose interaction constantly transformed the product even within the intellect of one person (see Shank's chapter) and certainly among a group of practitioners making use of one another's results. In this framework of thought, knowledge is "on the move" to begin with. Furthermore, both history in general and the intellectual history of the fifteenth and sixteenth centuries in particular show that the production of knowledge tends to be enhanced wherever and whenever circulation of knowledge across boundaries of languages, disciplines, and cultures occurs. This view of knowledge is at the heart of the intense preoccupation of many historians today – including us – with practices of circulation, transfer, and transmission of knowledge. Working within this framework, we are not inclined to presuppose that the ultimate response to Copernicus's problem is to be found through one correct derivation of a model that necessarily led to a coherent and true astronomical-cosmological picture. Rather, from our perspective, the Copernican system resulted from many practices that included attempts to deal, mathematically, with violations of physics found in Ptolemy's models, discussions of the relation of natural philosophy and mathematics, and epistemological forays into the "true" cosmology and the human capacity to discover it. Likewise, we tend to hypothesize that fifteenth-century astronomy was the outcome of multiple transformations along different paths that crystallized in the work of Copernicus into some kind of coherent whole that differed enough from the preceding astronomical discourse to open the door to additional, enhanced transformations.

Thus this volume focuses on the intersection of three kinds of transformations in the background to the Copernican system:

1 First, transformations in the body of knowledge are presented and discussed in most of the chapters. One example concerns the transformation of Ptolemaic two-dimensional circles into physical, three-dimensional orbs performed by Abū ʿAlī al-Ḥasan ibn al-Haytham, Naṣīr al-Dīn al-Ṭūsī, Georg Peurbach, Albert of Brudzewo, John of Głogów, Joseph ibn Naḥmias, Johannes Regiomontanus, and many others (see F.J. Ragep's, Sylla's, and Morrison's chapters). Another example touches on the importance of new types of models. One case is the transition from the epicyclic models for the second anomaly of the inferior planets to their eccentric models in the work of ʿAlī Qushjī and Regiomontanus (see Shank's chapter). A second case is the role played by the Ṭūsī-couple in constructing non-Ptolemaic models (see F.J. Ragep's and Morrison's chapters). Yet another example concerns conceptual transformations related to a moving Earth, "new ways of seeing," that would facilitate Copernicus's heliocentric choice by helping to explain away some of the irregularities observed in the heavens (see Sylla's, Shank's, and Chen-Morris and Feldhay's chapters).
2 But there were transformations of another kind that intersected with those in the body of knowledge. These transformations were related more to the image and status of astronomy and tended to intervene in the order of the disciplines that was more or less accepted in both Islamic and Christian environments for centuries. As early as Ibn al-Haytham's *On the Configuration of the World*, a

serious effort was made by Islamic astronomers to charge Ptolemaic mathematical devices with physical meanings – thus to transform Ptolemy's two-dimensional mathematical circles into a discourse on three-dimensional physical orbs (see S.P. Ragep's and F.J. Ragep's chapters). As a result, a "second-order" reflection on whether astronomy was to be seen as a mathematical or physical science, or both, was developed and fed back into the body of knowledge by attempts to return to pre-Ptolemaic concentric or homocentric models using better mathematical tools (see Shank's, Sylla's, and Morrison's chapters). Furthermore, the epistemic status of astronomy was questioned once the empirical-observational origins of astronomy's "first principles" had to be addressed following the "physicalization" of astronomy by Islamic astronomers. Thus new categories for catching the nature of astronomy, such as theoretical but nondemonstrative-narrative astronomy versus demonstrative-theoretical astronomy, emerged and enhanced reflection about the epistemic status of its procedures and conclusions (see Sylla's, Shank's, Chen-Morris and Feldhay's, and S.P. Ragep's chapters).

3 However, knowledge transformations did not occur only within the body of knowledge. Nor were they simply determined by the status and image of knowledge among its practitioners. Also decisive were the paths of the transmission of knowledge, which were obviously connected to its carriers and their identities. Thus one transformation was much connected to Cardinal Basilios Bessarion's identity as a refugee from Constantinople and a cardinal in the service of several popes. He recognized the need to find an expert on astronomy like Regiomontanus to retranslate Ptolemy's *Almagest* from Greek to Latin, while possibly also carrying translations of Islamic texts containing astronomical knowledge and practices that found their way to his library in Rome (see Shank's and F.J. Ragep's chapters). Whereas Sally P. Ragep's chapter tells the story of the transmission and dissemination of knowledge and practices (synchronically and diachronically) in Islamic lands related to subjects in Islamic astronomy, Robert Morrison's chapter throws light on cases of Jews expelled from the Iberian Peninsula who resettled in the eastern Mediterranean and travelled to Istanbul or Italy, thus becoming mediators between Islamic and Christian cultures. Transmission also occurred within Europe, as shown by Edith Dudley Sylla in her chapter's argument about the diffusion of the *Configuration* tradition in medieval Europe. Finally, Raz Chen-Morris and Rivka Feldhay's chapter points to the circulation of knowledge within informal, intellectual-artistic circles that associated around a site of

knowledge such as Bessarion's library in Rome, known to be visited by Nicholas of Cusa, Leon Battista Alberti, Paolo Toscanelli, and Johannes Regiomontanus.

Whereas the two last sections of the volume consist of chapters that point out the likelihood of certain paths of transmission, the first section is devoted to discussion of the political conditions that opened up some paths but closed others. Thus Christopher Celenza tends to stress the tradition of travelling scholars that stretched from medieval masters and students to Renaissance scholars in search of patronage, and Nancy Bisaha points out the constraints on transmission by political conflict that produced animosity, as mirrored in the negative images of self and other.

In the fifteenth century, astronomy was seen by many as the most antique field of scientific discussion; it was conceived as the most variegated scholarly tradition not only linguistically but also from the point of view of the cultural and religious identity of its carriers. In addition, the practice of astronomy as a scientific discipline involved crossing disciplinary boundaries between mathematics, philosophy, theology, medical astrology, and logic, as well as between theory and practice, astronomy and astrology, and models and instruments. Last, but not least, in the fifteenth century, the tradition of astronomical knowledge common to Europe and the vast Islamic civilizations surrounding it was dispersed among many different kinds of sites of knowledge – from universities and academies to observatories and madrasas; from princely courts to printing houses and instrument workshops.

In light of this state of affairs, our project and this book should be seen as an attempt to encourage a conversation on the multicultural, multireligious, multilingual, and multidisciplinary contexts of learning pertaining to fifteenth-century astronomy, on Copernicus's radical choices when other "sensible" alternatives existed, and on other radical transformations that in some ways existed parallel to his own. But like the Copernican revolution, the consequences of those transformations did not become apparent until later, sometimes much later. As historians of that period know well, it was a time of great transformation and equally great resistance to transformation, which is why Copernicus strikes us today as being so conservative (the "timid canon" in Arthur Koestler's inimitable turn of phrase) while paving the way for almost all the scientific upheavals that were to come.[26]

Thus this volume is the outcome of an exploration and an experiment to see whether scholars working in different fields and cultural areas could generate synergy that would allow for new insights and perspectives. Our hope is that it offers the reader a wide range of ideas, views, and narratives that provides substance to that elusive term "background to Copernicus."

CHAPTER SUMMARIES

Roughly speaking, the chapters fall into three main categories:

1 The fifteenth-century European social and political contexts (Celenza and Bisaha);
2 The fifteenth-century European intellectual and scientific contexts (Sylla, Shank, and Chen-Morris and Feldhay);
3 The multicultural astronomical background to the Copernican revolution (S.P. Ragep, F.J. Ragep, and Morrison).

The European cultural context as it relates to Copernicus is considered by both Christopher Celenza and Nancy Bisaha, who specialize in the intellectual history of Europe. Celenza focuses on everyday practices of scholars in the "long fifteenth century." He draws attention to the intense and interlinked form of intellectual life, which may have had an impact on how authorship was conceived. He emphasizes travel as a constitutive element in the lives of scholars searching for personal patronage and for a community of supporters. He also points out the collaborative approach to knowledge making that was then prevalent. This way of life may well be related to the view of authorship and to the lack of an institutionalized manner of quoting predecessors. Copernicus's reticence about his sources may have been part of the everyday practices of scholarly life in the fifteenth century, which are so different from the modern ones through which we tend to read our predecessors.

Bisaha reconstructs the complex relations that prevailed between western Europeans, the Ottoman Empire, and Byzantine refugees. By reading the humanist rhetoric on Europe and Asia, she is able to reconstruct the connections, but also the tensions, that characterized the exchanges that took place among European, Asian, and Byzantine scholars at the time. In particular, she draws attention to the images of *Europe* and *Asia* constructed in the two volumes written by Aeneas Silvius Piccolomini (Pope Pius II), which reflect, in her eyes, the crystallization of a European identity vis-à-vis the perception of Asia as "the other." This could well have influenced how someone like Copernicus identified with, acknowledged, or did not acknowledge his sources.

Copernicus's transformation involved important epistemological and conceptual shifts as well as astronomical and mathematical innovation.

These factors are explored by Edith Sylla, Michael Shank, and Rivka Feldhay and Raz Chen-Morris. Drawing on commentaries on Aristotle's *Posterior Analytics*, as well as those on Johannes de Sacrobosco's *On the Sphere of the World* and some Cracow commentaries on Peurbach's *Theoricae novae planetarum*, Sylla delineates how Ibn al-Haytham's *On the Configuration of the World* influenced the quest of European astronomers for a physical interpretation of the celestial orbs. She then analyzes the status of the science of astronomy as it was articulated by professional astronomers, philosophers, and logicians, which she maintains is an essential part of the story behind Copernicus's first articulation of his new system in the *Commentariolus*.

Shank takes up the life of Regiomontanus, who was arguably the most talented mathematical astronomer in Europe in the generation before Copernicus. He and his antecedents deserve special attention if one wants to understand the level that the science of the stars had attained in late-fifteenth-century Europe. Regiomontanus's *Epitome of the Almagest*, which he completed after the death of his teacher Peurbach, would constitute a foundational work for Copernicus's project, providing a competent summary of Ptolemaic astronomy and hints that would prove crucial for the transformation from a geocentric to heliocentric cosmology.

The status of the "appearances" in the traditional science of the stars has always been central in the eyes of mathematical astronomers; in their chapter on the European discourse on visibility in the fifteenth century, Feldhay and Chen-Morris reconstruct the shifting boundary between the visible and the invisible as part of the changing practices of observation in the fifteenth century, a topic that has recently been discussed by historical epistemologists. The discourse on visibility and invisibility, the authors show, touches upon the boundary between mathematical entities and their representations, between artificial perspective and natural vision, and between the human and divine ability to perceive minute mathematical differences. Feldhay and Chen-Morris consider the ways that the new discourse on visibility may throw light on Copernicus's intuition concerning the need to reconsider the point of view of the observer. This reflection may have had implications for his choice to put the invisible motion of the Earth at the centre of his new cosmological system.

The multicultural background to Copernicus and the importance of transmission of knowledge are explored in the chapters by Sally P. Ragep, F. Jamil Ragep, and Robert Morrison. S.P. Ragep's chapter depicts the existence of a staggeringly large number of assembled scholars engaged in studying the mathematical sciences in fifteenth-century

Samarqand. In an attempt to shed some light on what they were studying, she gives a quantitative estimation of the scope of production of astronomical texts in the Islamic world of that period. She focuses in particular on theoretical astronomy (i.e., the *hay'a* tradition) and provides us with a sweeping overview of the works that the Samarqand scholars inherited. In so doing, she asks us to consider the extent to which a deeply rooted, "naturalized" tradition of science may, paradoxically, inhibit the type of change that made a Copernicus possible.

F.J. Ragep's chapter focuses on the various versions of the Ṭūsī-couple – the main device invented by Naṣīr al-Dīn al-Ṭūsī (d. 1274) to replace Ptolemy's use of the equant. The development of alternative models allowed Islamic astronomers to seriously challenge Ptolemy's hegemonic astronomical models. Copernicus, too, used similar devices to replace Ptolemaic models and ultimately to justify his rejection of the main traditional astronomical presuppositions. F.J. Ragep discusses possible ways of transmission of the Ṭūsī-couple from East to West on the basis of new evidence and a reinterpretation of existing data.

Another aspect of the early modern circulation of scientific knowledge is taken up by Morrison, who examines the role of Jews as scientific intermediaries in the European Renaissance. In the first part of the chapter, he points out parallels between a text written by a Jewish astronomer originally in Judeo-Arabic and works of early modern European astronomers; he then delineates specific connections between Islamic, Jewish, and European scholars as well as important pathways by which Jews became intermediaries between Islamic astronomy and Renaissance intellectual communities.

PART ONE

The Fifteenth-Century European Social and Political Contexts

1

What Did It Mean to Live in the Long Fifteenth Century?

Christopher S. Celenza

THE FOCUS OF THIS BRIEF CHAPTER is the world in which Nicholas Copernicus moved and lived. Many of its features were shared alike by the learned and unlearned and by the rich and poor; other aspects served to distinguish Copernicus and other intellectuals as a class of people with certain unique sets of experiences, acquired skills, and mental horizons. Before I move to these features, it will be helpful to foreground the term "long fifteenth century." As I am using it, the term designates a time from the mid-fourteenth century to approximately the year 1525.

I first used this term in my 2004 book *The Lost Italian Renaissance.*[1] There, I intended to refocus attention on fifteenth-century, predominantly Latin-writing Italian intellectuals, that group of thinkers who inhabited a roughly five-generation span from when Francesco Petrarch (1304–74) was in his full maturity to the era of Angelo Poliziano (1454–94), with a sixth generation tacked on as a coda, taking the period to the middle of the 1520s, when Latin ceased, for the most part, to be a language of creative literature in Italy.[2] My argument in the book was that this period – which produced artists like Masaccio and Masolino da Panicale, Sandro Botticelli, and Leonardo da Vinci – was relatively lost when it came to intellectual history. Few educated people could name more than two or three Italian intellectuals from the long fifteenth century since canonical lists tend to skip from the vernacular Petrarch to Niccolò Machiavelli or, in the realm of philosophy, from William of Ockham to René Descartes (an even longer chronological hiatus).[3] The book addressed the questions of why these canons were created, what could be done about studying these "missing" intellectuals (relatively speaking, of course), and even whether doing so was worthwhile.

To a small extent, I intended a deliberate contrast with the better-known "long sixteenth century," associated with the brilliant and controversial work of the sociologist Immanuel Wallerstein, who himself had drawn, in historical terms, on the work of Fernand Braudel, among others.[4] Focused on global geopolitical and economic transformation, the work of practitioners of this kind of inquiry highlighted the growing importance to historians of a series of factors (especially in the 1970s and 1980s when Wallerstein published the influential, three-volume *The Modern World System*): economic conditions, global linkages, and the way politics fostered and, reciprocally, was created by the emerging global economy. My concern instead was to bring a socially and politically engaged intellectual history to the forefront in discussion of Italian Renaissance history. Thus, for reasons sketched out in the book and touched on in a series of subsequent articles, I suggested that the "long fifteenth century" was an appropriate and useful analytical tool for looking at Renaissance intellectual history.[5]

The term is also useful here, for it can help us to frame a series of interlinked factors that, considered together, had a fundamental, shaping effect on the way people lived and worked. For intellectuals like Copernicus, the most important of these factors was the way that information was gathered and transmitted: the practices of reading, writing, and publishing. These practices are left to this chapter's final section. First, at least three other areas of cultural life deserve attention since they, too, impacted Copernicus's life, intellectual activity, and mental horizons: political life, global economic conditions, and belief in the supernatural.

As to political life, most people in Copernicus's era lived in a culture within which "rule of law" assumptions did not represent an internalized way of looking at one's place in the world. By the late eighteenth century, the notion that there could be no universally agreed-upon set of pre-existing natural laws (and hence that all human beings are in principle free), coupled with the idea that there is no universal standard for determining relative value among individuals (and hence that all human beings are equal), had led to a historically unique theory of the human. This theory, which has never, as yet, been fully adhered to in practice, formed the basis for the development in the West of theories of civil and eventually human rights, which were then instantiated within legal systems wielded by modern states with coercive power. This entire process can be said to have its own internal history, with elements being either adumbrated or present in ancient Greek democracy, Roman republicanism, the Magna Carta, late-medieval legal theory, Italian humanism, and Huguenot thought. Still, it was not until the late eighteenth century that the full complex of these theories was both realized and made into a program for political action, itself supported by laws.[6] Copernicus's world, instead, was one in which personal patronage mattered greatly, not only for advancement but also for personal safety and for the parameters of intellectual activity. Personal contacts and webs of intellectual alliances, as opposed to radical individualism, formed the basis of intellectual life and intellectual productivity.[7] This intense, more interlinked form of intellectual life found expression also in the way that authorship was conceived and practised.

This sense of the importance of the local was reinforced by economic conditions. Again, it was not until the late eighteenth century that global capitalism found expression in both theory and practice. On the one hand, John Locke eventually succeeded in arguing that the maintenance of one's private property should be considered a "natural" right. Here, too, there were many antecedents, especially in early fifteenth-century Florence, where the protection of private property was considered important and where legal measures, including tax surveys, were created to monitor the holding of private property.[8] On the other hand, it was not until the coming together of two essential factors that the global economy coalesced. First, the establishment of Atlantic trading routes shifted the focus of trade from the Mediterranean to the Atlantic world in the seventeenth century. Second, by the end of the eighteenth century, these accumulated experiences were being observed, recorded, and theorized. The result was the theory of capitalism being joined to its practice, with the 1776 publication of Adam Smith's *Wealth of Nations* as the capstone of this phenomenon.

That these two developments coincided, broadly speaking, with the beginnings of industrialization reminds us of an obvious, if sometimes unremarked, fact: Copernicus and his contemporaries lived well before mass transportation. Given this piece of information, one can see Copernicus as a member of a relatively small minority: travellers. Itinerancy over long distances was rare, and Copernicus's membership in that group signalled him as a person to whom ties of sociability took on a colouration that was slightly different from the normal, close-knit patterns of the agricultural village or the neighbourhood-oriented premodern city. As someone who spent roughly seven years outside of his home studying and travelling, Copernicus can be grouped with mercenaries, travelling police professionals (like the *podestà* in late-medieval Italy), merchants, traders, clergy, university students, educators, and intellectuals who brought to their lives a sense of cosmopolitanism, naturally acquired through travel and exposure to diverse communities and customs.[9] In turn, these cultures of sociability engendered certain social rules, sometimes unspoken, that helped to govern behaviour and individual activity. The culture of honour and reputation (as opposed to a seemingly neutral culture of empiricism), which has become a subject of comment and controversy in the recent historiography of science, was partially a result of these patterns of contact.[10]

Finally, travel afforded those who partook in it the opportunity to experience different cultural traditions. In Copernicus's case, his first foreign experience was at the University of Bologna, which was distinctive, as were a number of other Italian universities, in several ways. First, at Bologna, as at other Italian universities, "theology was weak," as David Lines has recently pointed out, and not connected in the pedagogical sense to the arts faculty, as was the case at the University of Paris.[11] Students pursuing doctoral work in theology at Paris, for example, taught liberal arts to students striving for the baccalaureate degree. Yet at Bologna it was the advanced doctoral students of *medicine* who did such teaching, which means that at a ground level there did not exist the same organic link between theology and the arts.

This sense of institutional "secularism" (in the broad sense of a lack of formal allegiance to canons) even extended at Bologna to the teaching of Aristotle, according to one Jesuit observer, Antonio Possevino. He claimed that, in their zeal to advance to the study of career-advancing and lucrative medicine, students and professors alike rushed through their natural philosophy courses.[12] In other words, at Bologna the general curricular culture did not encourage traditional reverence toward long-respected authority in certain fields of research and intellectual activity, something that may well have been a factor in Copernicus's interactions with Domenico Maria Novara, his mentor at Bologna. The same lack of an organic link between the arts and theology obtained at Padua, the other Italian university where Copernicus spent a significant portion of his time. Curricular cultures, as Johannes Kepler noted a century later, are both profoundly shaping influences and inherently conservative, as members of one generation by and large continue to teach, semester by semester and year by year, what they have become accustomed to teaching.[13] At a number of Italian universities, part of that culture was represented by a willingness to question authority, a willingness inherited from the Italian humanism of the long fifteenth century, where the posture of the intellectual "outsider" gained prominence.[14] Indeed, in Italy there was far less distance between university culture and Italian humanism than certain caricatures, both from the epoch itself and from later eras, might have us believe.[15] Given this situation, Copernicus's willingness to entertain divergent techniques (like the Ṭūsī-couple) and possibly revolutionary viewpoints (like heliocentrism) becomes more understandable.

A third element in which Copernicus and his contemporaries in the long fifteenth century differed markedly from later thinkers was in the relation of individuals to nature. Copernicus lived at the beginning of a three-century period of transition within which the views of intellectuals regarding nature's boundaries were in flux.[16] Many thinkers from the middle of the fifteenth through the seventeenth centuries attempted to define the boundaries of the natural, focusing attention on "marvels" that seemed superficially to lie outside the normal order of nature.[17] These thinkers attempted to recalibrate the boundaries of the natural by arguing that, even if these phenomena were outside or beyond the normal order of nature, or preternatural (*praeter naturam*), they could still, if properly scrutinized, be explained by resorting to natural causes. One professor at the University of Bologna, Pietro Pomponazzi (1462–1524), as Lorraine Daston has brought into relief, did just this in his critical examination of a miraculous apparition, viewed by an entire town. Explaining the apparition, he "invoked astral influences" and thus reduced something apparently "miraculous" to something "natural."[18]

Still, one segment of belief in the supernatural went unresolved throughout the long period that extended from the fifteenth through the eighteenth centuries: demons.[19] Even the most hardnosed thinkers believed in the existence of airy beings, intermediate between humanity and divinity, who could have discernible effects on the natural world. There were many differences in the levels of belief involved. Here, for example, is a thinker at the tail end of Italy's long fifteenth century, Francesco Guicciardini (1483–1540), the historian of late-Renaissance Florence and godson of Marsilio Ficino (1433–99), in his *Ricordi* (maxim 211): "I am entitled to state that the spirits do exist. I mean those things we call spirits, that is, the aerial beings that speak with human beings in a direct and open way. My personal experience has convinced me of their existence." The maxim goes on: "But what the spirits are, and of what sort they are, is known as little to the person who believes he understands them as it is to someone who hasn't even given a thought to the matter. This and predicting the future, as you see people sometimes do either by design or by means of a frenzy [*furore*], are hidden powers of nature. Or rather, they derive from that superior power that guides all. To Him they are clear, to us, secret. As such, human minds cannot reach them."[20]

Four maxims earlier, at number 207, Guicciardini writes that it "is madness even to speak of astrology, that is, of that science that passes judgment about future things. Either the science is not true, or all the matters that are necessary to it cannot be known, or human capacity cannot reach it. In any case the conclusion is this: thinking to know the future this way is a dream."[21] Guicciardini does not deny that "spirits" with airy bodies, or demons, talk to human beings since, in his view, they are part of the order of nature.

Guicciardini stood at one end of a spectrum, practical as he was and thus disinclined to worry overmuch about demons. At the other end were those, like Marsilio Ficino, who believed in a much more viable possibility for human contact with and knowledge of demons and the supernatural. When Ficino warns, as he does in his *Three Books on Life*, against image worship, the reason is unsurprising and wholly in line with his core beliefs: people "are very often deceived in this matter by evil daemons encountering them under the pretense of being good divinities."[22] The existence and presence of demons is a fact taken for granted. From antiquity through the late eighteenth century, belief in the existence of demons on the part of intellectuals marks a fundamental dividing line. The loss of belief among intellectual elites in an activist supernatural realm was still centuries away when Copernicus was active at the tail end of the long fifteenth century. It seems artificial, from this perspective especially, to separate Copernicus the astronomer too emphatically from the astrological tradition.[23] His teacher at Bologna, Domenico Maria Novara, actively engaged in astrological prognostications, as did Georg Joachim Rheticus, Copernicus's pupil. In fact, Novara's prognostications are his only published remnants that have come to light.[24]

Certain cultures change slowly, and this slow pace of change is most evident in one of the cultures within which Copernicus, like other intellectuals, was deeply embedded: that of reading and writing. The first prominent factor is overall content. On the one hand, Copernicus lived at the beginning of a time notable for its sense of "information overload," as Ann Blair has dubbed the sense among intellectuals from about the middle of the sixteenth century through to the end of the seventeenth century that they were trying to find new ways to manage an increasingly overwhelming amount of printed information.[25] Special techniques for note taking and new forms of reference books are among the sixteenth-century phenomena that signal the onset of this burgeoning of available information.

On the other hand, historians of the book, most notably Adrian Johns and David McKitterick, have noted certain fundamental continuities in the ways that reading material was handled through to the beginnings of the nineteenth century and the advent and popularization of the steam press.[26] Undoubtedly, printing with moveable type in the fifteenth century increased exponentially the amount of material available.[27] Still, during the era of the hand press, reading communities tended to be smaller, for books, although far less costly than in the manuscript era, were still expensive and, compared to books after the steam press, relatively rare; and at times, texts that to an individual scholar might have been most important were also the most likely to be lost over time since they might have been transported with the scholar as unbound quires and lost or destroyed before ever having been bound.[28] Printing did not change this latter situation overnight.

One can also note that those staples of the eighteenth-century "public sphere" of Jürgen Habermas, newspapers and relatively widespread novel reading, were absent in the long fifteenth century.[29] In absolute terms, few people read, few people had regular access to books, and the books to which they did have access were comparatively rare.

Reading communities themselves remained heavily inflected by the sorts of politics of reputation that have become controversial in the history of science. Cultures of reading are conservative in that they change slowly; for a work to garner attention, it needed more than the interest of an individual. It needed community support. An example of this sort of community was the intellectual circle comprised of Giovanni Pico della Mirandola (1463–94) and Angelo Poliziano (1454–94), both of whom were active in Florence in the 1480s and early 1490s, together with their Venetian friend and correspondent Ermolao Barbaro (1454–93). Here is Pico writing to Barbaro upon receipt of one of Barbaro's letters: "I, and our own Poliziano, have often read whatever letters we had from you, whether they were directed to us or to others. What arrives always contends to such an extent with what there was previously, and new pleasures pop up so abundantly as we read, that because of our constant shouts of approbation we barely have time to breathe."[30] The point is that the approach to knowledge making was collaborative.

A new letter would arrive highlighting, for example, a text that its author had recently found challenging or interesting. The letter, although addressed to one member of the group, was understood by both writer and recipients to be part of collective discussions. Its arrival would prompt the other members of the group to read the text in question, comment on it, respond to the author's inquiries, and so on.[31] Conversation shaped reading, writing, and investigation. To understand this process fully, one needs to conceive of a sense of collective authorship, of numerous voices inflecting the one voice that shows up in the written text.

The terminology of Saint Bonaventure, born Giovanni di Fidanza (1221–74), is useful in making distinctions that held throughout the Middle Ages and in varying degrees through the long fifteenth century. The following passage occurs at the very beginning of Bonaventure's *Commentary* on Peter Lombard's *Sentences*, the twelfth century's greatest work of theological synthesis, a staple of medieval theological curricula, and a work on which many later thinkers wrote commentaries.[32] The passage in question occurs when Bonaventure wishes to determine, first of all, whether Lombard is in fact the author of the *Sentences*:

There is one [type of writer] who writes things of others [*aliena*] without adding or changing anything, and he is just called a 'scribe' [*scriptor*]. Another writes things of others, and he adds things, though they are not his own [*de suo*]; he is called a 'compiler' [*compilator*]. Yet another writes both things of others and of his own; but he writes the things of others as the main part of the work [*principalia*], adjoining his own just for clarity [*ad evidentiam*]; he is called a 'commentator' [*commentator*], not an 'author' [*auctor*]. Yet another writes both his own things and things of others, but mainly his own, adjoining the things of others just for support [*ad confirmationem*]. And it is this sort of writer who ought to be termed an 'author' [*auctor*].[33]

One can see what Bonaventure takes for granted: that even the most "individualistic" type of person who engaged in graphic culture, the "author," was perceived as someone who still added "things of others," or *aliena*, whether ideas, citations, passages, and so on, even if most of what he wrote was "his own." Medieval thinkers invested this latter type of writer with authenticity.[34] It troubled no one that *aliena* would be added to a text, provided it was done "for support" (*ad confirmationem*). Acknowledgment of having done so was not perceived as necessary for this type of authorial authenticity.[35]

Also important were diverse styles of reading.[36] One style can be characterized as slow and meditative, a method of reading that was available and important in Copernicus's day.[37] Copernicus was probably right on the mark when, at the outset of *De revolutionibus orbium coelestium*, in his dedication to Pope Paul III, he stressed that, dissatisfied with the uncertainty of the mathematics that had been handed down, "I took it as my task to reread whatsoever books I could get my hands on of all the philosophers, in order to find out whether anyone had ever arrived at the opinion that there were motions of the spheres of the world that were different from those that the people who teach mathematics in the schools put forth."[38] One style of reading in Copernicus's day was repetitive, closer to memorization than to the gathering of information, and intended to serve as a stimulus to intellectual production on the part of the reader.[39]

Practising this style of reading, which possessed fundamental continuities from at least the twelfth century to early modernity, meant that an author read in order to write: "he read to compose a text of his own that was largely made up of the citation of others; he read by writing, because he continuously annotated books in the margins."[40] Medieval scholars were aware of different types of reading. John of Salisbury (1115–80) differentiated between reading done in the context of a master teaching a student and hence a public and reading done in private as a form of meditation:

> Therefore, may whosoever aspires to philosophy come to grasp reading, learning, and meditation with the exercise of good work, lest he anger his Master, who may think that he is losing that over which he used to seem to have possession. But the word 'reading' [*legendi*] is equivocal: it refers as much to the work of the teacher and the student as it does to the business of carefully examining the Scriptures for oneself; in the first case, which is to say referring to what is communicated between a teacher and student (to use Quintilian's terminology), it is called 'reading aloud' [i.e., *praelectio*],[41] whereas in the second case, that is, in the case of what comes under the scrutiny of the person meditating, it is termed, simply, 'reading' [i.e., *lectio*].[42]

John of Salisbury's clarifications concerning various types of reading and the differentiations he makes among them alert us, in the first instance, to one medieval scholar's understanding of different types of reading practices, one at least somewhat public and the other private.

More broadly, they allow us to continue to account for other premodern styles of reading. One of these might better be termed "information gathering" rather than "reading." Robert Westman points out that Isaac Newton acquired his knowledge of Kepler's innovations not from reading Kepler's *Astronomia nova* but from intermediary sources.[43] One must assume that all sorts of intermediary sources, from anthologies and *florilegia* to oral transmission, formed part of Copernicus's toolkit as he was studying. Given this notion, it would be unsurprising if Copernicus came across a version of the Ṭūsī-couple in a way that is now difficult to trace definitively.

Finally, there is one more crucial area regarding the practices of reading and writing that, taken as a whole, allows us to see that Copernicus lived in a period when a great debate was coming to an end: that concerning the nature and purpose of the Latin language, which for him and others in his era was the primary language of intellectual life. To understand this debate, and its ramifications for the world in which Copernicus lived, one must step back and take a bird's-eye view of the "long fifteenth century" in Italy, which can be said to begin in Petrarch's maturity.[44]

Petrarch, like other medieval thinkers, believed that Latin was a *lingua artificialis*, or "artificial language." He and others meant by this conception that Latin was a language of *ars*, or "craft," and that by definition it therefore possessed rules, which implied order, institutions necessary to teach the rules, and uniformly trained people to perpetuate this international language. Medieval thinkers, and Petrarch was no exception, employed the term *grammatica* (grammar) as a synonym of "Latin." These associated notions about Latin also implied a kind of immutability. Yet for Petrarch, the language he called "the foundation of all our arts" evinced a sense of dissonance as well.[45] As he and others around him explored their passion for Roman antiquity, they could not help but notice that *grammatica*, the Latin then in use and a staple of both the institutional church and university life, did not seem equivalent to the Latin of the ancient authors whose work was increasingly in vogue.

Petrarch's main objective with respect to Latin was reform: to set the Latin language, and hence the culture around him, on a path toward greater authenticity. This goal contained within it the seeds of two ultimately disharmonious elements. First, a key element of Petrarch's approach had to do with exegetical imitation.[46] When Petrarch is trying to find, say, Lucius Annaeus Seneca's true meaning, he is not trying to uncover Seneca's intentions in a historicist fashion. Instead, he is accessing the truth contained within Seneca's work, the full extent of which Seneca himself might not even have been aware. One did not, as a rule, do philosophy (including natural philosophy) in the pre-Cartesian period with the intention of being strikingly original. Instead, one used the authorities with which one was surrounded and the traditions that one had inherited to express one's own original viewpoint. This perspective allows us to see Petrarch and the four generations of humanist thinkers who followed him not as a break from but as a part of a long, relatively continuous ancient and medieval tradition that continued long after their heyday.

Second, however, Petrarch's recognition of a gap between the language of his own time and that of the ancients introduced into humanist culture the notion of historical discontinuity, a fact that led to the demise of Latin as a language of creative philosophical literature in Italy in the generation after Angelo Poliziano. A debate ran throughout the fifteenth century: what sort of language did the ancient Romans speak? Did they possess both a vernacular and a polished speech endowed with rules that needed to be learned in school, like moderns? Or did they learn Latin (a complicated, inflected language) from birth?[47]

This debate began in the generation of Leonardo Bruni (1370–1444), effectively the third in this five-generation cultural experiment.[48] Bruni remained tied to the view that Latin was an "artificial" language.

For him, even the ancient Romans had to learn Latin's rules rather than coming to them naturally. By the end of the fifteenth century, most humanists, if they considered the question, believed the opposite: that ancient Latin had been a historically contingent natural language, with a trajectory of growth, glory, and decline. Poliziano, the last and fifth generation's leading figure, considered himself a *grammaticus*, literally "grammarian" but more aptly, in his case, "philologist." His strong point as a scholar was not, as it had been for many humanists before him, relatively lengthy dialogues or other prose treatises. Instead, small, philologically acute essays best captured his abilities and, in their form, encapsulated the state of the Latin language question at the time. Advanced as he was, Poliziano signalled the beginnings of a key transformation, one of the many that the Latin language has undergone in its long history.

Thereafter, two main trajectories presented themselves. The first, which was Poliziano's choice, was that of Latinity as code, whereby one created a scholarly community, both real and imagined, by weaving together a rich texture of classically appropriate but necessarily eclectic allusions, quotations, and intertextual gestures that could be recognized by the community's members. The second, which represented the majority tendency in the early modern period, reflected Latin's metamorphosis into a language of scholarship: one that was classicizing in a loosely Ciceronian fashion, relegated to the margins as a language of creative philosophical literature, but suitable for and translatable across the bureaucracies, universities, and academies that were woven into the fabric of a newly emerging order of European sovereign states.

In Italy the generation after Poliziano saw a move toward the vernacular, as the works of Niccolò Machiavelli and Baldassare Castiglione testify. Figures like Thomas More and Desiderius Erasmus indicate well the creative potential of Latin as Latinate humanist culture moved north. And one should add Copernicus himself, whose *De revolutionibus* has been judged by our era's foremost student of Neo-Latin style to be an example of "beautiful" Latinity, given how classicizing, flexible, and creative Copernicus's treatise was.[49]

On the one hand, and by way of conclusion, it is appropriate to situate Copernicus in the "long fifteenth century" simply because certain key features shaped his personal universe. The lack of well-developed civil and human rights theories and laws; the relative rarity of travel and the concomitant and unique status of the traveller, embedded in and yet different from the more local character of the majority of his fellows; the continuing belief in the possibility of supernatural intervention in the world; and the existence of many different varieties of reading and writing practices, some of which show significant continuities through the eighteenth century: these factors shaped his outlook, and a number of them make it likely that he may well have come across a theory like the "Ṭūsī-couple" without feeling the characteristically modern need to record precisely where, when, and in what format he encountered it.

On the other hand, Copernicus, in his own early sixteenth-century moment, stood at the end of one set of developments and at the beginning of another. A relatively standard, classicizing Latinity was adopted as an instrument of recording and expressing natural scientific conclusions, just as it had been in the fifteenth century for matters literary and historical. That Copernicus employed this means of expression, alert as he was like many in his era to finding the boundaries of the "natural," reminds us of an important fact: he stood at the beginning of precisely those developments that later issued forth in some of the habits of modernity that were attuned to a culture of scientific invention whose members were inclined to value individual achievement and citation over collective authorship.

2

European Cross-Cultural Contexts before Copernicus

Nancy Bisaha

FOR CENTURIES, Nicholas Copernicus was regarded as a lone visionary whose discoveries freed astronomers from the stifling bonds of the Ptolemaic system and changed the direction of scientific thought. A few decades ago, historians of science complicated this view by demonstrating the closeness between late-medieval Islamic treatises and Copernicus's findings. Others see strong signs of borrowing from recent European thinkers like Johannes Regiomontanus.[1] It is likely, then, that Copernicus, who is silent on these sources, drew on prior theories via some path of transmission that historians of science are attempting to unravel. My chapter considers the recent Islamic sources upon which Copernicus likely drew. Lacking a clear sense of how he may have acquired this learning from the East, we are left with many questions regarding the reception and repackaging of this information. Specifically, why did Copernicus and his contemporaries say nothing about recent Islamic astronomers if they were so heavily indebted to them? How or why did such precise astronomical knowledge travel great distances in the early modern era, only to have its origins vanish so effectively that scholars did not discover them until the last few decades? Here is where my research on decidedly nonscientific subjects, namely Renaissance humanist perceptions of Europe and the Ottoman Empire, fits into this volume.

To understand the transmission of knowledge across cultures and time, one must grapple more broadly with the ways that those cultures perceived one another and themselves. Any information that travelled between East and West was subject to layers of interpretation that could easily obscure and change the message; this situation was particularly true of the highly charged period of the Ottoman advance in the fifteenth and sixteenth centuries. This chapter examines one aspect of the cultural backdrop in western Europe that led up to and followed Copernicus's *De revolutionibus orbium coelestium* (1543). Specifically, it addresses the growing belief among Europeans in their cultural unity and their perception of stark differences between their society and that of Islamic Asia. These concepts of self and other can shed much-needed light on the ways that Islamic astronomical learning may have been presented to and used by Copernicus and, consequently, interpreted by his contemporaries and the next few generations.

When it comes to Copernicus's own reception and perceptions of post-twelfth-century Islamic astronomical learning, we have very little direct information to go on. Copernicus's writings make no mention of Islamic thinkers from the period after 1200. But Noel Swerdlow, Otto Neugebauer, and F. Jamil Ragep have pointed persuasively to the uncanny similarities between mathematical models and ideas found in the works of Copernicus and predecessors like Naṣīr al-Dīn al-Ṭūsī (d. 1274), ʿAlāʾ al-Dīn ibn al-Shāṭir (d. ca. 1375), and ʿAlī Qushjī (ca. 1400–74).[2] The likelihood of Copernicus coming up independently with the same complex models seen in these works – the cumulative knowledge of several centuries of challenges to Ptolemaic theories – is extremely slim.[3] These Islamic models, it has been argued, must have found their way westward via unknown intermediaries and/or translators. Jewish scholars and translators are a strong possibility.[4] Another plausible theory is that Greek scholars, possibly refugees of the Ottoman advance like Cardinal Basilios Bessarion, brought these ideas in some form to Western scholars.[5]

Taking this hypothesis further, once these ideas were transmitted, one can imagine a range of ways that they might have been received. At one extreme, they could have been seen by Copernicus and his predecessors as wisdom from the East. One can readily cite the examples of Giovanni Pico della Mirandola's *Oration on the Dignity of Man* (1486), with its enthusiastic embrace of Arab and Persian thought alongside Latin, Greek, and Jewish scholarship, and Guillaume Postel's respect for Islam and Arab learning and the rise of interest in Arabic among Western scholars.[6] Whereas Renaissance scholars more often admired earlier medieval Arab thinkers like Avicenna (Ibn Sīnā) and Averroes (Ibn Rushd),[7] one sees some appreciation for late-medieval Islamic learning in later Byzantine scholarship and can find intriguing suggestions of respect for – if not necessarily a clear knowledge of – learning in central Asia and Persia in the later fifteenth and sixteenth centuries.[8] The examples suggest the possibility that both transmitters of Islamic astronomical models and western European recipients happily embraced them and that

Copernicus, for any number of reasons, simply did not credit his more recent Islamic sources. After all, one finds other examples of contemporary scholars doing the same in the sciences and other disciplines.[9] Christopher Celenza's chapter in this volume speaks to the collaborative nature of Renaissance authorship and the long tradition of compilation that preceded it – processes where names of individual contributors were often either lost or not seen as requiring mention. Alternately, one could argue that Copernicus did not know the authors of the ideas that were transmitted to him, their identities having been lost in the process.

These theories are very plausible and must be given due consideration. At the same time, we would also do well to consider the political and cultural context of Renaissance Europe – specifically prevailing attitudes toward the Ottoman Empire and Islam in general. That one or more European nations were almost constantly at war or preparing crusades against the Ottoman Empire for three centuries is all too often ignored or only briefly mentioned. As I discuss below, war and religious conflict by no means cut off curiosity and transmission, but we should certainly expect that they complicated the process in a number of ways.[10]

The Ottoman Empire expanded rapidly and at a steady pace from the fourteenth century on. Numerous Christian defeats, like Nicopolis (1396), Varna (1444), Constantinople (1453), Venetian Negroponte (1470), Otranto (1480), Rhodes (1522), and Mohacs (1526), increased Western Christian fears, while bloody and dramatic accounts of battles and sacks added to their revulsion of the "barbaric Turks." What few victories Christians could claim, like Belgrade (1456), Vienna (1529), and Lepanto (1571), were celebrated with enthusiasm, but there was little long-term respite until the late seventeenth century. In between major battles, a frequent terrifyingly effective Ottoman practice was to send out irregular troops or raiders ahead of potential campaigns to harass and abduct poorly defended locals and essentially soften up the area for more organized conquest.[11] This context, of course, does not account for the sum total of European Christian exchanges with and perceptions of the Ottomans. In between wars and moments of heightened tensions, Italians traded in Ottoman ports, temporary truces and alliances were made, and a range of productive interactions took place in many areas of the Mediterranean with mixed populations.[12]

Although a growing number of studies have found nuanced European attitudes toward the Ottoman Empire, on balance, more written works in the Renaissance dealt harshly with it.[13] All these defeats or near defeats struck fear among Europeans and generated a wave of pamphlets, crusade calls, sermons, popular laments, and literature. Humanist rhetoric on the Turks was, for the most part, especially hostile.[14] Rhetoric may seem an unreliable representation of an individual's true feelings, but there are strong indications that humanist anti-Turk rhetoric reflected and/or increased hostility toward Ottomans and Muslims in general. For good reason, Muslims were extremely wary of travelling in Christian Europe, with the exception of Venice, throughout the period.[15] In the fifteenth century, Ottomans went from being cast as the already troubling "infidel" to also being characterized, thanks to the humanists, as "the new barbarian." One such humanist who arguably had the greatest influence in constructing both this highly charged notion of the Turks and what he perceived to be their polar opposite, "Europe," was Aeneas Silvius Piccolomini, elected Pope Pius II (1458–64). The youngest son of impoverished Sienese nobles, Aeneas struggled to make his way in the world, first as a student who had to catch up with wealthier and younger colleagues at the studio in Siena, then as a secretary and diplomat in the service of cardinals at the Council of Basel, and later in the more illustrious court of Holy Roman Emperor Frederick III. In his early years, he wrote poetry, a famous erotically charged love story, and at least one comedy; he also sired two children out of wedlock.[16] But by 1446 Aeneas's mood had grown more serious, and he turned increasingly toward religion. He took holy orders that year and shortly thereafter was made bishop of two sees and then made a cardinal. In 1458 he was elected pope. His stunning rise to power can be attributed to his great skill as a rhetorician, his shrewd political abilities, and his personal commitment to unite fellow Christians against the Ottoman advance. This last ambition, coupled with his demonstrated knowledge of crusading and the Turks, played a central role in winning him the papacy.

Aeneas's initial interest in the concept of Europe most likely sprang from his unusual circumstances as (1) a participant in the Council of Basel (1431–42) and (2) an imperial secretary and diplomat (1442–55). These unique opportunities afforded Aeneas years of close experience with two entities that claimed authority over all Latin Christians: council and empire. His work with the two, moreover, forced Aeneas to wrestle with the notion and nature of papal power from a variety of angles. Added to these experiences was his unique vantage point as a keenly observant outsider: he was a layman at the council and an Italian at the largely German imperial court.

Aeneas's thinking on the supranational role of empire can be seen most clearly in his 1446 treatise *De ortu et auctoritate imperii Romani*. As Cary Nederman has shown, this treatise is an underappreciated work of subtle Ciceronian criticism, which deserves a place alongside better known examples of Renaissance political thought, namely works of civic humanism.[17] In *De ortu*, Aeneas tries to resolve the inherent tension in Cicero's vision of ideal government, which upholds independent republics but fails to confront the problems that arise when all these independent states expand, butt up against one another, and ultimately go to war. The civic humanists knew this only too well from Florence's own expansionist wars. Aeneas's proposed solution to these disturbances was the stability and centrality that only empire could bring. In *De ortu*, he supports imperial rule by election but sees almost no excuse for revolt in the case of an inadequate ruler – a useful argument for an emperor like Frederick III, who was slow to act and demonstrated little vision or leadership. Aeneas also affirms the necessity of the proverbial two swords, papal and imperial power, averring that each should respect the province of the other and keep to its own respective spiritual or temporal realm.[18] Around the same time he composed this treatise, Aeneas rejected the Council of Basel, begged papal forgiveness, and became a key negotiator in reconciling Pope Eugenius IV and Emperor Frederick III, who had parted ways over Frederick's support, however weak, of the council.

Both *De ortu* and Aeneas's shift from conciliarism to the support of papal primacy have been unfairly characterized as self-serving and disingenuous. On closer examination, each fits perfectly with his later views on the papacy and his lifelong mistrust of popular government. Poor he may have been, but there is no evidence that Aeneas ever harboured populist sympathies.[19] His years at Basel, exhilarating at first but desultory by the end, brought him as close to representative government as this nobleman would ever come, and the bickering and irresolution clearly left a bad taste in his mouth.[20]

Aeneas's firsthand experience with the divisiveness of Basel and the constant disputes and warfare that beset the Holy Roman Empire already had him thinking about disunity within the church and empire, but it was only in 1453 that he began seriously to evoke the term or concept of "Europe." That year marked the fall of Constantinople to the Ottoman Turks and brought newfound urgency and emotion to what was already a long-held interest.[21] He wrote a flurry of letters and delivered three orations in the next two years dealing with the topic.[22] Perhaps his most vivid and well-known rhetoric on the Turks is seen in his letters to Pope Nicholas V and Nicholas of Cusa, written in July 1453. In these letters, he expresses righteous indignation at the wanton abuses of fellow Christians and their shrines, and he calls for action in defence of the Christian faith and Europe. Equal time, however, is given to the siege's impact on learning and culture. Repeating tales to the pope of the destruction of countless books in the siege, Aeneas calls the Turks "barbarians" and laments the event as "a second death of Homer and Plato."[23] He goes on at great length in his letter to Cusa about these losses. What sort of men, he wonders, would attack learning? Xerxes and Darius "waged war on men, not letters," and the ancient Romans held Greek learning in high regard despite their conquest of the land. But under the Turks, he asserts, Greek learning is sure to perish.[24] For Aeneas, as for many other humanists, tales of the Turks' sack of so rich a city, particularly its libraries, conjured up parallels to the fifth-century sacks of Rome and the subsequent "dark ages."[25] A critical point regarding Aeneas's rhetoric on Constantinople and a proposed crusade is that his letters and orations were widely circulated, first in manuscript and within two or three decades in print.[26] He had a large reading audience, indeed.

A corollary of Aeneas's rhetoric on the Turks was his evolving notion of "Europe" and the political, religious, and cultural destruction he believed they would visit upon the continent if left unchecked. Aeneas obviously did not invent the term "Europe," nor did he stop using other terms like "Christendom" (*Christianitas*) or "Latins" to describe this larger collective or region; in fact, like many contemporaries and later thinkers, he continued to use these terms interchangeably.[27] But it has been shown that Europe was not a widely used term before the fifteenth century and that the *idea* of Europe as a cultural collective was still a young one.[28] Aeneas's role in articulating that notion is undeniable. His most influential and widely read work dealing with the subject was *Europe* (1458), written within a few months of his election as pope.

Europe, in Aeneas's words, "records for posterity what, to my knowledge, were the most memorable deeds accomplished among the Europeans and the islanders who are counted as Christian during the reign of Emperor Frederick III." He adds, "I will also include earlier material from time to time, when the explanation of places and events seems to demand it."[29] It is a fascinating hodgepodge of mostly recent history, geography, cultural observations, and entertaining gossip or folklore. It begins in southeastern Europe with Hungary, circles around central Europe, moves up to the north in Scandinavia, and finally passes briefly over to France, England, and Iberia before finishing in Italy, with roughly one-third of the text devoted to his homeland. The work lacks a strong central core or single message; its attention to geography, moreover, slips by the time he reaches the outer edges of western Europe and is all but missing in Italy. Citing these inconsistencies, scholars have rightly noted that he seems to have taken a work he had composed on Italy in the time of King Alfonso of Aragon and Naples and tacked it onto this text devoted more broadly to Europe in the time of Frederick III, who began his reign in 1440.[30]

Despite its seeming lack of textual unity, there are several key themes that give the piece coherence and offer clues to Aeneas's goals. Since the work opens with Hungary, Transylvania, Serbia, and other southeastern regions, the Ottoman Turks loom large from the outset. We read about the Ottomans' dynastic origins and history, their military progress across the region, and their steady takeover of what was once Christian territory and the home of Greek and Roman heroes and thinkers. Occasional Christian victories are celebrated but shown to be all too rare given the dynastic struggles for the throne of Hungary and squabbling between local rulers and noble families. The work opens, then, with the region of Europe most directly in danger of being subsumed into a non-Christian and seemingly "non-European" empire. Ironically, the amount of ink Aeneas spills on the Ottomans shows their integral role in recent politics in Europe, especially eastern Europe, from dynastic intermarriage to various short-term treaties and alliances. Thus Aeneas ends up discussing the Ottomans as a powerful, *European* dynasty, even if his goal was to portray them as interlopers to be expelled.

Another dominant theme in *Europe* is the role of the papacy as a stabilizing, unifying force in European society. Aeneas rehashes the fallout from the Papal Schism and its troubled resolution with conciliarism – a movement that ironically culminated in the Council of Basel's deposition of Pope Eugenius IV and election of anti-pope Felix V in 1440.[31] Other clergy, mostly high-ranking, appear frequently as arbiters or enablers of political disputes, one key resolution being the efforts of Pope Nicholas V, Fra Simonetto da Camerino, and Cardinal Domenico Capranica to broker the Peace of Lodi between Italian powers in 1454. Glances back to the ancient and medieval past also note the clergy's role in spreading Christianity across the continent and, like the Roman Empire, bringing "civilization" in its wake. With all of these examples, one is presented with the clear impression that "Europe" was defined, in large part, by the reach and authority of the Roman Catholic Church.

The authority of secular princes and their supposed progress toward greater cooperation and unity parallel the authority and progress of the church. Rulers like the holy Roman emperors and King Alfonso of Aragon and Naples figure even more prominently in the work than do the clergy. Several lords in the Holy Roman Empire and kings like Charles VII of France also appear numerous times. Although Aeneas does not spare any of these figures his criticism, he is largely on their side and seems to support their goals. Even the power-hungry and vainglorious Alfonso is described as having "become the keeper of peace in Italy."[32] Object lessons to those who do not take peace and cooperation between Christians seriously are shown in catastrophic losses to the Turks, like those of Varna and Constantinople. The reverse can be seen at the Battle of Tannenberg (1410), where Poland and Lithuania united against the land-hungry Teutonic Knights, although Aeneas clearly regrets the loss of Christian life. The overall effect of these mounting resolutions and peace pacts is all the stronger when compared to the Ottomans' seeming inability to maintain peace pacts.[33] The structure of *Europe*, however, seems designed to lead the reader away from the depressing fragmentation of eastern Europe and toward some assurance that the currently strong, unthreatened portions of the continent will eventually unite against the menace with which he has so gloomily opened.

The other defining aspect of Europe, as Aeneas has drawn it, is more elusive. Perhaps more than any other factor, the interconnectedness of the entities he describes makes them appear to be a single unit. No sooner does he leave one kingdom or province than he must refer back to it as he moves on to its neighbour and the following region. The same goes for rulers: several figures, major and minor by our standards, appear again and again, literally all over Europe by reach of their influence, if not their physical presence. Scholars and clergy, especially, show how easily a "European" traversed the internal borders of the continent, finding a home in any number of places. In short, the modern reader will not find simple, discrete histories of one city, region, or ruler. It is the quality of constant interplay of persons and interests that makes it easier to discuss "Europe" than to firmly define the limits of Saxony or France.

Perhaps what defined Europe most clearly for Aeneas was its opposition to Asia. Three years after completing *Europe*, he composed *Asia* (1461). The two were often printed together and read as one piece called the *Cosmographia*.[34] As they do in *Europe*, the Turks in *Asia* play a prominent role. In *Asia*, however, Aeneas does not begin with the Turks but ends with them. He opens *Asia* with a shadowy picture of China, known by only a few Western travellers and surrounded by much myth. As Aeneas moves westward, the Turks come increasingly into play. He builds dramatically to, and ends upon, an ominous picture of the Ottomans' growing power, unbridled violence, and savagery. It is also worth noting that Aeneas spends a disproportionate amount of the work discussing the Turks: Asia Minor takes up about two-thirds of the text. It clearly served Aeneas's rhetorical purposes to make the Turks and Asia almost synonymous. One can truly see, then, how prominently the Ottomans featured for Aeneas and his readers when they tried to visualize "Asia."

Also unlike *Europe, Asia* spends much more time on the ancient past. As a result, Asia appears shrouded by the haze of history and legend – frequent references are made to Alexander the Great, Homer, and tribes that had not been heard from for centuries. Aeneas's general approach is to first discuss a city or area geographically, then the great men and women of antiquity who may be connected with it, and then the area's conversion to Christianity. Thus far it is a reassuring story of progress from a humanist's point of view. The narrative soon darkens, however, as he skips over the medieval period – presumably to obliterate the role of the Byzantines who rejected the Latin Church – and then abruptly describes the way the Turks took over each area and blotted out all these cultural and religious achievements.[35] This pattern marks most of *Asia*.

If we read *Europe* and *Asia* side by side, as most readers in the generations following Aeneas would do, two starkly different portraits emerge: *Europe* presents an image of growing unity, albeit an incomplete, complicated, and tenuous sense of unity. Again and again in *Europe*, we see how various areas were civilized by the Roman Empire and converted to Christianity; in Aeneas's era, we see them striving toward peace and other common goals. *Asia* presents a completely opposite picture: it is a land in a state of cultural and religious disintegration, mostly of course at the hands of the Ottoman Turks, although the Eastern Christians are handed some responsibility for not uniting with Rome. Hence we see an exaggeratedly rosy view of Europe coming together while Asia is falling apart, in the thrall of its powerful conquerors, with its greatest achievements in the distant past.

This harsh dichotomy may have softened at times even for Aeneas. As Pope Pius II, he briefly entertained hopes for a political alliance against the Ottomans with central Asian ruler Uzun Ḥasan.[36] Aeneas's famous letter to Mehmed II (1461) has also been viewed as a change of heart. It is easy to see why many have viewed the letter as a genuine attempt to convert the sultan. Aeneas calls him an "excellent man" whose "nature is good"; he possesses "many natural gifts" but, alas, was raised in ignorance.[37] The letter invites Mehmed to accept baptism, to be loved and followed by all, and to be supported by Aeneas himself as a legitimate ruler of all his domains and, potentially, all of "Europe."[38] Perhaps Aeneas had received news since 1453 of Mehmed that challenged his previous image of him as a "most terrible beast."[39]

That is not likely. As I have argued elsewhere, this letter, which was almost certainly never sent, should not be read as a genuine conversion piece. On the contrary, it was a work of rhetoric or even deliberate propaganda to be read by Europeans alone; the letter was first printed around 1470 and received wide circulation in Europe.[40] If nothing else, Aeneas's decision personally to lead the crusade he had spent his papacy planning (he died before it could depart) shows his commitment to destroying the Turks.[41] Indeed, for a letter that was supposed to welcome Mehmed and the Turks to the fold, it is a chauvinistic boast of European superiority and stands as perhaps the strongest example of Aeneas's use of language to polarize Europe and Asia, or West and East. For Denys Hay, the letter reads as "a catalogue not of Christian peoples, but of European Christian peoples," in which Eastern Christians under Ottoman rule are "written off as not true Christians."[42] One of Aeneas's arguments to Mehmed in support of conversion is a warning that Mehmed may have to battle more masculine Western Christians: "You will not fight against women if you invade Italy, Hungary, or other occidental areas."[43] Later he evokes classical stereotypes of Eastern decadence, arguing that the Turks would find western Europeans "better companions than the effeminate Egyptians or unwarlike Arabs."[44]

In addition to faith and manliness, the learning of Westerners is compared to that of Muslims. Only "among us [western Europeans] does the study of liberal arts flourish ... No branch of learning is ignored. Famous literary schools are found in many of Italy's cities. Across the Alps, in Spain, in France, in Germany, in Britain, faculties of excellent men are not lacking who teach wisdom to those who lack it."[45] By way of contrast, Aeneas boldly asserts, "There was once a great and flourishing school of philosophers in Alexandria; many of its learned men whose names have come down to us were known throughout Syria and Asia. But ever since the law of Mohammed won the day, few have attained renown for revealing the secrets of nature." From there, he recites a list of commonplaces about Islam as mired in ignorance and luxury.[46]

This perspective resonates with earlier quotations about 1453 and Aeneas's view of the Turks as inimical to learning. *Europe* and *Asia* also convey Aeneas's sense of the contrast in learning between the two continents. He celebrates schools and scholars in every corner of Europe, but in *Asia* learning is treated as an achievement of the distant, mostly Greek, past. Interestingly, Aeneas mentions Samarqand, praising the city as a symbol of the "Parthian" Tīmūr Lang's power and the opulence of his empire. But instead of noting the great observatory patronized and used by the scholar-prince Ulugh Beg (1394–1449), Tīmūr's grandson, Aeneas ends his account with division and decline under Tīmūr's warring sons.[47] Either Aeneas was unaware of the observatory or chose not to call attention to it. It bears mentioning that some contemporaries, like Francesco Filelfo, uttered brief and intriguing statements about the civility of central Asians, whom they saw as descendants of the

Persians, in contrast to the supposedly rough and uncultivated Turks.[48] But such statements tended to be motivated by hopes for an alliance against the Ottomans and do not stand out boldly against the growing discourse of Europe's pride in its enormous scholarly achievements.

For Aeneas, Europe and the West were more than words: they represented the sense of unity he sought to articulate and increase among fellow Christians of different nations. This unity might radiate out from the Holy Roman Empire or other Christian kingdoms; it should certainly, at the very least, derive from common allegiance to the Roman Catholic Church and the pope. But his hopes for papal and imperial primacy would prove unattainable with the approach of the Reformation and multifront wars against the Habsburgs. Yet in one way, Aeneas's vision took root and survived him for centuries. His notion of Europe as a cultural entity, where similarities outweighed differences, became a concept that appealed to Christians, who increasingly referred to themselves as Europeans.[49] Not only did Aeneas's notions of European identity survive, but so did the sense of opposition to Asia, with Europe always in the position of superiority.[50]

What connections might be drawn between Aeneas's rhetoric on Europe and Asia and the questions I opened with about Copernicus? We need not imagine that hostile attitudes toward the Turks and even open warfare stopped transmission. On the contrary, studies have shown that war throughout the centuries created a zone of cross-cultural exchange and that dismissive cultural stereotyping often masked or compensated for borrowings.[51] Ideas certainly flowed between East and West, but in what form? According to A.I. Sabra, "As for cross cultural transmission, it is clear that its presentation in isolation from cultural factors would remain an incomplete description, one which cannot by itself explain large transformations that frequently occur when cultural boundaries are crossed."[52] In other words, cultural perceptions during any exchange of information can greatly affect reception and alter the concepts along the way. The belief that transmission of ideas denotes respect and admiration for the "donor" culture proves to be too facile, at least in this case, and I suspect in many others.[53] My surmise is that a much more complex transmission took place in western Europe before Copernicus.

Take the example of Cardinal Bessarion, who has been suggested as one possible purveyor of Islamic ideas or manuscripts to western Europe. His visit to Vienna in 1460 inspired Johannes Regiomontanus and Georg Peurbach to take on a new epitome of Claudius Ptolemy's *Almagest*; Bessarion may also have shared ideas with them from central Asian astronomers. Regiomontanus travelled back to Italy with Bessarion, where he had access to his enormous library of Greek and Latin texts.[54] The question we should ask is what other ideas Bessarion may have brought in his "suitcase" along with those theoretical manuscripts or models? He had been sent to Vienna as papal legate by none other than Pope Pius II specifically to promote a crusade against the Ottoman Turks. He, no doubt, had diplomatic orders and prepared speeches for negotiating military and monetary commitments. He also brought his concern for protecting the Greek corpus of learning, which began to fragment as the Ottomans swept over Byzantium.[55] What, then, might his conversations have been like with Regiomontanus and Peurbach if he introduced recent Islamic theories that fell outside the canon of earlier, church-approved authorities? These theories, I might add, likely came through and were added to in the Ottoman Empire. Years earlier, Bessarion reacted emotionally to the fall of Constantinople by describing the Ottomans as the "fiercest of wild beasts" who "consumed the public treasure, destroyed private wealth, stripped temples of gold, silver, and saintly relics."[56] Yet a closer look at Bessarion and other Greek writers such as Doukas and Michael Kritoboulos on 1453 shows that they did not engage in western European tropes of the Ottomans as being anti-learning, suggesting that they knew such characterizations were untrue.[57]

It is difficult to say whether any of the interlocutors in Vienna would have treated recent Islamic astronomical learning with complete urbanity, some measure of discomfort, or even disbelief. It seems possible, given the context, that later Islamic origins of Copernicus's ideas were obscured at some point by Greek refugees who found the provenance a sensitive subject given their adamant calls for crusade and the rhetoric of Ottoman barbarism that was so fashionable in western Europe. It is also possible that Copernicus himself knew the origins and chose not to note them for fear of unpleasantness or a harsh reaction from the papacy. It is equally plausible that this lack of provenance was simply due to an innocent omission at some point in the transmission. But there is one common denominator that emerges despite all of this uncertainty.

Somewhere between the fourteenth and seventeenth centuries, someone or several individuals either deliberately omitted or simply failed to note the origins of later Islamic astronomical models that made their way westward – in striking contrast to the reverence in which earlier medieval Arab authorities were held. Even if it was an innocent slip, it is noteworthy that none of Copernicus's contemporaries or later readers noticed or bothered to mention what many scholars today view as blatant borrowing from recent Islamic sources. To me this lack of comment suggests a kind of transformation, such as Sabra notes above. The ideas travelled westward and were used, but they were changed or cloaked, consciously or unconsciously, perhaps to make them fit with the growing belief among Europeans that their current scholarship had surpassed that of the East. The reality may have been far from the case, but the confidence of this belief was very much in line with Aeneas Silvius Piccolomini's rhetoric.[58]

One might well ask just how influential this humanist rhetoric could have been to scientists. Today we often think of science and the humanities as separate camps and are surprised to find a scholar who shows expertise in both fields, but this was hardly the case in the Renaissance, when it was common to see individuals who excelled in math and science as well as literary studies, such as Marsilio Ficino and Niccolò Tignosi.[59] Greek émigrés, like John Argyropoulos, Cardinal Bessarion, and George of Trebizond, who trained in Latin rhetoric at Italian universities, also provide useful cases. Peurbach, Regiomontanus, and Copernicus had rhetorical as well as mathematical training.[60] All of these scientific thinkers were certainly exposed to humanist diatribes against the Turks, and several of them personally added to that rhetorical corpus.

What I am suggesting is a cultural context that helps to explain how Copernicus's ideas came to be seen as harbingers of a "European scientific revolution," not part of a broader tradition of exchange between Islamic, Latin, Byzantine, and Jewish scholars. As Walter Andrews and Mehmet Kalpaklı argue in a recent study on Ottoman poetry, the sheer military might of the Ottoman Empire often obscured other aspects of this creative and complex society, allowing contemporaries and later historians to view them as "better at war and less good at … culture."[61] The growing cultural discourse of European superiority seen in Aeneas's works may also help to explain why scholars began to give up on the study of Arabic as a catalyst for scientific discovery in the seventeenth century; several scholars who devoted their lives to the study of Arabic appear to have bought into the notion that it was pointless.[62] Aeneas's portraits of Europe and Asia, Christendom and Islam, and civilized versus barbaric can easily be taken apart for their errors, bombast, and oversimplifications, but to contemporaries and several generations of later readers, he was probably the ranking expert on these concepts. Long after his work was surpassed by newer geographical, ethnographical, and historical works, Aeneas's value-laden designations of Europe as an advanced civilization, superior to backward Asia, continued on in scholarly and more popular discourses of identity. This lasting influence, I believe, was an important factor in the evolving perception that Copernicus's theories, and modern science in general, were wholly European inventions.

PART TWO

The Fifteenth-Century European Intellectual and Scientific Contexts

3

The Status of Astronomy as a Science in Fifteenth-Century Cracow: Ibn al-Haytham, Peurbach, and Copernicus

Edith Dudley Sylla

NICHOLAS COPERNICUS'S FIRST KNOWN astronomical writing is *Nicolai Copernici de hypothesibus motuum caelestium a se constitutis commentariolus* (Nicholas Copernicus's Small Commentary on the Hypotheses of Celestial Motions Constructed by Him).[1] In the *Commentariolus,* Copernicus expounds his hypotheses, among others, that the Earth has three motions, whereas the Sun and fixed stars have none.[2] The *Commentariolus,* which is the best evidence we have of Copernicus's conception of astronomy when he first proposed heliocentrism, is a work of the same type as Georg Peurbach's *Theoricae novae planetarum* – that is, theoretical rather than practical, narrative rather than demonstrative, and based on the assertion of hypotheses or principles.[3]

In his very thorough and foundational 1973 article on the *Commentariolus,* Noel Swerdlow writes:

> Now, the principal textbook of planetary theory in Copernicus's time, the *Theoricae Novae Planetarum* by Georg Peurbach, contains descriptions, and very elaborate descriptions they are too, of spherical representations of Ptolemy's planetary models. Peurbach gives exceedingly careful attention to the proper alignment of the eccentric sphere that carries the epicyclic sphere through its proper path, to the inclinations of the axes about which the spheres rotate, and to all the different motions of the spheres and axes required to produce the apparent planetary motions in longitude and latitude, and the precession of the apsidal and nodal lines along with the sphere of the fixed stars. This entire apparatus of spheres and axes, rotations and inclinations, is taken over by Copernicus in the *Commentariolus,* although his description is not as thorough

since the reader's familiarity with such models is taken for granted. Copernicus usually describes planetary motions in terms of rotations of spheres and inclinations of axes.[4]

The genre of Peurbach's *Theoricae novae planetarum* is made explicit by Albert of Brudzewo, master at Cracow University in the late fifteenth century. In his commentary on Peurbach's *Theoricae novae planetarum*, which in its closing statement is also called a *commentariolus*, Brudzewo describes the various genres of astronomical writing.[5] According to the definition of Isidore of Seville, Brudzewo reports, astronomy provides the laws of motion of the heavenly bodies.[6] According to Claudius Ptolemy and Haly (ʿAlī ibn Riḍwān), astronomy has two principal parts, the first concerning the motions of the heavenly bodies and the second concerning their effects on Earth. The first principal part is theoretical or speculative and the second is practical.[7] Speculative or theoretical astronomy is divided into narrative and demonstrative. The *Theoricae novae planetarum* is theoretical and narrative, not demonstrative, he continues, and is intended as an introduction. The *Theorica planetarum* of Campanus of Novara is likewise narrative, Brudzewo goes on, as is the *XXXI Differentiis* of Alfraganus (al-Farghānī). In contrast, Ptolemy in the *Almagest*, Geber (Jābir ibn Aflaḥ), and Albategnius (al-Battānī) provide demonstrative presentations.[8] Works on tables and instruments are parts of practical astronomy. It is worth noticing here how many works Brudzewo includes in his survey that are translated from Arabic.[9]

The earliest recognized source of the configurations of celestial orbs (*theoricae orbium*) in Peurbach's *Theoricae novae planetarum* is Abū ʿAlī al-Ḥasan ibn al-Haytham's *On the Configuration of the World*, a work that was transmitted to Latin-speaking Europe at the latest by the end of the thirteenth century and that could have reached Peurbach by so many alternative routes that it does not make sense to look for a primary route of transmission. The conception of multipart physical orbs carrying around the planets was widely familiar in Europe already by the fourteenth century. It is distinctly different from conceptions found in Ptolemy's *Planetary Hypotheses*, which, in any case, was not known in Europe. But the European descendants of Ibn al-Haytham's configurations also included material that postdated *On the Configuration*, such as inclusion of the apsides of the Sun among astronomical variables affected by the one degree per century motion of the eighth orb, so there was not a single early transmission from Ibn al-Haytham and then further development in Europe, but there must have been later transmissions from the Islamic science of configuration (*ʿilm al-hayʾa*). In this chapter, I argue that Ibn al-Haytham's *On the Configuration of the World* and

Peurbach's *Theoricae novae planetarum* are genealogically linked, while not attempting to trace the descent of ideas within Islamic *hay'a* or the likely multiple migrations of the ideas to Europe beginning in the late thirteenth century before they appear in Peurbach's *Theoricae novae planetarum.* Without attempting further to contribute to the study of transmission, one might immediately suggest the possibility of oral rather than written transmission and consider the many possible intercultural connections over time between those with interests in astronomical or astrological subjects and methods. That the Europeans at some point seem to have stopped making translations from the Arabic to Latin does not mean that other modes of transmission had to end, let alone that astronomical activity in Islamic areas ceased to progress because European interest in it declined.

When the idea of physical orbs spread from Islamic to European areas, the use of Ptolemy's mathematical methods was not abandoned, for the astronomy of physical orbs was incomplete, compelling astronomers to use Ptolemaic mathematics in order to do their work (e.g., compiling almanacs and ephemerides, predicting eclipses, and so forth). So in his commentary on Peurbach, Brudzewo quotes Richard of Wallingford as explaining that, in order to do their work, astronomers were forced to use imaginary mathematical devices, which they in no way thought really existed in the heavens.[10]

It followed that Peurbach's *Theoricae novae planetarum* presented alternate *complementary*, neither unified nor mutually exclusive, physical and mathematical approaches for understanding the motions of the planets. The physical approaches involved real three-dimensional orbs, while at the same time there were mathematical approaches represented in *theoricae* figures as two-dimensional geometrical circles and lines, which were understood not to be real things existing in the external world but to be the products of mathematical construction or imagination. Thus astronomical research could proceed along two simultaneous paths, one mathematical and the other physical.

Islamic astronomers had long followed a program of trying to propose physical bodies that might lie behind the observed motions described mathematically in Ptolemaic astronomy. They had not, however, completed the job of finding a physical configuration consistent with every mathematical regularity. As a result, there was a continuing felt need for new and better physical hypotheses. Astronomers, therefore, could seek to advance independently both mathematical astronomy, on the one hand, and possible physical configurations, on the other.

Existing at the same time as this situation in the discipline of astronomy, there was a wider Averroistic-Aristotelian conception of science according to which there were many autonomous scientific disciplines, none of which needed to be subservient to another, at least at that time, since scientific knowledge was still unfinished. In relation to astronomy in particular, it was possible to accept a situation in which mathematical astronomy, such as contained in the *Almagest*, was considered a tool with proved success and in which physical astronomy involving real, three-dimensional, uniformly rotating aethereal orbs was also considered worthy of attention, despite the fact that existing mathematical astronomy and physical astronomy were, at that given moment, not fully consistent with one another. No one thought that bare mathematical lines were in the heavens, but astronomers had to use mathematical approaches lacking physical support because they had no other way to do their work.

This picture of the components of the conception of astronomy as a science present in Cracow when Copernicus was a student is what I have arrived at after looking for the last several years at the major sources relevant to the background of the *Commentariolus* at Cracow. The *Commentariolus* was composed at a time when Copernicus had not worked out significant parts of the mathematics that would be included in *De revolutionibus orbium coelestium.*[11] At the time of the *Commentariolus*, Copernicus assumed that there were real, three-dimensional orbs made out of aether. He recognized that the existing physical astronomy was a work in progress. He seems to have believed that he had caught a glimpse of a more satisfactory configuration or system of orbs than the ones that had been devised for an Earth-centred system. So he decided to communicate his ideas to some others, short of publication, to test the waters and see what reception his ideas might receive. For the rest of his life, he continued to pursue his ideas until, finally, Georg Joachim Rheticus persuaded him that it was time for him to publish his large work.

COPERNICUS'S *COMMENTARIOLUS* AND PEURBACH'S *THEORICAE NOVAE PLANETARUM*

The *Commentariolus* mirrors the *Theoricae novae planetarum* in starting with a statement of principles. In Copernicus's work these principles are stated postulates (*petitiones*), and in Peurbach's work they are the *theoricae* (figures) themselves, together with their descriptions. Each work then elaborates a configuration consistent with these principles. Copernicus's principles (or postulates) are as follows:

1. There is no one centre of all the celestial orbs or spheres.
2. The centre of the Earth is not the centre of the universe, but only the centre towards which heavy things move and the centre of the lunar sphere.

3. The orbs surround the Sun as though it were in the middle of all of them, and therefore the centre of the universe is near the Sun.
4. … [T]he distance between the Sun and the Earth is imperceptible compared to the great height of the sphere of the fixed stars.
5. Whatever motion appears in the sphere of the fixed stars belongs not to it but to the Earth …
6. Whatever motions appear to us to belong to the Sun are not due to the [motion] of the Sun but [to the motion] of the Earth and our orb with which we revolve around the Sun just as any other planet …
7. The retrograde and direct motion that appears in the planets belongs not to them but to the [motion] of the Earth.[12]

These *petitiones* state guiding principles for constructing a configuration of the world that could account for what astronomers had observed in the heavens and described mathematically. In his *Letter against Werner* (1524) concerning Johannes Werner's treatise *On the Motion of the Eighth Sphere* (1522), Copernicus describes how astronomers worked from measured positions to arrive at their theories:

The science of the stars is one of those subjects which we learn in the order opposite to the natural order. For example, in the natural order it is first known that the planets are nearer than the fixed stars to the earth, and then as a consequence that the planets do not twinkle. We, on the contrary, first see that they do not twinkle, and then we know that they are nearer to the earth. In like manner, first we learn that the apparent motions of the planets are unequal, and subsequently we conclude that there are epicycles, eccentrics, or other circles by which the planets are carried unequally. I should therefore like to state that it was necessary for the ancient philosophers, first to mark with the aid of instruments the positions of the planets and the intervals of time, and then with this information as their guide, lest the inquiry into the motion of heaven remain interminable, to work out some definite planetary theory, which they were seen to have found [*quam tum visi sunt invenisse*] when the theory agreed in some harmonious manner with all the observed and noted positions of the planets.[13]

Thus in the *Commentariolus* Copernicus's *petitiones* represent hypotheses derived from experience, which are to be accepted as true, even though they could be wrong given that astronomy is a science still in the process of development. The currently existing theories may not be perfect, but they have been adopted for the present "lest the inquiry into the motion of heaven remain interminable."

In Peurbach's *Theoricae novae planetarum*, the parallel to Copernicus's postulates are Peurbach's statements about the orbs of the planets and the figures (*theoricae*) that represent them, starting with the Sun:

The sun has three orbs, separated from one another on all sides and also contiguous to one another. The highest of them is concentric with the world on its convex surface, but is eccentric on its concave surface. The lowest, on the other hand, is concentric on its concave but eccentric on its convex surface. The third, however, situated in the middle of these, is eccentric to the world on both its convex surface and its concave surface … Therefore, the first two [orbs] are eccentric relatively, and they are called the deferent orbs of the apogee of the sun. The apogee of the sun varies according to their motion. The third is eccentric absolutely and is called the deferent orb of the sun. The body of the sun is attached to it and moves indeed according to its motion. These three orbs take two centres. For the convex surface of the highest and the concave of the lowest have the same centre, which is the centre of the world. From that fact the whole sphere [*tota sphaera*] of the sun, just as the whole sphere of any other planet, is said to be concentric with the world. But the concave surface of the highest orb and the convex of the lowest, together with the surfaces of each side of the middle orb, share another center, which is called the centre of the eccentric.[14]

Peurbach then goes on to describe and to represent by a figure (i.e., the *theorica orbium*) the basic configuration for the Sun, which is repeated with variations for the other planets (see figure 6.2). These whole three-dimensional, three-part spherical shells, or orbs, are the identifying DNA, so to speak, of the configuration that Peurbach's *Theoricae novae planetarum* shares with Ibn al-Haytham's *On the Configuration of the World*, a configuration that by Peurbach's time had been shared between European authors since at least the late thirteenth century – so, for instance, a passage very similar to this long and detailed paragraph about the three component orbs of the whole orb of the Sun can be found in John of Sicily's *Scriptum super canones Azarchelis de tabulis Toletanis* (ca. 1290).[15] John of Sicily's description of the orbs obviously belongs to the family tree linking Ibn al-Haytham and Peurbach even if it is not in the direct line linking the two.

Within each whole thick spherical shell of this configuration, there is an eccentric deferent just thick enough to contain the Sun (or the epicycle of a planet), which touches the convex and concave surfaces of the whole orb at two opposing points. Inside and outside of this eccentric deferent, there are unevenly thick orbs that not only fill out the rest of the volume of the whole orb but also carry around the apsides (i.e., apogee and perigee) of the planet west to east, around an axis parallel to the axis of the ecliptic, at a rate of one degree every hundred years. These outer shells are called the deferents of the apogee or the deferents of the apsides of the eccentric of the planet. They are not simply there filling up space but have a function of moving the apsides west to east at the very slow rate of one degree per century. Between these two outer suborbs, there is a fixed cavity within which the eccentric deferent of the Sun (or of the epicycle and planet) rotates continuously at a uniform velocity into itself, never impinging upon the deferents moving its apsides.[16]

These orbs are rigid, not fluid. Edward Grant is in error in his *Planets, Stars, and Orbs* when he argues that orbs were not thought to be rigid until the sixteenth century.[17] If Grant had followed the development of configuration astronomy, or *ʿilm al-hayʾa*, from Ibn al-Haytham through its transmission into Europe, he would not have made this mistake. For instance, in his *Al-Tadhkira fī ʿilm al-hayʾa*, which built upon Ibn al-Haytham's *On the Configuration of the World*, Naṣīr al-Dīn al-Ṭūsī wrote, "Nothing having the principle of circular motion can undergo any rectilinear motion at all, and conversely, except by compulsion. Thus the celestial bodies neither tear nor mend, grow nor diminish, expand nor contract: neither does their motion intensify nor weaken. They do not reverse direction, turn, stop, depart from their confines, nor undergo any change of state except for their circular motion, which is uniform at all times."[18] Instead of paying attention to Ibn al-Haytham, Grant credits the idea of three-part orbs to Ptolemy's *Planetary Hypotheses* and makes little use of works with the title *Theorica planetarum* in his research:

> The old *Theorica planetarum*, probably composed in the latter half of the thirteenth century, includes nothing relevant to cosmology, omitting even discussion of the orbs. Apart from a discussion of the order and distances of the planets, there is little that is cosmological in Campanus of Novara's similarly named *Theorica planetarum*. The same judgment applies to Georg Peurbach's *Theoricae novae planetarum* of 1460–61, which is virtually devoid of cosmological content, although Peurbach does discuss the orbs as if they were real physical bodies. Bernard of Verdun chose to include more cosmology than most astronomers. In his *Tractatus super totam astrologiam*, he devoted the first eleven brief chapters to cosmological themes, approximately 4 percent of the entire treatise.[19]

When he comes to the introduction of what should be recognized as Ibn al-Haytham's configuration, Grant writes,

> References to epicycles and eccentrics appear in widely used thirteenth-century works like [Johannes de] Sacrobosco's *On the Sphere* [*of the World*] and in the anonymous *Theorica planetarum*, although neither author implies or suggests that they might be real, material, solid orbs. If Roger Bacon (ca. 1219–ca. 1292) was not the first to mention material eccentrics and epicycles in the Latin West, he may well have been the first scholastic natural philosopher to have presented a

serious evaluation of their cosmological utility. After some hesitation and ambivalence, Bacon rejected physical eccentrics and epicycles and opted for Aristotle's system of concentric spheres.[20] Ironically, it was his description of the system of eccentrics and epicycles that was most widely adopted by medieval natural philosophers and which still found defenders well into the seventeenth century.[21]

Actually, Roger Bacon did not grasp all of the features of the three-orb model – for instance, that the orbs surrounding the eccentric deferents of the planets move one degree per century with the motion of the eighth sphere. Bacon called the three-orb model the "imaginatio modernorum," leaving it open to speculation whom he meant by moderns. John of Sicily's more organized and complete statement of the theory in his *Scriptum super canones* is a much more likely route for the spread of *'ilm al-hay'a* into Europe.[22]

After a long list of writers who gave the view described by Bacon "more than a cursory glance," Grant goes on to describe the three-orb view in detail: "I shall frequently refer to the 'modern' theory as the 'three-orb system,' but I shall also refer to it as the 'Aristotelian-Ptolemaic system,' since Aristotle's concentric spheres were assigned a significant role within the system of eccentrics … Bacon introduces another interpretation – 'a certain conception of the moderns,' as he put it – in which the external surfaces of each planetary orb are concentric but which contain at least three eccentric orbs."[23] Despite Grant's Latin-language perspective, the authors that Bacon called "the moderns" should be understood to be Islamic astronomers in the *'ilm al-hay'a* tradition or those authors writing in Latin who took up their views.[24] Among scholastics who accepted the physicalized three-orb system, Grant lists Albertus Magnus, Duns Scotus, Aegidius Romanus, and Durandus de Sancto Porciano.[25] He then describes in more detail how the system was defended by Pierre d'Ailly in his *14 Questions on the Sphere of Sacrobosco* and by many others in succeeding centuries.

It is clearly the case in Ibn al-Haytham's *On the Configuration* and in the Islamic *hay'a* tradition generally that the orbs are rigid bodies and that the included deferents, epicycles, and planets are held tightly in place, except that they can rotate uniformly, never exceeding the place or cavity they are in. Moreover, these orbs spin. The Latin *orb* is a good translation of the Arabic *falak*. Not only are these orbs and suborbs geometric spherical shells, but they are also physical: they uniformly spin, and there are movers (i.e., separate or immaterial substances or intelligences) that cause their spinning.

Recently, historians such as Peter Barker and Michela Malpangotto have supported the view that Peurbach and Copernicus accepted the existence of orbs.[26] In the meanwhile, however, the connection of Peurbach's orbs to Ibn al-Haytham's orbs, known perfectly well in 1976 to Swerdlow, has been neglected, undermining the correct interpretation of the place of orbs in sixteenth-century astronomy.

In the *Commentariolus*, Copernicus is commenting on the hypotheses that he himself has set out (*de hypothesibus motuum caelestium a se constitutis*), whereas in commenting on Peurbach, Brudzewo states what he plausibly understands to be the principles of Peurbach's *Theoricae novae planetarum.* According to Brudzewo, they are as follows:

1. The heaven is a simple body.
2. Of any simple body there is only one simple motion proper to it naturally.
3. A motion belonging to one body unnaturally, necessarily is proper to another body naturally.
4. One orb is not moved with several motions by the same intelligence, nor is the same orb moved by several intelligences equally proper to it.
5. To this may be added that a lower sphere does not influence the motion of a sphere above it, but rather the opposite, the superior influences the inferior. Not everything in a lower orb that is derived from a higher orb is natural to it, but it may belong to the lower sphere beyond [*praeter*] nature, because the lower orb has a nature obedient [*oboedientialem*] in everything, by which it obeys the higher spheres. Therefore, the higher orbs may influence by their motion the lower orbs and carry them around, whereas the opposite does not happen.[27]

These principles have a relation to Peurbach's *Theoricae novae planetarum* similar to the relation of Copernicus's *petitiones* to his *Commentariolus.* They derive ultimately from thinking about observations and how they could be explained by underlying reality. They are physical principles rather than mathematical ones.

Thus an idea about the nature of astronomy as a science that Copernicus was exposed to when he was a student at Cracow was that at least one genre of astronomical writing (works that are theoretical but not demonstrative) can start by stating principles or postulates upon which the following exposition will be based. This format could also be found within the determinations of scholastic questions in which, after stating principal arguments for one conclusion, the author mentions an opposite solution and then starts a determination of the question with definitions, suppositions, and the like.

In fact, the fourth and fifth of the principles that Brudzewo states for Peurbach's *Theoricae novae planetarum* are principles found earlier in Albert of Saxony's *Quaestiones subtilissime in libros de caelo et mundo*, where, as is typical, he states principles or suppositions before answering questions. In this case, Albert of Saxony's question is how many celestial spheres or orbs there are.[28] After some discussion, he proposes the opinion that there are nine orbs, starting with the suppositions that lead to this conclusion:

For the proof of this position, first it is supposed that the eighth sphere or orb is moved with more than one motion: it has one motion from east to west on the poles of the world [once a day], and another motion from west to east on poles of the Zodiac, one degree in a hundred years. The second supposition is that a single orb is not moved with several motions by the same intelligence, nor is the same orb moved by several intelligences equally proper to it. Aristotle proves this in the *Metaphysics*, Book XII.[29] The third supposition is that a lower sphere does not move with itself a higher sphere, but a higher sphere does move a lower sphere.[30]

Later, in *14 Questions on the Sphere* by Pierre d'Ailly (1350–1420), also in a question about the number of the orbs, a similar set of suppositions appears.[31]

Thus Albert of Saxony, Pierre d'Ailly, and others before Brudzewo put their theories or parts of their theories into a structure in which there are suppositions, principles, or premises (i.e., hypotheses) on which conclusions are based. These principles are typically physical rather than mathematical, and they are thought to derive from experience. They are not demonstrated but are the foundations of demonstrations.

Knowing that the principles are not proved and that the processes by which they are arrived at are not logically rigorous, practitioners of the discipline of astronomy could in exceptional circumstances think that a reformation of principles was called for. This is what Copernicus says at the beginning of the *Commentariolus*:

I understand that our predecessors assumed a large number of celestial orbs (*multitudinem orbium coelestium*) principally in order to account for the apparent motion of the planets through uniform motion … the stronger opinion (*potior sententia*), in which the majority of experts finally concurred, seemed to be that it is done by means of eccentrics and epicycles. Nevertheless, the theories concerning these matters that have been put forth far and wide by Ptolemy and most others, although they correspond numerically [with the apparent motions], also seemed quite doubtful, for these theories were inadequate until they also envisioned certain equant circles, by which it appeared that the planet never moves with uniform velocity either in its deferent sphere or with respect to its proper center. Therefore, an opinion of this kind seemed neither perfect (i.e., complete – *absoluta*) enough nor sufficiently in accordance with reason. Therefore, when I noticed these [difficulties], I often pondered whether perhaps a more reasonable way [*modus*] composed of circles could be found from which every apparent irregularity would follow while everything in itself moved uniformly, just as the principle of perfect motion requires. After I had attacked this exceedingly difficult and nearly insoluble problem, it at last occurred to me how it could be done with fewer and far more suitable things [*rebus*] than had formerly been put forth if some postulates called axioms are conceded to us which follow in this order.[32]

And then Copernicus lists the seven postulates (*petitiones*) quoted earlier in this chapter. That Copernicus, like the authors of *theoricae planetarum*, starts with physical principles supports the contention that he conceived his research program within the *theorica planetarum* genre.

Thus I claim that as a student at Cracow where Peurbach's *Theoricae novae planetarum* was a model for the status of astronomy as a science, Copernicus would have learned that astronomy was both mathematical and physical and that, although it had many real achievements, it might still be improved by new insight into the hidden physical structures behind the appearances. Whether or not Copernicus was aware of it, the conception of astronomy as a science reflected in Peurbach's *Theoricae novae planetarum* derived from Islamic *hay'a* astronomy going back to Ibn al-Haytham's *On the Configuration of the World*, to which I now turn.

PEURBACH'S *THEORICAE NOVAE PLANETARUM* AND IBN AL-HAYTHAM'S *'ILM AL-HAY'A*

The *Theoricae novae planetarum* and the *Commentariolus* not only presuppose the research tradition based on Ptolemy's *Almagest* but also presuppose the Islamic tradition of *hay'a* astronomy. An ancestor of Peurbach's *Theoricae novae planetarum* (and hence of Copernicus's *Commentariolus*) is Ibn al-Haytham's *On the Configuration of the World*.[33] In one Arabic manuscript of Ibn al-Haytham's text, although not in the Latin or Hebrew translations, there are principles like those in fourteenth-century authors and in Peurbach. These principles may in some way have affected the statements of principles in Brudzewo and other Latin authors.

Historians of astronomy looking for the origins of the concept of thick physical celestial orbs have typically suggested that it derives from the orbs of Ptolemy's *Planetary Hypotheses*, but Ptolemy himself, having first described three-dimensional orbs, then decided instead in favour of rings, tambourines, or "sawed-off-pieces," and it is for the latter that he was known in the Islamic world. Moreover, Ptolemy suggested that these rings could interpenetrate and that the planets were the movers of their orbs rather than vice versa, both rejected by Peurbach.[34]

Unlike Ptolemy's *Planetary Hypotheses*, Ibn al-Haytham's *On the Configuration of the World* was translated and assimilated over time, so evidence of its reception in the West is widespread, although it is difficult to distinguish what comes from Ibn al-Haytham and what comes from other parts of the Islamic tradition of physical representations of celestial orbs, or *hay'a*. As F. Jamil Ragep defines it, "The Arabic term *hay'a* had several distinct significations when used in the medieval astronomical context. The basic meaning is 'structure' or 'configuration' … An astronomer writing in the *hay'a* tradition was charged with transforming mathematical models of celestial motion, usually those of Ptolemy's *Almagest*, into physical bodies that could be nested, along with the sublunar levels of the four elements, into a coherent cosmography (*hay'a*)."[35] As to the influence of Ibn al-Haytham's work and the *hay'a* tradition in Europe, Ragep states, "Given that the *Planetary Hypotheses* was unknown during the Latin Middle Ages, it is clear that the main source for the European *theorica* tradition was Islamic *hay'a*; Ibn al-Haytham's *On the Configuration of the World* was certainly quite influential."[36] Robert Morrison writes, "The best known achievement of Islamic astronomers from the thirteenth century onward was the creation of physical models that could represent available observations. In fact, the genre in which these astronomers wrote, *'ilm al-hay'a* (astronomy, literally 'science of the configuration') was a product of Islamic civilization."[37]

What about evidence for the connection of Peurbach's *Theoricae novae planetarum* to the longstanding history of *'ilm al-hay'a* in Islamic astronomy? In Peurbach's *Theoricae novae planetarum*, the word "theoricae" may be understood to refer to the figures in the book representing astronomical hypotheses or parts of theories, many of which include the word *theorica* in their titles. Some of Peurbach's *theoricae* represent physical bodies called orbs, whereas other *theoricae* represent mathematical lines or motions with no pretense of being bodies. Typically, between the first *theorica* representing coloured physical orbs of unequal thickness and the third *theorica* representing circles and the motions on them in two dimensions, there is a second *theorica axium et polorum*, which by means of lines and circles shows the axes and directions of rotation of the physical orbs. The distinction between the physical and the mathematical may have been made clearer in the printed versions of the

Theoricae novae planetarum by their coloured representations of the partial orbs, in contrast to the narrow black lines used for the mathematical *theoricae*, but the distinction was not new.

In any case, the relation of the physical and the mathematical in this sort of astronomy can be appreciated better by seeing its roots in Ibn al-Haytham's *On the Configuration of the World* and subsequent *hay'a* astronomy. Peurbach's *theoricae novae* have many features linking them to the configurations found in Ibn al-Haytham's *On the Configuration of the World*, especially the postulation that the "complementary" orbs surrounding each planet's eccentric deferent move the apsides of the deferent with the slow motion of the eighth sphere, most commonly one degree per century. Importantly, the three-part orbs are concentric with the world on their outside and inside surfaces, thus avoiding the problem that epicycles might be thought either to stick out and conflict with nearby deferents or to require that there be a vacuum or compressible material between celestial spheres in order to allow room for rotations of spheres with epicycles protruding (see figure 6.2).

In the introduction to the *On the Configuration of the World*, Ibn al-Haytham states the motivation of his work, which is to find physical configurations that might lie behind mathematical astronomy in the Ptolemaic tradition. The introduction is not included in the only known complete translation of the *On the Configuration of the World* into Latin,[38] but it is present in thirteen of the fifteen manuscripts of the work in Hebrew found in European libraries, so it may well have entered the European context through Jewish intermediaries.[39] Ibn al-Haytham's goal is to associate each of the motions that Ptolemy describes with a real, solid, uniformly rotating orb that does not conflict with any other orbs that may exist. Like others in this genre, Ibn al-Haytham is going to describe the bodies and motions in the heavens without attempting to demonstrate them. The audience for his book is people who desire a rapid way of learning the basic facts of astronomy. Ibn al-Haytham carries his program out most completely for the Sun. As far as I can judge, he completely succeeds in carrying out his program only for the Sun, for which Ptolemy had not resorted to the equant. For the three outer planets and Venus, both Ibn al-Haytham and Peurbach succeed in embedding the planet within the epicycle and embedding the epicycle within the deferent – making a consistent combination of orbs explaining a good part of the motions, as was the case for the Sun, with the addition of the epicycle containing the planet. Nevertheless, Ibn al-Haytham has something corresponding to Ptolemy's equant for the planets, and Peurbach likewise has equants in the mathematical *theorica linearum et motuum*. Thus astronomers working in the *hay'a*, configuration, or *theorica* traditions had arrived at a view of astronomy that included both physical configurations and mathematics functioning in a complementary way. Late-fifteenth-century conceptions of scientific disciplines, for theoretical rather than practical reasons, supported this conception of a diversity of approaches within astronomy.

CONCEPTS OF SCIENCE AS A DISCIPLINE IN THE LATE FIFTEENTH CENTURY: *ANTIQUI* AND *MODERNI*

Much of historians' efforts to understand Copernicus's position on the status of astronomy as a science has concerned his *De revolutionibus orbium coelestium* and its reception. A common conception is that medieval astronomers following Ptolemy saw their task as being limited to saving the phenomena by making use of deferents, eccentrics, epicycles, and even equants or other mathematical devices for which they did not claim physical reality.[40] Then (Andreas Osiander's introduction notwithstanding) Copernicus in *De revolutionibus* asserted that his heliocentric (or heliostatic) planetary system represented the real system of the world. In the years that followed the publication of *De revolutionibus* in 1543, according to this commonly received view, many astronomers following what is called the "Wittenburg interpretation" made use of Copernicus's mathematical techniques, while still following the old conception of astronomy as a science, claiming only to save the phenomena, not to assert the reality of a Sun-centred system.[41] Only with Johannes Kepler, according to this view, was a unified physical and mathematical astronomy proposed.[42]

This chapter, however, seeks not to track the conceptions of astronomy as a science that existed after the publication of *De revolutionibus* in 1543 but to examine the conceptions of astronomy that were common in Cracow University at the time that Copernicus was a student, considering how these conceptions may have shaped Copernicus's *Commentariolus*. At that time, as already indicated, Peurbach's *Theoricae novae planetarum*, the main astronomical textbook of the day, already attempted to provide a physical basis for as much as possible of Ptolemaic astronomy. Although late-fifteenth-century celestial physics was not the same as the physics of sublunar bodies – it described the uniform rotations of bodies made of aether rather than the rectilinear motions of sublunar elements and compounds – it was a physics with movers as well as moved bodies. Astronomy in the late fifteenth century did include equant circles, which were purely mathematical and not paired with physical bodies, but this could be understood to be the case because the program of supplying a physical basis for Ptolemaic astronomy had yet to be brought to a successful completion.

One reason why many historians continue to assume that the astronomers who received and reacted to Copernicus's *De revolutionibus* in the way that they did were typical of the whole later medieval period is that they find similar positions among Aristotelians of the thirteenth century and among those of the later sixteenth century. It was the case, however, that medieval Aristotelianism went through alternating phases of conservativism or traditionalism, following the so-called *via antiqua*, and progressivism or renovation, following what was called the *via moderna*. The dominant form of Aristotelianism present during the decades of reception of *De revolutionibus* reflected the decisions of the Council of Trent and the Catholic Counter-Reformation, which advocated a return to the Christian Aristotelian synthesis found in the work of Thomas Aquinas and of other thirteenth-century thinkers such as Albertus Magnus and Aegidius Romanus. The advocates of this form of Aristotelianism were later called *antiqui*, as contrasted with fourteenth-century Aristotelians such as William of Ockham, John Buridan, Albert of Saxony, Marsilius of Inghen, and others who were called *moderni*.[43]

On the basis of the evidence described here, I argue that the conception of astronomy as a science that Copernicus encountered as a student at Cracow University, the one reflected in the *Commentariolus*, was closer to the attitudes of the *moderni* than to those of the *antiqui*. From the fourteenth century up through Copernicus's time at Cracow, the conception of celestial orbs as physical rather than only mathematical was widely held. More important, astronomy was conceived as a progressive scientific discipline in which principles were derived a posteriori from experience and hence could be rederived from new or added experience.[44] The conception of science in general and of astronomy in particular can be seen not only in works of astronomy but also in commentaries on the works of Aristotle, particularly in commentaries on the *Posterior Analytics*. Although the adoption of physical orbs as well as purely mathematical methods of calculation may have developed for internal reasons within the discipline of astronomy proper, corresponding attitudes toward the relation of mathematics and physics were supported by what was written in commentaries on Aristotle's *Posterior Analytics*.[45] Although I think the conception of astronomy in the *Theoricae novae planetarum* was probably of greater influence on Copernicus's *Commentariolus* than were commentaries on Aristotle, the background of Aristotelian philosophy at Cracow also helps to explain why Copernicus might have proposed a new configuration of the world in the *Commentariolus*.[46]

Aristotle had understood astronomy to be intrinsically both mathematical and mechanical, similar to what was understood to be the case for terrestrial mechanics in the pseudo-Aristotelian *Mechanica*, where the properties of a rotating rigid body are used to explain the law of the lever.[47] In the *Mechanics*, the mathematical and the physical are inextricably bound together in the analysis of simple machines. The same could be the case in astronomy, where, rather than planets moving on elliptical orbits through space, one has the planets carried around by rigid orbs or aether shells, constraining their motions to the possibilities for a rigid orb rotating in place with uniform velocity.

If late-medieval astronomy was both mathematical and physical or mechanical, at the same time mathematics was conceived to be a product of the human mind rather than referring to quantitative forms really existing in the external world. In book 6 of the *Physics*, Aristotle had argued that all kinds of continua – geometric, corporeal, temporal, and so forth – are isomorphic. Thomas Bradwardine, in the fourteenth century, still assumed the same isomorphism in his *De Continuo*.[48] Averroes (Ibn Rushd) had argued, however, that physical continua were not necessarily isomorphic to geometric continua. The latter were assumed to be divisible *ad infinitum*, but physical continua might be composed of atoms, or *minima naturalia*, and therefore might have a limit to their divisibility.[49] Like Averroes, the late-medieval *moderni* did not assume that geometrical truths must always be consistent with physical truths or vice versa.[50]

Aristotle's conception of demonstrative science, as set out in his *Posterior Analytics*, posits that demonstrative sciences have structures similar to Euclidean geometry. They have principles (i.e., axioms, postulates, definitions, suppositions, or hypotheses) on the basis of which conclusions are demonstrated. In some cases, a scientific principle may be known in itself (*per se nota*), such as the principle that the whole is greater than the proper part, but in other cases a principle may be known from experience, such as the principle that fire is hot or the principle that the stars are on a sphere rotating around the Earth once a day. For Aristotle, there are many distinct and autonomous scientific disciplines, but in a few cases, including astronomy, one scientific discipline may be "subalternate" to another. Astronomy is subalternate to geometry, and music or harmonics is subalternate to arithmetic, in the sense that the subalternated science makes use of arguments from the scientific discipline subalternating it. For the *moderni* as a rule, astronomy was not subalternate to natural philosophy.[51]

FOUR POLARITIES BETWEEN THE *ANTIQUI* AND *MODERNI*

Aristotelian commentators over time developed many different interpretations of the Aristotelian conception of scientific knowledge. A first polarity within medieval views on demonstrative science concerns the emphasis on necessity and certainty in science as opposed to a conception of science as developing over time. According to a taxonomy of senses of "science" frequently quoted from Robert Grosseteste's commentary on Aristotle's *Posterior Analytics*, there are four levels of scientific knowledge. Science, in the fourth and most proper sense, is the apprehension of a necessary truth, of which the necessity is known by its cause, and in this way we know only the conclusions of demonstration. Science, in a broader sense, however, includes apprehension of contingent truths or of principles of demonstration.[52]

In the thirteenth century and again in the sixteenth, the *antiqui* paid attention to Aristotle's view that, in the most proper sense, science is demonstrative, universal, and necessary. In commenting on Aristotle's *Physics*, Averroes had, however, modified what Aristotle had said by proposing a distinction between mathematics and empirical science, saying that physics differs from geometry because in geometry the same thing (e.g., an axiom) is better known in itself and to us, but in physics, what is better known in itself or to nature (e.g., the causes) is not better known to us because we observe the effects before knowing their causes. For example, astronomers first know by observation that lunar eclipses occur (this is the effect); then they reason to causes of eclipses, namely that the Moon shines by reflected sunlight and that an eclipse occurs when the Moon moves into the Earth's shadow (this is the cause). Once the cause of eclipses is known, an astronomer can predict or demonstrate that a lunar eclipse will occur when, by the relative positions of the Sun, Moon, and Earth, it follows that the Moon will be in the Earth's shadow.[53]

What does such a history of the discovery of the causes of eclipses mean with regard to the status of the science of eclipses? A commentator following the *via antiqua* would likely hold that the demonstration that an eclipse will occur based on its causes (i.e., a demonstration *propter quid*) is certain in Grosseteste's fourth, most proper sense. He would understand this to mean that the principles or premises of such a demonstration are known for certain to be true; some would even say that they are known per se or a priori. In contrast, a commentator following the *via moderna* might explain that astronomers do not doubt the principles of such a demonstration because the practitioners of any demonstrative science do not doubt their principles – there are no more certain principles in the science by which the principles at issue could be demonstrated – but, in fact, the principles of eclipse science could be false. For the *moderni*, most of the principles of the natural sciences were in fact thought to be based on experience, not on a priori or innate knowledge. Even with respect to a mathematical science like geometry, John Buridan, a quintessential *modernus*, went so far as to say that the principles or postulates of geometry itself are not certain. Thus for Buridan, it was not necessarily true in all sciences that a continuum is divisible in infinitum (as a principle of geometry states) since physicists might discover that the world is made up of indivisible atoms. Nevertheless, geometers do not doubt divisibility in infinitum because, if it were not true, many of their theorems would be false.[54]

Consistent with this first polarity, a second, related polarity between the views of the *antiqui* and *moderni* concerns the certainty of the method by which the principles of an empirical science come to be known. The *moderni* followed Averroes in believing that, in contrast to mathematics, in empirical sciences, there is a two-fold process in which scientists first work a posteriori from observed effects to causes and then later, once the causes have been established, work *a priori* from causes to effects. They call a demonstration from effects to causes a demonstration *that* or *of fact* (*quia*). They call a demonstration from cause to effect a demonstration of the reasoned fact (*propter quid*). But must one always begin from sense, memory, and experience, or might there be arguments entirely a priori? Might there be a third kind of demonstration (*potissima*, most powerful) that could at the same time demonstrate the fact and the reason for the fact? Avicenna (Ibn Sīnā) is said to have held that there is only one species of demonstration, the *propter quid* demonstration; Averroes is said to have held that there are three species of demonstration: *quia*, *propter quid*, and *potissima*, the last of which proves both the cause and the existence of the effect.[55]

Related to these questions is the question of how, in fact, demonstrations *that* or *quia* leading to principles are supposed to work. Clearly, principles cannot be demonstrated syllogistically within the given discipline; otherwise, they would not be principles or there would be circular demonstrations. Walter Burley argued for confidence in science derived from experience on the grounds that humans have the natural ability to understand rationally what they observe, just as fish have the ability to swim and birds to fly. If such an ability is denied to humans, Burley said, what justifies their high place in the cosmos?[56] By the time of Giacomo Zabarella (1533–89), among the most prominent of the Renaissance Aristotelians, who in this respect belongs among the *antiqui*, there was great concern about what was called *regressus*, or a multistage process by which, after causes have been found a posteriori by a demonstration *quia*, they can somehow be stabilized or proved true before being used as the premises in a *propter quid* demonstration.

Buridan, for his part, responded to arguments that the process of deriving principles from experience is not certain by saying that anyone who was not willing to accept theories that were true only "for the most part," rather than always and necessarily, did not deserve to take part in natural philosophy, where this uncertainty is a condition of the work.[57] In the second question of book 2 of his *Metaphysicen Aristotelis*, Buridan argued against the view (defended by Nicholas of Autrecourt) that all principles have to be traced back to the principle of noncontradiction.[58] In empirical sciences, Buridan said, many principles are based on sense, memory, and experience, and there may be as many principles as there are conclusions. We do not doubt these principles based on sense experience, but they cannot be demonstrated. Rather, we assent to a principle such as "fire is hot" because we have always observed it to be the case and have never observed an exception. We have developed a habit of assenting to the principle. Nevertheless, we could begin to worry that a principle we have thought was certain is not certain. Here, Buridan gave an example found in more than one place in his work, where old women are led to doubt a first principle. Buridan first asks them whether it is possible for them to sit and not sit at the same time. They say it is impossible. Then he asks whether they believe that God could do it, and they answer that they do not know.[59]

In the *Prior Analytics*, Aristotle himself had said that most of the principles of natural science are based on experience: "But in each science the principles which are peculiar are the most numerous. Consequently, it is the business of experience to give the principles which belong to each subject. I mean for example that astronomical experience supplies the principles of astronomical science; for once the phenomena were adequately apprehended, the demonstrations of astronomy were discovered. Similarly, with any other art or science. Consequently, if the attributes of the things are apprehended, our business will then be to exhibit readily the demonstrations."[60] Generally speaking, in discussions of learning from experience in science, we now think of induction, where by induction we have in mind collecting a body of data and then reasoning from the data to the generalization that encompasses these data. In Aristotle and in the Middle Ages, however, the process was described as moving from sense, to memory, and then to "experiment," where experiment means grasping clearly in the mind what has been sensed. An astronomical example would be observing the positions and motions of the stars, noticing that they all move together without changing their positions relative to one another, and then concluding that the stars are attached to or embedded in a rigid rotating sphere. Thus the process involved insight, not the collection and methodical analysis of sets of data.

Thus, on this second point, the *antiqui* differed from *moderni* in their estimation of the certainty of the principles of natural science. The *antiqui* concentrated on the ideal demonstrative science, whereas the *moderni* had in mind the scientific disciplines as they currently existed. This focus does not mean, however, that the *moderni* were "skeptics"; this label – which, to me, has negative connotations – was placed on them by *antiqui*, whose goal was certain knowledge. They are better understood as critical realists; they aimed for true theories and they had a high opinion of the success of natural philosophy, but they were not dogmatic.[61] The current theories, although the best available, might be wrong.

A third polarity between *antiqui* and *moderni* has to do with their conceptions of the relations of scientific disciplines to each other. Aristotle generally held that there are many scientific disciplines, each with its own principles, subject matter, and conclusions, and he argued that demonstrations should not mix concepts or terms from different disciplines. Geometry as a demonstrative science is separate from the demonstrative science of physics. It follows that Aristotle would have rejected René Descartes's later claim that he could deduce all of physics a priori from mathematics, Aristotle's grounds being that this deduction would involve an illegitimate transgression of disciplinary boundaries (*metabasis*).[62] The *antiqui*, by contrast, tended to the view that scientific disciplines could and perhaps should be synthesized or at least coordinated under metaphysics into a unified worldview, one that is sometimes called by historians the "Christian-Aristotelian synthesis."

In how many subject matters is demonstrative scientific knowledge possible? In the *Nicomachean Ethics*, Aristotle had said that it was foolish to expect scientific demonstrations in ethics, the subject matter being far too complex; at best, only probable arguments could be obtained.[63] In other sciences, however, it might be difficult but not impossible in the long run to develop a demonstrative science. In the case of such a complex subject matter as meteorology, Aristotle said that a scientist may have to be satisfied with a hypothesis that saves the phenomena and is *at least possible*, containing nothing impossible or self-contradictory.[64] At the time of Copernicus, eccentrics and epicycles had been shown to be possible if they were embedded within thick concentric orbs. Equants, however, had not been shown to be physically possible. Because of the observed regularity of celestial motions over long periods, from which it had been inferred that celestial bodies are incorruptible and celestial movers infatigable, it was taken to be more likely that astronomy could reach high levels of scientific certainty than meteorology.

Taken together, what did these opinions mean about the present state of astronomy for the given practitioner around 1500? In the Aristotelian scheme, the practitioner of a given science is expected to take the principles of his science as true; they are the ultimate basis of proofs and cannot be proved themselves. When, however, a natural science is in the process of being established on the basis of observation, the most obvious facts of observation – on the basis of which the scientist reasons a posteriori to causes – should not be denied. In this sense, a physicist does not need to dispute with a follower of Parmenides who denies that anything moves because the most obvious empirical truth of physics is that all or some things move.[65] These are Aristotelian ideals of demonstrative science, but a given science at a given point in time may not have reached the ideal. Since this is the case, it is always possible that in the future someone will be able to propose better or more satisfactory principles or hypotheses for the given science. In the 1270s and 1280s, at the University of Paris and at Oxford University, there were lists of condemned theses, many of which seemed to assert that God could not contravene the truths of Aristotelian physics – saying, for instance, that God could not cause a vacuum to exist because vacuums are physically impossible. After the condemnations, it became normal to distinguish between what is naturally possible and what would be supernaturally possible if God so chose. Thus it became common to discuss conceptions of the world differing from Aristotle's.

On the one hand, then, the *antiqui* and *moderni* generally agreed with Aristotle that ideally scientific disciplines should demonstrate conclusions on the basis of principles and that our ideas of the world originate in sensation. On the other hand, *antiqui* and *moderni* tended to differ in that the *antiqui* emphasized that science should be universal, necessary, and certain, as well as that the disciplines should be consistent with each other, coordinated under the discipline of metaphysics. In contrast, the *moderni* took scientific disciplines to be distinct rather than tied together in a single overarching worldview, and they took them to be not absolutely certain but still in progress, true for the most part, but not perfected.

What would happen if a conflict arose between an established theory and observation? In the thirteenth century, Roger Bacon considered one possibility. In responding to Averroes's rejection of epicycles and eccentrics on the grounds that they conflict with truths of natural philosophy, although the observation that planets are sometimes nearer and sometimes farther seems to require them, Bacon supposed that those like Averroes would choose established theory over observation. So Bacon wrote that the naturalists who say that the heavens must contain only uniform circular motion and who reject epicycles and eccentrics say that it is better to contradict sense than the order of nature because sense is known to fail at a great distance.[66] But then Bacon seems to go off on a different tack, saying that if the natural mathematicians succeed in saving the appearances just as well as the pure mathematicians do, it is better to follow the natural mathematicians, for the principles are in natural things and mathematicians accept principles from natural scientists.[67] Moreover, Bacon had determined earlier that there is solid evidence from eclipses that the Sun and Moon are sometimes nearer to the Earth and sometimes farther away.[68] Everything visible that sometimes appears under a larger angle and sometimes a smaller one, although the visible thing itself is not changed, nor the medium, nor the vision, is sometimes nearer and sometimes farther away. But this is true of planets, notably the outer planets, which always appear larger when they are in opposition to the Sun.[69] From this it follows that no theory that does not allow the distances to the planets to vary can be correct. In the end, Bacon concluded in a conciliatory way, saying that it should be known that although the pure method of mathematicians differs from that of the one who knows natural things with regard to saving appearances in the heavens, everyone has the same goal, albeit approaching it by different routes, namely to find the places of the planets and stars with respect to the zodiac. Thus, however much pure mathematicians and mathematical astronomers may diverge along the way, one and the same goal terminates their endeavours.[70]

Another response to a situation in which theory and observation disagreed within a given discipline or in which there were competing theories concerning the same subject would be to reconsider the principles that had been assumed. In his *Optics*, Ibn al-Haytham addresses the lack of unanimity in previous physical and mathematical optics, one camp holding that rays come from objects into the eye and the other camp holding that rays go out from the eye.[71] He says that when there are conflicting theories, both theories may be false, one may be true and the other false, or the two theories could be shown to be consistent if the inquiry were taken farther.[72] To resolve the disagreement between the intromission and extromission theories in optics, Ibn al-Haytham proposes to "recommence the inquiry into the principles and premises, beginning our investigation with an inspection of the things that exist and a survey of the conditions of visible objects."[73] Ibn al-Haytham's suggestion was known in the Latin Middle Ages.[74] On a smaller scale, in giving their answers to or determinations of scholastic questions, the *moderni* often began by listing several suppositions on which they would base their conclusions. This habit of listing suppositions as so-to-speak lower-level principles corresponded to the practice of authors of *theoricae planetarum*.[75]

A fourth, more particular polarity between *antiqui*, exemplified in the person of Thomas Aquinas, and *moderni*, exemplified in the person of William of Ockham, has to do with their conceptions of mathematics. Aquinas thought that there are quantitative forms existing in bodies, which may be abstracted from bodies and studied in themselves by mathematicians, although they never really exist separately as Platonic forms are supposed to do. The resulting abstract mathematics is the most certain of any science, but it is not empirical. In contrast, Ockham thought that there are no quantitative forms; everything that exists is either a substance (i.e., a substance consisting of matter and a substantial form) or a qualitative form existing in a substance. Following Ockham, the fourteenth-century *moderni* treated mathematical quantities as concepts existing in the minds of mathematicians, not in the external world.[76] Rather than abstracting quantitative forms from the bodies in which they actually exist, mathematicians, thinking creatively, come up with mathematical concepts. Sometimes the concept might be based on a "phantasm" existing in the mind as a result of seeing a body in the external world. The concept of a geometric line might be based on the vestige or trace of the motion of a body, reduced to a single dimension. Or the mathematician could imagine a "latitude of heat" by analogy to a line, without there being any such qualitative line in the external world.[77] At Cracow in the late fifteenth century, John of Głogów and Albert of Brudzewo tacitly assumed what Ruth Glasner calls a "divorce" between mathematics and physics, together with the belief that mathematical entities are conceptual or imaginary.[78] In relation to astronomy, it is worth noting that the *moderni* treated indivisibles, such as points, lines, and surfaces, as mathematical, not physical, entities. Already, this means that Peurbach's *theoricae orbium* will be understood very differently from his *theoricae linearum et motuum*. In the next section, then, I examine what Głogów and Brudzewo had to say about mathematics and physics in relation to the polarities I have sketched between the *antiqui* and *moderni*.

JOHN OF GŁOGÓW ON ARISTOTLE'S *POSTERIOR ANALYTICS* AND ON SACROBOSCO'S *SPHERE*

Copernicus almost certainly became acquainted with Aristotle's *Posterior Analytics* when he was a student at Cracow.[79] He might have seen Albert of Saxony's questions on the *Posterior Analytics*, which are typical of the *via moderna*, and/or he might have heard more eclectic lectures by John of Głogów or one of his students. The publication dates of the questions on the *Posterior Analytics* of Albert of Saxony (1497) and of John of Głogów (1499) are both after Copernicus left Cracow for Bologna, so if Copernicus knew of these works or others like them, it was from manuscripts or from the lectures of his teachers.[80]

John of Głogów's commentary is in effect an edition of, or supercommentary on, John Versor's commentary on Aristotle's *Posterior Analytics*, and it takes some of its colouration from Versor. Basing himself on Versor's questions, Głogów says he will also draw upon the commentaries of Aegidius Romanus, Thomas Aquinas, Albertus Magnus, and Paul of Venice, among others.[81] These are names mainly associated with the *antiqui*.

In answer to the question of whether it is possible to know something de novo (*utrum possibile sit aliquid scire de novo?*), Głogów says that we can have scientific knowledge and that it can be new rather than, as Plato said, always something that we knew previously but forgot. First of all, this new knowledge comes by way of sense experience. Our senses are actualized by sensible species, and our intellect is actualized by intelligible species poured into it by the agent intellect.[82] Beyond new knowledge (*intellectus*) from experience, Głogów describes how we demonstrate scientific conclusions by reasoning and argumentation. Here, Głogów seems to have in mind a science like geometry in which mathematicians gradually prove new theorems on the basis of the definitions, axioms, and postulates. Not everyone who is presented with the principles of geometry immediately sees the truth of all possible theorems.[83]

In Aristotle's opinion as Głogów reports it, everything we come to know originates in sense. Once we have concepts in mind, the intellect reasons with phantasmata.[84] Correspondingly, as Albertus Magnus wrote in the thirteenth century, in physical or mathematical reasoning phantasmata are required.[85] Would Głogów have thought that humans can, in this way, gain certain scientific knowledge about the world? Or, given the role of human sensation and cogitation, is all putative scientific knowledge fallible? It appears that Głogów, like most fourteenth- and fifteenth-century scholastic Aristotelians, holds that scientific knowledge is fallible.[86] This view would put Głogów more in the camp of the *moderni*.

With regard to the relation of the special sciences to the more general sciences, such as metaphysics or logic, and with regard to the relation of subalternate sciences to those to which they are subalternated (e.g., astronomy to geometry), Głogów notes Aristotle's argument that scientific demonstrations must be based on proper rather than common principles.[87] A geometer would not use the common principle "every whole is greater than its part," he says, except only insofar as it is contracted or limited to the subject matter of geometry, namely magnitude.[88] In particular, demonstrations *propter quid*, or the most powerful demonstrations, should be from the proper principles of the given science. If special sciences use common principles, it is only as they are contracted or limited to the particular subject matter.[89] Głogów does say that special sciences might in some way be corroborated or strengthened by a common science such as metaphysics,[90] but he qualifies this effect by repeating Aristotle's statement that it is difficult to know whether we have scientific knowledge, explaining why this is the case, namely that we often do not know whether we have proper principles for our demonstrations.[91]

In response to the question of whether sciences such as mathematics are the most certain, Głogów says that they are because in them one is less likely to err, given that they are universal rather than particular and given that the arguments are often convertible. But the first argument he makes is more surprising, namely that mathematical sciences are more certain because they offer more to sense and nothing is in the intellect that was not previously in the sense. Thus, he says, mathematics points to a circle drawn in the sand.[92] For such reasons, demonstrative mathematical sciences are the most certain and rarely err.[93] Here, Głogów cites Euclid, Aegidius Romanus, Averroes (Ibn Rushd), and Paul of Venice, as well as other books of Aristotle. From Paul of Venice, Głogów quotes an answer to the argument that mathematics assumes something false when it says a line may be extended infinitely. According to Paul of Venice, Głogów says, this assertion is to be understood according to mathematical imagination, not physical reality. For any given line, one can imagine a larger one.[94]

What Głogów writes in his commentary on Sacrobosco's *On the Sphere of the World* is also consistent with the views of the *moderni*. Commenting on the different types of scientific proof, Głogów says that the daily rotation of the *primum mobile* can be proved with an argument "from a sign," namely by the paths of the fixed stars rising in the east and setting in the west. This is a weak form of argument, he says, but of a type that Aristotle frequently uses in his writings. Such an argument from a sign shows *that* something is the case but not why. The same conclusion can be shown a priori, he says, by referring to the purpose (or final cause) of such a situation in nature. The reason why there is daily rotation is to cause the generation and corruption observed in the sublunar realm. In other words, causing generation and corruption is the purpose for the rotations of the heavenly bodies.[95]

Later in his commentary on *On the Sphere*, Głogów repeatedly distinguishes between the parts of astronomical theories that are supposed to represent physical bodies and the parts that are merely mathematical. First, the various circles drawn on the celestial sphere are not real, unless perhaps the zodiac, understood as being twelve degrees wide and containing constellations, is real.[96] Sacrobosco had said that every planet except the Sun has three circles, namely the equant, the deferent, and the epicycle. The equant of the Moon, he said, is a circle concentric with the Earth.[97] Głogów explains that equants for the three outer planets are imagined circles, which astronomers proposed because the planets did not move with constant velocity either around the centre of the world or around the centre of their own deferents.[98] In contrast, the main three-part orbs of the planets are real and thick even if sometimes they are referred to as circles. They are only represented as projected on a plane because otherwise they cannot be painted or shown in a figure.[99] Głogów observes that Gerard of Cremona writes about the eccentric and equant circles of the superior planets in the same terms, but Georg Peurbach and Johannes Regiomontanus often criticize him.[100] Epicycles should be understood to be embedded within the depth of the deferent.[101]

So in his commentary on *On the Sphere,* Głogów distinguishes between what is mathematical (and hence imaginary) in astronomical theories and what is physical. He includes final causes or purposes as belonging to the theory. Although commenting on *On the Sphere,* which deals mainly with circles, not solid orbs, Głogów imports into his commentary the solid orbs of Peurbach's *Theoricae novae planetarum.* Thus, in harmony with the views of the *moderni,* Głogów distinguishes in *On the Sphere* between what is to be understood as real or physically existing and what is mathematical and hence dependent on human thought.

ALBERT OF BRUDZEWO'S COMMENTARY ON PEURBACH'S *THEORICAE NOVAE PLANETARUM*

Finally, clear evidence concerning the conceptions of astronomy as a science present at Cracow when Copernicus was a student is contained in Albert of Brudzewo's commentary on Peurbach's *Theoricae novae planetarum.*[102] If Brudzewo did not lecture on Peurbach's *Theoricae novae planetarum* with Copernicus in attendance, he had taught such a course in previous years, and whoever taught Peurbach with Copernicus in attendance had likely heard Brudzewo himself earlier. Before the commentary was printed in 1495, there were likely manuscript versions in circulation in Cracow. Brudzewo has a clear conception of astronomy as a science in part physical and in part mathematical; the commentary makes this view abundantly clear, even if it were not obvious from Peurbach's work itself.

Near the start of his introduction, Brudzewo raises two questions. First, he asks what moved the wise (*sapientes*) to posit several celestial orbs. He replies that the movement of the fixed stars together in an unchanging configuration led them to posit that there is a single sphere to which the fixed stars are attached. Then they noticed that there were certain stars (the planets) that changed positions relative to the fixed stars, so they posited separate spheres for each of them.[103] Second, he asks how many mobile orbs there are. In reply to the second question, he cites Averroes's questioning of the existence of any sphere not containing a visible star and then cites the *theoricae*, which posit three or more orbs for an individual planet. He is not going to discuss every opinion on the number of orbs, he says, but only the more probable theories. He describes three different senses in which the word "orb" is used – taking this from Albert of Saxony or some other previous author – and then lists the suppositions or principles discussed above.[104]

If one looks in the commentaries of Averroes and Albert of Saxony in connection with Brudzewo's reference to Averroes or with his silent borrowing from Albert of Saxony, two features of these sources emerge. First of all, Averroes repeatedly analyzes the logical structure of Aristotle's arguments and describes many and various astronomical opinions. On the basis of Averroes's report of Aristotle's views, one easily gets the impression that cosmology and astronomy are works in progress. With regard to the number of orbs, for instance, Averroes writes, "What [Aristotle] says, that the Eighth Orb is near the first orb, we find written thus. The opinion of the ancients is that the Eighth Orb, that is the starry orb, is the first orb. Ptolemy, however, posited a ninth, because he said that he himself found in the fixed stars a slow motion in the order of the signs [i.e., west to east], and the modern Arabs say that this motion is a motion of access and recess, and that it is not perfected, and they say that this motion was spoken of by the ancient Babylonians."[105] The fundamental motions of the celestial orbs are uniform, so if planets appear to move with changing velocities, Averroes writes, it must be because they are moved by several uniform motions, which combine in varying ways to produce the observed effect.[106] Here, Averroes says, Aristotle does not have in mind epicycles and eccentrics, but Averroes himself is motivated on this point to remark that there is no way to demonstrate whether a given irregular motion is caused by an eccentric or by an epicycle. Just because, if an epicycle is assumed, the observed positions can be calculated or just because, if an eccentric is assumed, the observed positions can be calculated, does not mean that from the agreement between the observations and the assumptions, one can infer that there is an epicycle or alternatively that there is an eccentric. Some people later argue that you might infer that there is one or the other but not which one.[107]

In the conclusion of his introduction, Brudzewo provides an alternate explanation of the relations of the three-dimensional orbs of the *Theoricae novae planetarum* to the two-dimensional eccentrics and epicycles of standard Ptolemaic astronomy.[108] The major two-dimensional figures are simply the projections of the three-dimensional orbs or, alternately, what results if the three-dimensional figures are collapsed into two dimensions. More than one projection of the three-dimensional figures is possible. If the celestial sphere is collapsed along its north-south axis, the celestial equator will be the largest circle, and the Tropic of Cancer will be superimposed on the Tropic of Capricorn. According to another projection, however, the various circles are represented as they would appear from a particular point of view; the nearest tropic may appear larger because it is nearer to the viewer. Astronomers use plane figures to represent astronomical theories because the theories are much easier to understand if they are represented visually. But it is the three-dimensional, thick orbs that are physically real, whereas the two-dimensional eccentric circles and epicycles are mathematical fictions, ones that play an important role in aiding the human intellect to understand the theory.[109] The figures in the printed editions of Peurbach's *Theoricae novae planetarum* no doubt reinforced in the minds of his readers the distinction between physical orbs and fictional lines by colouring the unequally thick real orbs. Here, Brudzewo increases the potential impact of the *Theoricae novae planetarum* in this respect by emphasizing the artistic or artificial properties of the figures. Brudzewo further clarifies the conditions of the art, or professional work, of the astronomer – why, for instance, astronomers make use of mathematical devices that are not related to real bodies – by contrasting astronomy and natural philosophy. The latter presents a less detailed view and tries to explain why the stars and planets display the positions and motions that have been observed. Whereas the astronomer considers partial orbs, the philosopher considers only total orbs. Whereas astronomy measures motions by the angles passed through by the planets, philosophy measures motions by the distances traversed. Thus, whereas the Moon may move through angles more quickly in the astronomers' view, being nearer, it traverses shorter distances in the philosophers' view.[110] When Averroes objects to Ptolemaic astronomy, Brudzewo says, he is writing as a natural philosopher, not an astronomer, and he cares only about total orbs.[111]

After his introduction, Brudzewo begins his comments on individual passages of Peurbach, marking those passages by the opening words. The commentary is too detailed to be summarized here, but there are some notable points worth reporting. For instance, in agreement with Aristotle's doctrine that the practitioners of a given scientific discipline do not doubt the principles of their science, Brudzewo argues that uniform circular rotation of the celestial bodies is a basic principle of astronomy, not to be disputed by astronomers. Thus he writes, "And if it is a first principle of astronomy that the Sun moves regularly in its eccentric (and therefore there is no further dispute in astronomy with someone denying it), nevertheless such a principle can be demonstrated by a subalternating science, namely mathematically."[112] Contrary to what might be thought at first, Brudzewo here has in mind an a posteriori, or *quia*, demonstration, based on observed positions of the Sun at different times: "The Sun in equal times describes equal angles upon its center and cuts off equal arcs. Therefore it is moved uniformly [*aequaliter*]. The consequence holds by the definition of 'to be moved uniformly,' which is taken from Book VI of the *Physics*. But the antecedent is clear in this figure."[113] In a related use of mathematics, astronomy can determine the eccentricity of the Sun's deferent by determining how much longer it takes the Sun to move from the spring equinox to the fall equinox than to move from the fall equinox to the spring equinox. Thus mathematics provides a method of reasoning *quia* from observed positions of the stars and planets over time to suppositions or principles of astronomy – as astronomers infer from observation that the stars are on a rotating sphere. This process establishes *that* something is the case.

A difficulty might arise, however, when more than one theory, such as an eccentric or an epicycle, produces results that fit the appearances equally well. This difficulty had been recognized from the time of Apollonius of Perga (ca. 262–190 BCE). Concerning alternative theories of the motion of the Sun, Ptolemy had said in the *Almagest* that there is no way to demonstrate that one of the two alternative theories is the correct one, eccentric or epicycle, so he is limited to probable rather than demonstrative arguments. However, it would seem more reasonable to associate it with the eccentric hypothesis since that is simpler and is performed by means of one motion instead of two.[114] Later, Naṣīr al-Dīn al-Ṭūsī made this same characterization of Ptolemy's situation.[115] Thus in this conception of Ptolemaic astronomy, many of the principles come from observation, memory, and experience, but there may be no unique set of principles corresponding to experience. The same observations could be explained by an eccentric and by a corresponding concentric deferent together with an epicycle.

Like Ptolemy, Peurbach in the *Theoricae novae planetarum* chose to use an eccentric to explain the motion of the Sun. Unlike Ptolemy, but following Arabic predecessors in the configuration or *hay'a* tradition, Peurbach embedded the Sun's eccentric within a thick concentric shell (see figure 6.2). For Venus, Mars, Jupiter, and Saturn, Peurbach proposed that within the eccentric there was embedded a solid epicycle containing the planet. Although, mathematically speaking, some motions produced by a concentric deferent and epicycle could equally well be produced by an eccentric, physical factors might favour one configuration over another. If Peurbach never raises the question of any possible interference of small circles near the centre, this is no doubt because they are not real. Where he has such a small circle for Mercury, he provides a second set of unequally thick nested orbs that cause the imagined centre of the deferent to trace the small (mathematical or imaginary) circle to which no body corresponds.[116]

In addition, there were some aspects of the planets' motions that had not been explained using uniformly rotating concentrics, eccentrics, and epicycles. Like previous astronomers, Peurbach tracked these other motions mathematically using geometric lines or circles rather than physical orbs (they could also be tracked arithmetically using tables). In his commentary on Peurbach, Brudzewo carefully notes where lines rather than physical orbs are being described. In particular, Brudzewo does not consider the equant to be a real physical thing in the celestial realm because it does not correspond to any aether sphere. So he inserts a note that does not correspond to any particular text in Peurbach:

> It should be noted that there is no work [*opus*] for the equant insofar as the motion in itself of the orbs is concerned. For the equant does nothing with regard to the regular motion of the orb since it is an imaginary circle. But it has a function with regard to the astronomical work and the calculation of tables. Tables are calculated using mathematical principles and conclusions, which conclusions indeed often cannot be accommodated and applied to the motions either in themselves or as they appear to us. Therefore, these mathematicians sometimes take the motions of celestial bodies other than they are in their nature or other than they appear to us, and they consider them in a way that serves their art and their operation, since it is certain that otherwise there is no way that they can arrive at their intended work correctly and precisely. Therefore, they imagine that the motion is uniform, which does not appear uniform in itself, on account of their work, that it be done more correctly. And from this they are convinced and compelled to posit imaginary equant circles, on which they consider the diverse and unequal motions of the orbs to be equal, and they reduce those diverse motions first to uniformity, as to that on the basis of which the

irregular motions should be taken. "The regular is the judge of itself and of the irregular" [as Aristotle says in book 1 of *De anima*].[117]

Thus Brudzewo repeatedly makes clear the distinction in scientific status between physical orbs and mathematical or imaginary circles.

To repeat: how does the use of equants fit into the proper role of an astronomer? Since it is a goal of practical astronomy (e.g., astrology) to support prognostications concerning the effects of the heavenly bodies on Earth, it should be understood that astronomers have adopted mathematical methods that go beyond the limits of their confidence in theoretical physical astronomy in order to make those predictions. Thus Peurbach and Brudzewo posit equant points in order to make the tables and predictions that are part of their task as practical astronomers. They do not think that equant points – or the circles that may be drawn around them – correspond to real physical bodies. Nevertheless, they posit and use them because they see no other way to carry out their craft. What Brudzewo says here is related to what Copernicus wrote in his *Letter against Werner*, saying that astronomers have settled on a chosen theory, "lest the inquiry into the motions of heaven remain interminable."[118] Astronomers are confident that their mathematical descriptions are correct in connecting the observational data in a systematic way, even though they have not yet been able to devise a system of real celestial orbs embodying that motion because the results of their mathematical calculations agree with what is observed.

Thus, in Brudzewo's interpretation of Peurbach, we are left with an astronomy partly grounded in what are supposed to be real, uniformly rotating aether spheres in addition to which astronomers imagine other mathematical devices, which are not based on physical bodies but are necessary to make the predictions called for. The physical orbs provide a pretty good theory, so they are taught, although the theory has not been perfected. As far as the real physical orbs in the heavens are concerned, it should be possible to explain the relation of rotations to their causes, including intelligences or final causes. If the heavenly bodies are the causes of being and life of the terrestrial bodies, how, Brudzewo asks, can one explain the existence of empty spheres, which presumably would have no terrestrial influence? And why do seven orbs have one body each, whereas the eighth orb contains all the fixed stars? These questions and their answers are taken, nearly verbatim, from the paraphrase of Aristotle's *Metaphysics* by Albertus Magnus. The questions are answered in terms of the intended effects of the stars and planets on Earth.[119]

But there is another aspect of the work of the theoretical astronomer, namely to fine-tune the celestial theories in accordance with the observed positions of the stars and planets. If there remain open questions about astronomical theory, new data about celestial positions may help to decide between alternatives. In this connection, Brudzewo has more to say than Peurbach about the importance of accurate observations, especially as aided by instruments, and he notes that the measurements of earlier astronomers may have been inaccurate.[120] In the course of his commentary, Brudzewo reports differences of opinion between past astronomers, such as with regard to the apparent sizes of the Sun and Moon.[121] To determine how far a star or planet has moved, it is necessary to have a fixed point of reference, he says, and if there is trepidation of the equinoxes, the point of intersection of the celestial equator and the ecliptic may itself be uncertain.[122] So another method of measurement is to use the angles between a planet and one or more fixed stars.[123]

CONCLUSION

When Ibn al-Haytham's *On the Configuration of the World* was translated into Latin, Europeans recognized his embedding of eccentrics and epicycles within concentric total orbs as showing how eccentrics and epicycles could be reconciled with real, nonoverlapping physical orbs. *On the Configuration of the World* became known in Europe in multiple ways. A Latin translation (or reworking) found in Oxford was "part of the corpus of Arabic scientific writings that were made available in Latin under the patronage of Alfonso X (the Wise) of Castile (1221–1284)."[124] Translations of the work into Hebrew were even more numerous.[125] By whatever route, the idea that Ptolemy's mathematics could and should be matched with thick physical orbs rotating uniformly spread widely in Europe beginning in the later thirteenth century. However the ideas spread, works called *theoricae planetarum* evolved from describing the motions of planets using mathematical lines to describing them, at least in part, using physical orbs.[126]

Thus from the late thirteenth century, a growing number of Western astronomers working in the *theorica planetarum* or configuration tradition explained as much as they could using solid orbs or aether shells but then presented other *theoricae* or figures containing lines and motions that did not correspond to bodies, these other *theoricae* of lines, motions, axes, and so on being necessary to ground the calculations of positions required in astronomical or astrological practice.[127] In his *Theorica planetarum*, Campanus of Novara begins to move from lines to physical orbs insofar as he describes how to draw three-dimensional orbs extending from the nearest possible approach of the planet to the

Earth to the greatest possible distance away.[128] In his *Scriptum super canones Azarchelis de tabulis Toletanis* (ca. 1290), John of Sicily notes that words like "eccentric" are ambiguous, sometimes referring to physical orbs and other times to mathematical lines, such as the path of the centre of a planet. Whereas he takes the three-part orbs descended from *On the Configuration of the World* to be real, he says that the equant is an imaginary circle (*circulus imaginarius*).[129]

From the later thirteenth century through the fourteenth and fifteenth centuries, the embodiment of as much as possible of Ptolemy's mathematical astronomy within physical orbs was accepted by European astronomers, but for purposes of astronomical prediction, they continued to use astronomical tables based on Ptolemaic mathematics that lacked a physical counterpart. In fourteenth-century Aristotelian commentaries, a typical treatment of epicycles first described Ptolemy's use of eccentrics and epicycles, then noted Averroes's rejection of them on the grounds that they would require either vacua or more than one body in the same place, and finally went on to say that this objection can be overcome by embedding epicycles and eccentrics within the depth of concentric shells such that any given orb remains within the same cavity in the body surrounding it, always rotating into itself. This, for instance, is what John of Jandun writes in his commentary on the *Metaphysics.*[130] John Marsilius Inguen (pseudo Marsilius of Inghen) makes a similar remark in his commentary on the *Physics* to refute the view that if there are epicycles, there must be vacua or fluids in the heavens.[131]

In this chapter, I have not tried to provide detail of possible routes of transmission between Ibn al-Haytham's *On the Configuration of the World* in the late tenth or early eleventh centuries, the *hay'a* tradition of Islamic astronomy, and the appearance of similar configurations in Europe at the end of the thirteenth century; there were many routes, of which John of Sicily's *Scriptum super canones* is only one. Works on configuration or *hay'a* continued to be produced in Islamic areas, and I hope that historians of Arabic astronomy can to a great degree trace the history of the science of configuration forward.[132] On the European receiving end, it is clear that European astronomers took it for granted that they had many Islamic predecessors. Albert of Brudzewo cites by name Alfraganus (al-Farghānī), Albategnius (al-Battānī), Averroes (Ibn Rushd), Geber (Jābir ibn Aflaḥ), Haly ('Alī ibn Riḍwān), and Thābit ibn Qurra.[133]

If I have shown that Copernicus wrote his *Commentariolus* within the conception of astronomy as a science exemplified by Peurbach's *Theoricae novae planetarum* and, moreover, that this conception descended from Ibn al-Haytham's *On the Configuration of the World,* the transition from *Theoricae novae planetarum* to Copernicus's *Commentariolus* can be more correctly understood. In several papers in recent years, Peter Barker and Michela Malpangotto have described in detail the place of orbs in Peurbach's *Theoricae novae planetarum* and in Brudzewo's commentary on that work. Barker has argued that what was new in Peurbach's *Theoricae novae planetarum* may have stimulated Copernicus's innovations in astronomy, and Malpangotto has given even more credit to Brudzewo, whom she sees as critical of Peurbach's failure to replace Ptolemy's equant and mean motion of the apogee of the epicycle with uniformly rotating orbs or circles.[134] I argue that both Peurbach and Brudzewo were carrying on a research program initiated by Ibn al-Haytham and carried on by Islamic *hay'a* astronomy. All three – Ibn al-Haytham, Peurbach, and Brudzewo – describe complementary figures or *theoricae.* Some of these figures portray physical orbs, but others represent Ptolemaic mathematical astronomy not matched with bodies because so far no one had succeeded in matching those parts of mathematical astronomy to physical bodies. Before Copernicus, Regiomontanus had tried to advance the program of finding homocentric astronomical configurations, but he had failed to complete this work.[135] Much earlier, in the twelfth century, Averroes had hoped to reconcile Ptolemaic mathematics with bodies but came up short. Thus Copernicus worked within a context in which astronomers were hoping for reform of the physical side of astronomy, a context having its roots in Islamic astronomy. Contrary to Malpangotto, I think that Peurbach and Brudzewo both accept the idea that there are some physical orbs uniformly rotating and other, purely mathematical methods that do not correspond to bodies. Brudzewo is not disappointed with Peurbach but is elucidating positions with which Peurbach would have agreed.

Like Regiomontanus, who took an interest in homocentric theories in astronomy as well as in theories of eccentrics and epicycles, Copernicus could see himself as within the current program of astronomy in proposing an alternate and more satisfactory *theorica* of the planets. Copernicus was not the first to try a new configuration, but by the time he published *De revolutionibus orbium coelestium,* he was able to make clear in mathematical detail that a configuration of the world with the planets rotating around the Sun could fit appearances as well as the Ptolemaic Earth-centred system.

4

Regiomontanus and Astronomical Controversy in the Background of Copernicus

Michael H. Shank

WITH NICHOLAS COPERNICUS, as with other figures deemed revolutionary, the more we learn about his predecessors and context, the less original he seems. It is therefore important at the outset to avoid misunderstanding the point of this chapter. To Copernicus belongs the credit not merely for his heliocentric insight but also especially for the hard labour of transforming it into a full-blown, predictive mathematical theory that could replace Claudius Ptolemy's *Almagest.* The background discussed below subtracts nothing from this achievement. Despite past claims to the contrary, there is no evidence that Copernicus found heliocentrism in Johannes Regiomontanus (1436–76). Recent research makes it exceedingly unlikely that any will turn up, for reasons that become clearer below.

That said, it is difficult to understate the significance of Regiomontanus's work and his larger context for the emergence of Copernicus's new astronomy. First, Regiomontanus was arguably the most talented mathematical astronomer in Europe in the generation before Copernicus. Both he and his immediate predecessors deserve special attention if one seeks to understand the level that the "science of the stars" had reached in late-fifteenth-century Europe.[1] Second, Regiomontanus is more than a milestone of fifteenth-century intellectual achievement; Copernicus would use Regiomontanus's work heavily and build directly upon it. Any explanation of Copernicus's competence and development must confront his extensive reliance on the *Epitome of the Almagest* (1496), which Georg Peurbach[2] had begun and Regiomontanus had completed (and edited) by about 1462. Most notably, one of the latter's equivalence proofs provided the geometrical foundation for Copernicus's shift to an Earth in motion around a (nearly) stationary Sun. Third, Copernicus was working not in a solipsistic bubble but against the background of thriving institutional developments in astronomy. Two generations before Copernicus's birth, the universities of Bologna and Cracow had pioneered chairs in the science of the stars – the first time specialization was recognized in a discipline associated with the Faculty of Arts. In addition, a major mid-fifteenth-century controversy about Ptolemy's *Almagest* motivated both the *Epitome* and several other works by Regiomontanus. Indeed, we still do not fully appreciate the extent to which Regiomontanus's participation in this controversy on behalf of his patron, Cardinal Basilios Bessarion, shaped the last fifteen years of his career. All of these elements framed the context of Copernicus's achievement.

As the leading author of the *Epitome of the Almagest*, Regiomontanus has long been read as advocating a "traditional" Ptolemaic mathematical astronomy. On this account, he valued the utility of the *Almagest*'s astronomical hypotheses, however imaginary epicycles or equant points might be. Characterized as a "fictionalist," he could serve as a foil to the "realist" Copernicus, for whom the fixed stars and the Sun were truly stationary. In fact, the outlook of Copernicus had much more in common with that of Regiomontanus and his immediate predecessors in mathematical astronomy than historians have usually suspected. Although he found much to admire in the mathematical acumen and predictive power of the *Almagest*, Regiomontanus was also a critical reader of Ptolemy's program as he understood it. In his letters and formal works, he criticized Ptolemy's theory for failing both to conform to some observations and to take the physical aspects of astronomy seriously.[3]

In contrast, Regiomontanus's contemporaries and immediate successors saw him pre-eminently as a reformer of astronomy. This designation acknowledged a range of activities; not only his mastery of astronomy and his critical attitudes toward aspects of received theory but also elements of the university textbook and commentary tradition, including the teaching and diffusion of astronomy. When he wanted to praise Copernicus, Georg Joachim Rheticus would put his teacher on a par with Regiomontanus.[4] As we twenty-first-century readers try to understand the place that Regiomontanus occupied in his own day, we inevitably find it hard to put Copernicus out of our minds. This chapter seeks to tread as gingerly as possible on the thin line between historical context and salutary anachronism. Its main burden is to argue that astronomy in the generation before Copernicus was far from being a period in the doldrums of stagnation. On the contrary, a bitter controversy about the interpretation and proper understanding of Ptolemy's *Almagest* erupted in the early 1450s and dragged on into the 1480s. It drew in Regiomontanus and shaped his best-known work, the *Epitome of the Almagest*.

EARLY YEARS

The man we call Regiomontanus came from the small Franconian town of Königsberg, near Bamberg. In 1450 he matriculated at the University of Vienna as Joannes Molitoris de Künigsperg, literally "John of the miller (i.e., the miller's son) from Königsberg."[5] His youthful exposure to cogs, gears, and shafts goes some way toward explaining his craftsman's skills in the manipulation of materials and in thinking mechanically. These tendencies would later serve him well when he designed planetary models, made instruments, tried new printing techniques, and built astronomical clockwork.[6]

Even by the standards of the day, Regiomontanus was intellectually precocious. As an eleven year old, he studied at the university in Leipzig, matriculating in 1447. Whereas most other young students in the faculty of arts would have heard lectures on Johannes de Sacrobosco's *On the Sphere of the World*, Regiomontanus was computing his first almanac – a genuine ephemeris that gave not only new and full Moons but also daily planetary positions.[7] These skills and computational abilities, and the sheer endurance to use them so repetitiously, already put the preteen Regiomontanus in a league of his own.[8]

Why did Regiomontanus matriculate at the University of Vienna in 1450? At the time, he was probably still unaware of Georg Peurbach, who was travelling in Italy and not yet a master of arts. Like many an institution since then, Vienna perhaps basked posthumously in the reputations of its dead – in this case, John of Gmunden (active ca. 1420–42), whose astronomical works had radiated strongly throughout central Europe. More charitably, Regiomontanus may also have known that John of Gmunden's students were teaching there. They enjoyed a high reputation in their day, even if their works have yet to be studied in ours.[9]

One thing is certain: Vienna was a major political centre. In 1450 the Habsburg duke who ruled the territory would soon be crowned German king and elected emperor as Frederick III. Since astrology was one of the chief political tools of the age, practitioners of the art were prized in the leading courts of Europe, including those of the Habsburgs for at least a generation. Both its university and its political importance thus made Vienna and its vicinity a very promising place for a precocious mathematician with exceptional computational skills.[10] Coincidentally,

Frederick III's astrologer at the time was Johann Nihil (Nitzka), a fellow Leipzig alumnus.[11] By making a similar trek to Vienna, Regiomontanus would join not merely the largest university in German-speaking territory but also one both renowned in astronomy and adjacent to a major court that appreciated specialized astrological skills.[12] At the time, the Habsburg court was also raising the young Ladislaus Posthumus, the heir to the crowns of Hungary and Bohemia. While still a teenager, Ladislaus claimed his thrones and required an astrologer. On Nihil's recommendation, Peurbach got the position, with which Regiomontanus evidently assisted until the seventeen-year-old king died very suddenly in 1457.[13]

As his 1448 ephemerides demonstrate, Regiomontanus's mathematical skills upon arriving in Vienna as a twelve year old already exceeded those required for the master of arts degree. To receive that, however, he would have to wait until 1457, when he reached the statutory age of twenty-one. In the meantime, his education in the science of the stars undoubtedly was already proceeding outside the bounds of the official Viennese curriculum, thanks to the well-stocked astronomical libraries of both the university and the nearby Augustinian convent in Klosterneuburg.

Peurbach's return from Italy around 1451 gave Regiomontanus a mentor, an intellectual companion, and a kindred critic devoted to astronomy. One of the first traces of their interaction is textual. Regiomontanus's earliest dated astronomical manuscripts include the copy he made of Peurbach's 1454 lectures on his own *Theoricae novae planetarum* at the "citizens' school" (*Bürgerschule*) near St Stephen's.[14] Regiomontanus's engagement with Peurbach's *Theoricae novae planetarum* runs like a thread throughout his career, from his early manuscript to his printing of the work almost two decades later. During this period, he would extol the virtues of the *Theoricae novae planetarum* as superior to the widely used thirteenth-century anonymous *Theorica planetarum communis*. He attacked the latter as full of errors in the *Disputationes contra deliramenta cremonensia*, a dialogue that he set (and probably wrote) in 1464 and published on his Nuremberg press around 1475, immediately before his last trip to Italy and sudden death there in 1476. These criticisms made a lasting impression on astronomical pedagogy for the next century and a half. Even so, Regiomontanus's promotion of Peurbach's *Theoricae novae planetarum* may not have been wholehearted, for he still hoped, perhaps against hope, for a mathematical astronomy based on concentric spheres, without the epicycles and eccentrics that pervaded the *Almagest* and remained fundamental to Peurbach's models.

Why was Peurbach's *Theoricae novae planetarum* new? Not on account of the title, which had been used several times since the fourteenth century.[15] A leading hint appears in Regiomontanus's 1454 manuscript, which entitles the work "New theorica presenting the *real* configuration and motion of the spheres, with the terms of tables" (my italics).[16] In this intriguing qualification, Peurbach illuminates his own understanding of the novelty of his work, here explicitly linked to its emphasis on physical considerations – "the real configuration and motion of the spheres" – within the framework of the traditional *theorica* genre. Justifications for the novelty in the title also include Peurbach's descriptions of separate, partial orbs constituting the total, concentrically circumscribed orb of each planet (see figure 6.2). This concern, which had antecedents in Ptolemy's *Planetary Hypotheses* and Abū ʿAlī al-Ḥasan ibn al-Haytham's *On the Configuration of the World*, somehow got a foothold in late-fourteenth- and early fifteenth-century Viennese astronomy.[17]

This interpretation of Peurbach is consistent with Regiomontanus's later criticisms of the old *Theorica planetarum communis* and Ptolemy's *Almagest.* In his *Disputationes*, Regiomontanus chastised the former for underreporting the number of spheres required by the lunar theory and thus implying contrary motions in the same sphere. He put forward Peurbach's *Theoricae novae planetarum* as the proper way to handle the problem. In the *Defensio Theonis contra Georgium Trapezuntium*, however, Regiomontanus criticized Ptolemy for implying that mere circles somehow carried the planets or were responsible for their motions.[18] In each case, Regiomontanus treated the three-dimensionality of Peurbach's models as an improvement.[19] Concurrently, however, around 1460 Regiomontanus hoped to find a "concentric astronomy" without epicycles or eccentrics, a criterion that Peurbach's work obviously did not meet. Strikingly, when he printed his mentor's book on his Nuremberg press around 1472, he called it simply *Theoricae novae planetarum*, dropping "the real configuration and motion of the spheres."[20] This omission is consistent with his ongoing search for a fully concentric astronomy, on which more below.

Although Peurbach had said nothing about the physical properties of the spheres themselves, he had described them as three-dimensional bodies. His language was predominantly geometrical, but he occasionally applied physical terminology to the spheres. Thus the *Theoricae novae planetarum* discusses orbs with contiguous surfaces, to which things like the "body of the Sun is attached (*infixum*)."[21] The point of such language was clearly physical since the three-dimensionality of the orbs contributed nothing to the mathematical theory or its ability to generate tables useful for predicting planetary positions. Consistent with Peurbach's program, if not the title of his book, Regiomontanus distinguished the mathematical from the physical in the large, striking illustrations he devised for his *editio princeps* of the *Theoricae novae planetarum.* Whereas the execution was new, the concept followed a graphic tradition already evident in late-medieval manuscripts in the generations before Peurbach.[22] In Regiomontanus's edition, the physical partial spheres of the planetary models are filled in with black ink or striking colours. The latter contrast sharply with the thin black-on-white lines of the purely geometrical diagrams. Most subsequent editions of the text, beginning with those of Erhard Ratdolt in Venice in 1482 and 1485, adopted the same graphic convention.[23]

During his Viennese years, both Regiomontanus's correspondence and the texts that he copied show that he was examining critically a variety of astronomical systems. Alongside Peurbach's *Theoricae novae planetarum,* Regiomontanus copied into his notebook Henry of Langenstein's *De reprobatione ecentricorum et epicyclorum* (1364).[24] In his Parisian critique of the thirteenth-century *Theorica planetarum communis,* the future reviver of the University of Vienna dismissed Ptolemaic epicycles and eccentrics on empirical and theoretical grounds. Instead, Langenstein proposed a homocentric system for the Sun and Moon in which their concentric spheres moved nonuniformly.

Langenstein's jarring proposal dropped the age-old linkage of concentric planetary spheres with uniform motions, a longstanding ideal anchored in Aristotle's *De Caelo* (e.g., book 2, chapter 6). Langenstein's acceptance of nonuniform spherical motion thus sidestepped not only the problem that had led Ptolemy to invent the equant circle but also all the objections associated with it. To many of his readers, however, Langenstein's solution must have seemed even less palatable than the equant itself, but it at least had the merit of reminding them of the dilemma they faced. His treatise offered additional criticisms pertaining to planetary distances and sizes, as well as to the distortions that latitude theory introduced into predictions of longitude. Some of his views intriguingly echo Ibn al-Haytham's homocentric alternatives to Ptolemy.[25] Yet other criticisms resonate with those found in the *Astronomy* of Levi ben Gerson (Gersonides, d. 1344). In one of the most striking of these, Langenstein notes that the Ptolemaic lunar model implies a consequence that is never observed: a four-fold variation in apparent size as the Moon moves around the Earth (he made analogous arguments about the sizes of other planets).[26] These criticisms suggest why Langenstein found the homocentric approach appealing: whatever its other problems, it was more consistent with observation than were the enormous variations in lunar size or the planetary brightnesses that should follow from Ptolemy's epicycles. Langenstein's discussion evidently stimulated Regiomontanus's openness to homocentric possibilities and his formulation of similar proposals and objections when criticizing Ptolemy's approach in the *Almagest*.[27]

In addition, Regiomontanus was aware of the Spaniard Nūr al-Dīn al-Biṭrūjī's twelfth-century *De motibus celorum*, an earlier homocentric system that Michael Scot had translated from the Arabic in 1217. After stressing his study of the *Almagest* and conceding the latter's predictive success, Biṭrūjī bemoaned Ptolemy's inconsistency not with Aristotle but with nature. Indeed, his explicit goal was to replicate the predictions of the *Almagest* with physically consistent astronomical models. In Biṭrūjī's view, eccentrics, epicycles, and contrary celestial motions all violated sound physical principles and should therefore be eliminated.

Although Biṭrūjī is often considered an Aristotelian, his work in fact abandoned some fundamental features of Aristotle's approach. Crucially, Biṭrūjī rejected models that took the celestial phenomena to be the results of two contrary motions: the rapid east-to-west daily motion of the fixed stars (and the entire heavens) and the slower west-to-east (hence contrary) proper motions of the seven classical planets against the background of those fixed stars. Aristotle had built on the three- or four-sphere planetary models of Eudoxus of Cnidus, who had treated each planet separately. However, to fit all planetary motions into one unified system, Aristotle had inserted interfaces between each of Eudoxus's planetary models. Between each of the separate Eudoxean models, Aristotle's "buffers" consisted of two or three additional spheres, the sole purpose of which was to neutralize the motions of the two or three spheres above them.[28] Aristotle had maintained that circular motion had no contrary: regardless of direction, around is always around. Rejecting this account and the contrariety that followed from that rejection, Biṭrūjī argued instead that the celestial spheres all moved in the same direction (east to west). The sphere of the fixed stars was the fastest, with each lower sphere exhibiting ever more retardation, right down to the spheres of the elements, leaving the sphere of the central Earth motionless.

After trying very hard, however, Biṭrūjī conceded in the end that he could not both duplicate Ptolemy's results and eliminate epicycles and eccentrics.[29] This was a serious flaw. Nevertheless, Biṭrūjī's homocentric system embodied the ideal of a physical "third way" critical of both Aristotle's contrary motions and Ptolemy's nonuniform devices. Interest in his work kept alive the ideal of a physically satisfying alternative to the dominant systems, even when it served merely as an articulation of the dilemma and as a target for criticism. In the thirteenth century, Biṭrūjī's work stimulated pro and contra discussions among mathematically inclined Franciscans in particular. Roger Bacon is the best known of four Oxford Franciscans who favoured it (initially at least), whereas his brothers Bernard of Verdun and Guido de Marchia, among others, opposed it (as did John Peckham and several others who changed their minds).[30]

Regiomontanus himself owned a manuscript of Biṭrūjī's *De motibus celorum.* We do not know when he acquired it, but the notes in his Viennese notebook show his familiarity with some of Biṭrūjī's views, including the unorthodox arrangement of the inferior planets according to the synodic period (Venus above the Sun and Mercury below it).[31] These notes, copied from a third party, both summarize Marchia's attacks on Biṭrūjī and also sketch Marchia's unusual cosmology. It consists of planetary epicycles located within eccentric rings immersed in a single fluid medium that extends from the Moon to the stars and rotates once a day.[32]

During his Viennese years, then, Regiomontanus was clearly exposed both to various cosmological alternatives to the mainstream and to the lack of consensus about fundamental questions in astronomy and cosmology. The latter included such central issues as the arrangement of the heavens and uncertainties in the order of the Sun, Venus, and Mercury, as well as doubts about the tensions between what was observed and predicted about planetary sizes and brightnesses.

Contrary to our stereotype of "*the* medieval cosmos" as a coherent, unitary, and therefore unassailable fortress, a striking plurality of alternative views was available to inquiring minds in fifteenth-century Vienna. Regiomontanus's confrontation with them evidently shaped his critical outlook and picture of astronomy as full of unsettled questions. Not surprisingly, then, toward the end of his years in Vienna, while still teaching at the university, Regiomontanus sketched the beginnings of his own homocentric system, aimed at integrating the physical and mathematical aspects of astronomy. In a 1460 letter to a fellow alumnus, the Hungarian bishop János Vitéz, he articulated his goal of eliminating eccentrics and epicycles, outlined concentric-sphere models for the Sun and Moon, and expressed the hope of finding suitable ones for the other planets as well. This work was not merely an exercise in physical astronomy, for he expected that this new foundation would eventually lead to the computation of new and better tables. He evidently already believed that physically consistent models would yield the best predictions. Noel Swerdlow has shown that Regiomontanus's two homocentric models successfully (and silently) corrected a flaw in Biṭrūjī's scheme, thus making it workable (but only for the Sun and Moon).[33] Although unattainable,

Regiomontanus's goal of producing a complete homocentric astronomy is notable for its documentation of what he held dear. It is also less far-fetched than it seems to us. The chief objection to a homocentric schema around a central Earth is its inability to model variation in distance; by definition, the planets are driven by combinations of spheres that all have the same centre, the Earth. To Regiomontanus, however, this would have been an improvement. Assuming that one could suitably adjust speeds, a fixed distance accounted far better for the observed sizes of the Sun and Moon – and even for the modest variations in brightness of some other planets – than did the large variations one would expect from Ptolemy's models.[34] This ideal clearly displays a meld of physical and mathematical concerns that, despite their differences, resonates with those of Biṭrūjī in Spain, the eastern Islamic tradition, Gersonides in Provence, and scattered Latin astronomers of the fifteenth century.

In Vienna, Regiomontanus did more than explore a variety of cosmological options. His university years show him deepening his understanding of Ptolemy's *Almagest* in collaboration with Peurbach and offering mathematically oriented lectures, notably on the *Perspectiva communis* and Euclid. In the fall of 1461, however, he abruptly left Vienna for good in the company of the Greek cardinal Bessarion.[35] This turning point in his life was intimately connected with a half-finished work and a decade-long controversy between two Greek émigrés in Italy.

THE CONTROVERSY BEHIND THE *EPITOME OF THE ALMAGEST*

During Regiomontanus's decade of studies and teaching in Vienna (1450–61), the *Almagest* had moved to the centre of an acrimonious philosophical and astronomical controversy between two expatriate Greeks in Italy. Intellectual historians are familiar with George of Trebizond's attacks on Plato and Cardinal Bessarion's defences of the latter, but the astronomical and astrological dimensions of that conflict are poorly integrated into the history of astronomy. Since this part of the controversy not only shaped Regiomontanus's career but also changed the course of astronomy, it deserves close attention, beginning with some background on the protagonists.

As his name fails to suggest, George of Trebizond (1396–1472) came from Crete, where his Trebizond ancestors had settled several generations earlier. George himself moved to Italy as a twenty-something, learned Latin with Guarino da Verona, and translated Greek works for Pope Nicholas V. He also taught rhetoric; one of his students was Pietro Barbo, the future Pope Paul II. In contrast, Cardinal Bessarion was a patriotic native of Trebizond itself. A student of the Platonist George Gemistos Pletho, Bessarion had become a Basilian monk and risen rapidly in the Orthodox Church hierarchy. In 1438 he travelled to Italy as the youngest member of the Byzantine imperial delegation to the Council of Florence, which had convened to unify Eastern and Western Christendom in the face of growing Ottoman pressure. Bessarion helped to negotiate a historical reconciliation. This settlement, however, barely outlasted the council. The elites could not bring to their side an Orthodox world still seething from the Latins' plunder and thirteenth-century occupation of Constantinople. The aftermath of the council turned Bessarion's life upside down. Considered a traitor at home, he was appointed a cardinal of the Latin Church and returned to Italy permanently. His library, the result of a multidecade effort, had assembled one of the best collections of Greek and Latin manuscripts in Europe, including many Greek mathematical and astronomical works. Bessarion eventually willed his library to Venice, where it remains the crown jewel of the Biblioteca Nazionale Marciana.

Although cordial at first, the relationship between Bessarion and George of Trebizond soured in the early 1450s. To secure papal patronage, George translated the *Almagest* from the Greek as a much-needed replacement of Gerard of Cremona's twelfth-century Latin translation from the Arabic version. Bessarion lent him his best Greek manuscript of the *Almagest*, recommending Theon of Alexandria's commentary as a guide. By December 1451, after nine months' work, George had completed both his Latin *Almagest* and a 300-folio commentary on it. These labours immediately drew fire. The pope's referees rejected the commentary as error-filled, and Bessarion was angered by George's attacks on Theon.[36] Although temporarily stymied, the cardinal's search for a worthy introduction to Ptolemy eventually succeeded and changed the course of Latin astronomy in the process.

After Constantinople fell to the Ottomans, the relations between the two men deteriorated further. In politics and religion, George began to see Mehmed II as a chastiser of Byzantine corruption who might become the Christian world emperor anticipated by prophecy. In philosophy, George's *Comparatio philosophorum Aristotelis et Platonis* (1455) was an apologia of Aristotle that attacked Plato and his followers, especially Pletho and Bessarion. George's animosity toward Plato was not so much philosophical in the narrow sense as biographical, "nationalistic," ethical, religious, and even eschatological, as we shall see. It was a measure of George's hatred for Plato that he saw the Prophet Muhammad as the second Plato and deemed him better than the first. Since Bessarion considered Aristotle and Plato complementary and esteemed both highly, he leapt to the defence of Plato.[37] The original intellectual and personal conflict between the two men widened and intensified as political events and ecclesiastical factors exacerbated their differences.

For years, Bessarion had advocated a crusade against the Ottomans, and he would do so until his death. George of Trebizond had initially favoured a crusade as well. As the fall of Constantinople became imminent, however, he secretly adopted another approach. A few weeks after the Ottomans took Constantinople in late June 1453, George wrote a treatise addressed to Mehmed. In it, he warned (in John Monfasani's paraphrase) that God "had given him [Mehmed] the city of the universal Christian monarch Constantine in order that he might emulate Constantine and unite all men under one scepter and one religion." George then demonstrated how Islam and Christianity could be reconciled. He concluded by calling Mehmed "'the king of all the earth and heavens' and by attributing to him the title of 'King of Kings,' and by divine right no less."[38] George had already been accused of consorting with Mehmed in the 1450s. His enemies now found serious new grounds for suspecting him of treason.

Meanwhile, Bessarion's attempts to garner imperial support for a crusade led him to Vienna. During his long diplomatic visit (1460–61), Peurbach and Regiomontanus impressed him by their competence in astronomy. Bessarion accordingly convinced Peurbach to write an epitome of the *Almagest* that would displace George's work on the subject. Peurbach set to work promptly, drawing on Gerard of Cremona's translation, which he allegedly knew "almost by heart," and on the *Almagestum parvum* (a synopsis of the first six books).[39] He had completed half of the *Epitome of the Almagest* when he died suddenly in April 1461. Political unrest trapped Bessarion in Vienna until September, when he finally left for Italy. Regiomontanus accompanied him and remained a member of the cardinal's *familia* until at least 1465 (in 1467, he turns up in Bratislava/Pressburg). In the Bessarion circle, he significantly improved his Greek, which he had begun to study in 1454.[40] He not only revised Peurbach's first half but also wrote the remainder of the *Epitome of the Almagest*, completed in about 1462, which he dedicated to Bessarion. The *Epitome* remained a limited-circulation manuscript until Johannes Hamann printed it in Venice in 1496, twenty years after Regiomontanus's death. Its existence, however, was no secret. Regiomontanus had advertised it in his printing prospectus and cross-referenced it ("in breviario Almaiesti") in his *Disputationes* (ca. 1475), which circulated even more widely after Erhard Ratdolt included it in his successful compendium of introductory astronomical texts of 1482 and 1485, a genre much imitated by other printers.

THE *EPITOME OF THE ALMAGEST*

Although Peurbach had drafted half the book, it is not unfair to see it, as the publisher did, as the work of Regiomontanus, who wrote the crucial second half and revised the whole. The *Epitome* is neither a translation (an oft-repeated error) nor a commentary but a detailed, sometimes updated, overview of the *Almagest*. Swerdlow once called it "the finest textbook of Ptolemaic astronomy ever written."[41] It granted Bessarion's wish for a "condensed and clearer" exposition of Ptolemy, a constraint that explains why the work omits some of Regiomontanus's own cherished views. Even so, his "digest" ran to more than two hundred folio pages in print.

The *Epitome* follows the thirteen-book structure of the *Almagest* but with a more Euclidean layout than the original. Each book is organized into propositions, many followed by proofs, a structure that Peurbach had set up along the lines of the *Almagestum parvum*. As the work of a practising astronomer, however, the *Epitome* sometimes updates the *Almagest* by commenting on post-Ptolemaic developments. Regiomontanus discusses improved parameters and brings newer observations and theoretical work from the Islamic astronomical tradition to bear on his exposition, making extensive use of Albategnius (al-Battānī) and Geber (Jābir ibn Aflaḥ) in particular.

Peurbach may have been more mindful of Ptolemy's voice, which is unmistakable at the beginning of book 1 of the *Epitome* (even in the first person, a rarity). The summary of book 1, the most natural-philosophical part of the *Almagest*, does not hint at the late-medieval natural-philosophical debates about the possible rotation of the Earth. Indeed, the list of six "conclusions" about the centrality and immobility of the Earth leave the discussion in the second century. A fascinating disclaimer follows: "It seemed agreeable to express the preface of Ptolemy literally, both because of the many opinions in it that are eminently worth knowing, and especially because of the authority of Ptolemy, to which our *imitatio* might be rendered more faithful."[42] These words are almost certainly Regiomontanus's, for they are not in the earliest extant version of the *Epitome*.[43] The reader was to understand that much more might have been said about these "opinions." Whatever the ambiguities of the term *imitatio*, the *Epitome* was not a commentary.[44] The critical judgment and corrections that Regiomontanus expected in commentaries – and that he found wanting in those on the old *Theorica planetarum* – were therefore inappropriate in his "abbreviation" or "epitome," which was the task at hand.[45]

Accordingly, the skepticism about epicycles and eccentrics that Regiomontanus expressed in the nearly contemporary "Letter to Vitéz" (1460) does not surface in the *Epitome*.[46] Even so, Regiomontanus could not refrain from noting problems with Ptolemy's lunar theory. In book 5, proposition 12, of the *Epitome*, he points out, as Gersonides and Langenstein had done before him, the contradiction between observation and the four-fold growth in the Moon's apparent size implied by Ptolemy's model.[47]

Note briefly here (more below) that in book 12, propositions 1–2, of the *Epitome*, Regiomontanus proved the equivalence of epicyclic and eccentric models for the second anomaly of the planets. These proofs once seemed to be silent corrections (hence implicit criticisms) of Ptolemy's claim in the opening of book 12, chapter 1, of the *Almagest* that the eccentric model was not possible for the inferior planets. The *Defensio Theonis contra Georgium Trapezuntium*, however, shows that Regiomontanus considered them to be – odd though this seems to us – expressions of Ptolemy's own intention (see below).[48]

Once he had finished the *Epitome*, Regiomontanus took full advantage of his sojourn in Italy. Bessarion's library gave him access to an exceptional collection of about one thousand Greek and Latin manuscripts, including several Greek *Almagests*, Proclus's *Hypotyposis astronomicarum positionum*, Theon of Alexandria's *Commentary on the Almagest*, and other hard-to-find Greek works. One such was the only surviving manuscript of the astronomical portion of Theon of Smyrna's *Mathematical Knowledge Useful for Reading Plato* (early second century; slightly pre-Ptolemaic).[49] A title of this sort is likely to have drawn Regiomontanus's interest, all the more so as Bessarion thought highly of the work.[50] Intriguingly, Theon of Smyrna treats at some length the differences and equivalences between the eccentric and the epicyclic models of the planets. After proving their equivalence for the Sun, Theon of Smyrna asserts that the same demonstration applies to the other planets.[51] He favours the epicyclic model as being in greater accordance with nature, whereas the eccentric is an "accident" – effectively the trajectory of the planet moving on the epicyclic sphere. Theon of Smyrna even chides the second-century BCE astronomer Hipparchus of Nicaea for being insufficiently attuned to the science of nature (*physiologia*) and thus for failing to understand the difference between the two models.[52] This ontological difference between the epicyclic and eccentric models, despite their mathematical equivalence for Theon of Smyrna, may help to elucidate Ptolemy's context, even if it does not help us to understand Ptolemy's own position. Although Ptolemy stated that the two models were equivalent for the second anomaly of the superior planets, he famously and puzzlingly denied their equivalence for the inferior planets in book 12, chapter 1, of the *Almagest.*

Yet another intriguing angle on book 12 of the *Epitome* is Proclus's reference in his *Hypotyposis astronomicarum positionum* to one of his students' proof of the equivalence between the eccentric and epicyclic models.[53] As shown by both his printing prospectus and the autograph of his *Defensio* (on which more below), Regiomontanus appreciated the *Hypotyposis,* which he titled *De sufformationibus* and which he had planned to print.[54]

Regiomontanus's access to new books was complemented by encounters with new people and experiences, more difficult to detail but nevertheless very important. In 1463 he entered into correspondence with Giovanni Bianchini, the pre-eminent mathematician of the Italian Peninsula. Bianchini had worked for the House of Este in Ferrara, had been a publicist for better knowledge of Ptolemy's *Almagest,* and had computed new tables, which he had dedicated to Emperor Frederick III (Regiomontanus had copied these tables in 1460, during his Viennese years).[55] The surviving correspondence with Bianchini is a remarkable testimony to Regiomontanus's mathematical skills, to his dissatisfactions with existing tables and mathematical models, and to his expectations about the consistency of physical and mathematical predictions. Regiomontanus wrote one particularly rich letter to Bianchini in 1464, in which, after praising the work of Battānī and others on the luminaries, he notes that "they left the other five planets nearly untouched." He goes on to address the inconsistencies between large variations in the predicted planetary distances of models and the failure to observe anything close to such changes for the areas of the Moon (4-fold), Venus (45-fold), and Mars (52-fold), implicitly blaming the eccentric and epicycle models.[56] This criticism is consistent with his hopes for the advent of a concentric astronomy.

Partly alerted by reading some of his predecessors and partly stimulated by the completion of the *Epitome,* Regiomontanus had been working closely with the *Almagest.* He chose to mention only a few of the issues in that book but had evidently uncovered enough to require another medium to air them. In his correspondence with Bianchini, he informed the latter in late 1463 or early 1464 that he had already completed two books of his projected *Problemata Almagesti,* mostly on mathematical problems.[57] More were underway, intended to cover the whole *Almagest.* In the mid-1470s, Regiomontanus's printing prospectus promised an edition of the *Problemata.* In 1496 so did Giambattista Abioso's preface to the *Epitome.* Neither materialized. The manuscript, which

Regiomontanus had in Nuremberg, apparently passed into the possession of Johannes Schöner, who also evidently intended to print it. The work disappeared in the sixteenth century. Only a few draft fragments survive in several of Bessarion's manuscripts, as do scattered references elsewhere.[58]

In 1464 Regiomontanus visited Padua, where he lectured on Alfraganus (al-Farghānī). Only his April inaugural lecture (the *Oratio*) survives, but it offers a grand historical vision of astronomy as the apex of the mathematical sciences – a high valuation that is very much in line with the Ptolemaic schema in book 1, chapter 1, of the *Almagest*. The *Oratio* reiterates the goal of constructing a concentric astronomy as a worthy one – Averroes (Ibn Rushd) would have earned glory if he had achieved it – and expresses Regiomontanus's excitement about seeing Giovanni Dondi's astrarium in Pavia.[59] In passing, we also learn that he knew the *Persian Tables*, the Trebizondian Gregory Chioniades's fourteenth-century Byzantine version of twelfth-century, pre–Marāgha School tables. Nicholas of Cusa had translated some parts into Latin, but Regiomontanus also had access to four Greek copies in Cardinal Bessarion's library.[60]

In the late summer of 1464, Pope Pius II (Aeneas Silvius Piccolomini) was both at death's door and in the Adriatic port of Ancona to support the imminent departure of a crusading fleet for Constantinople. Nothing came of these efforts. Pius's death dashed once again the hopes for the crusade that Bessarion had long advocated. On this low point followed a brief moment of great anticipation for the Bessarion circle. In the upcoming papal election, Bessarion would be not only an elector but also a possible candidate. Soon thereafter, Regiomontanus wrote a short dialogue, set in Rome, just outside St Peter's, as the conclave was about to start. In its earliest version, the protagonists are Ioannes and Martinus, none other than Regiomontanus himself and Martin Bylica of Ilkusch, astronomer-astrologers attached respectively to Cardinal Bessarion and perhaps Cardinal Pietro Barbo.[61] As the two men wait for the conclave to start, Martin fetches a copy of the *Theorica planetarum communis*, a text that he trusts. Regiomontanus, however, attacks its errors and instead extols the virtues of Peurbach's *Theoricae novae planetarum*, emphasizing both mathematical and physical considerations. He appeals notably to Peurbach's solution of such problems as vacua and collisions of spheres, very much in the tradition of homocentric critiques of Ptolemy. The dialogue ends just as the cardinals enter the conclave. Thanks to this artifice, the protagonists do not know the election result, so profoundly disappointing to the Bessarion circle: the new pope was not their cardinal, but George of Trebizond's former student

Pietro Barbo (Pope Paul II). Regiomontanus would publish the dialogue around 1475, more than a decade later, three years after his patron, his enemy, and Paul II had all died.

In the intervening years, the election of Paul II evidently gave George an extra measure of audacity. The ill will between himself and Bessarion culminated soon thereafter in a potent mix of geopolitics, religion, and astronomy. In 1466 the cardinal learned that George had recently written two letters to Mehmed II dedicating his *Commentary on the Almagest*, among other works, to the conqueror of Constantinople. Beyond his plea for patronage, George exalted the sultan as the most pre-eminent of all world emperors, urged him to convert to Christianity, and suggested that he conquer Rome.[62]

In his first letter to Mehmed, George concluded by mentioning his attacks on Geber and Theon of Alexandria in his *Commentary on the Almagest*. In the second, he touted again his works on the *Almagest* and denigrated George Gemistos Pletho (arch-Platonist and Bessarion's teacher). After exposing the letters, Bessarion convinced the reluctant Pope Paul II to imprison George for four months in the Castel Sant'Angelo in 1467.[63]

This episode set up the Hungarian phase of Regiomontanus's career. The move is usually explained by the involvement of Regiomontanus and Martin Bylica in the new university that Matthias Corvinus founded in 1467 in Bratislava (Poszony/Pressburg), today the capital of Slovakia.[64] The likelier rationale for Regiomontanus's trip was that, touting his astrological skills, the recently freed George of Trebizond had now decided to dedicate his *Almagest* translation and commentary, among other works, to King Matthias Corvinus of Hungary.[65]

This latest of George's quests for patronage and geopolitical influence gives genuine urgency to Regiomontanus's move to Hungary. Bessarion would have had excellent reasons for assigning this task to Regiomontanus. First, the astronomer had access to the highest levels of power. János Vitéz, for whom Regiomontanus had sketched his homocentric solar and lunar theories in 1460, was now archbishop of Esztergom, the king's chancellor, and the other object of George's entreaties. Second, Regiomontanus was in the midst of writing the *Defensio Theonis contra Georgium Trapezuntium*, a work intent on demolishing George's *Commentary on the Almagest*, one of the two works that George planned to dedicate to Corvinus. The goal of finishing the *Defensio*, which Regiomontanus had probably begun while writing the *Epitome*, became more urgent. George now sought to influence the Hungarian king, a quest that had serious repercussions for the Ottoman question as well as for himself.[66] Regiomontanus obtained George's dedication of his commentary to Corvinus, a copy of which is bound into his *Defensio* autograph. In the manuscript, it follows Regiomontanus's own dedication of this work to Corvinus, drafted well before he completed his demolition of George's commentary.[67] In the *Defensio*, he set out to do for Corvinus, and in much greater detail, what Jacobus Cremonensis had done so successfully to George's *Almagest* commentary when the latter had sought papal patronage almost two decades earlier.

For Bessarion's benefit, the *Defensio* was also a vindication of Theon of Alexandria's *Commentary on the Almagest* (Regiomontanus had accused George of plagiarizing it when he was not attacking it).[68] Corvinus apparently never saw the final version of that work, which Regiomontanus completed in Nuremberg, after he had left the Hungarian court. It survives in a single autograph with no colophon, but it seems complete, ending in book 13 with a discussion of the penultimate chapter of George's *Almagest* commentary. In 1474 Regiomontanus still planned a wide diffusion for it: he listed the *Defensio* in his printing advertisement. Although George of Trebizond had died in 1472, his writings, still promoted by his sons, deserved a refutation. Regiomontanus's own death in 1476 ended those printing plans, on which no subsequent printer followed up either.

Even though the *Defensio* did not circulate, its significance for understanding Regiomontanus's outlook is considerable. As a very negative supercommentary, it follows the structure of the *Almagest*, focusing only on what George got wrong. Regiomontanus's autograph runs to 302 folios, which transcribed to more than 950 typescript pages. The *Epitome* had summarized Regiomontanus's best reading of Ptolemy's intentions, presented in an expository, nonpolemical vein. By contrast, the sharp and often personal attacks in the *Defensio* make explicit Regiomontanus's assumptions, predilections, and intellectual antipathies, which are usually deeply hidden between the lines of his formal writings. Despite its geometrical demonstrations, the *Defensio* reads like a giant letter, variously addressed in the second person to the reader, to George himself, and rhetorically to a tribunal of Ciceronian judges who are hearing his prosecution ("o iudices"). Accordingly, Regiomontanus's asides and flashes of anger are a still unexplored source for a fuller grasp of his astronomical commitments, including many technical details. In terms of the "big picture," Regiomontanus continues to hope for an astronomy that integrates physical and mathematical considerations. The *Defensio* thus shows his conflicting sympathies – Ptolemaic, homocentric, and Peurbachian – pulling him in several directions that I examine in turn.

First, despite its title, the *Defensio Theonis contra Georgium Trapezuntium* is also occasionally a defense of Ptolemy: Regiomontanus is unrelentingly eager to correct every one of George's misunderstandings and misinterpretations.[69] That said, the task of "getting the *Almagest* right" for both intellectual and polemical purposes does not prevent Regiomontanus from criticizing what he takes to be Ptolemy's approach. In particular, he points to the latter's failure to take the physical aspects of astronomy seriously enough, a flaw illustrated by many models that rely on mere circles rather than spheres. Instead, Regiomontanus promotes the vision of a complete astronomy "that not only accommodates computation to the appearances, but also truly imparts a complete knowledge of the figures of the celestial bodies." For Regiomontanus, this integration of computations with physical considerations went hand in hand with the explicit rejection of astronomy as a "fictitious art [or discipline]."[70]

In addition, the *Defensio* shows that, consistent with this criticism, Regiomontanus still hoped to find homocentric solutions to astronomical modelling, probably to his dying day. First discussed in his "Letter to Vitéz" (1460), this ideal reappears in his correspondence of 1463–64 with Giovanni Bianchini and in scattered passages of the *Defensio.* Although Regiomontanus sharply criticized Biṭrūjī's homocentric system for its flaws and its ordering of the planets, he did so as a fellow labourer in the homocentric camp. He drew out and modified Biṭrūjī to a surprising extent.[71] Despite their differences in approach and skill, the two shared a vision of astronomy that insisted on joining mathematical computation to physical coherence. Indeed, even when he was not pressing for the value of concentrics, Regiomontanus echoed Biṭrūjī in criticizing Ptolemy for being too two-dimensional.

Finally, Regiomontanus remained an advocate for Peurbach. There is no question that Regiomontanus rated Peurbach's combinations of the *Theoricae novae planetarum*'s partial orbs as an improvement on the two-dimensionality he saw in both the *Almagest* and the old *Theorica planetarum communis.* Equally clearly, the *Theoricae novae planetarum* was not close to Regiomontanus's dream of a complete homocentric astronomy in the *Letter to Vitéz.* Far from eliminating epicycles and eccentrics, Peurbach's work physicalized them without, however, saying anything about their physical characteristics. Given Regiomontanus's homocentric propensities, then, his endorsement of Peurbach's *Theoricae novae planetarum* must have been more pragmatic, comparative, and perhaps even reverential than absolute.[72]

In short, Regiomontanus faced a trilemma that left unresolved the tensions between the pros and cons of his three options. This crisis was an individual one, not – as far as we know – shared by the community of astronomers.[73] What were its consequences for Regiomontanus's critical outlook? One of the most striking was his attempt to keep open issues that many of his contemporaries considered settled and to demand evidence for what they were willing to take for granted. Regiomontanus's polemical stance toward George of Trebizond certainly exacerbated this skeptical outlook. The latter's commentary on book 9 of the *Almagest*, for example, treats as secure the order that Ptolemy was assumed to have assigned to the planets.[74] Where George saw obviousness, necessity, and consistency, Regiomontanus saw contingency, doubts, and other possible alternatives. Of course, he had had many predecessors on this score.[75] Without casting aspersions on Ptolemy, Regiomontanus thus pointed out that grouping the planets according to the solar linkage of either their epicycles (i.e., the superior planets) or their longitudinal motion (i.e., the inferior planets and the Moon) was a rhetorical, or nondemonstrative, argument for their location and order.[76] In other words, it did not compel assent, and the apparent rationale for a grouping seemed odd: "The three superior planets are tied to the Sun by their epicycle; the three inferior, not by their epicycle, but by their motion in longitude." Regiomontanus went on to ask why nature did not group together the planets with retrograde motions. The criteria for a ranking were not obvious. Behind his question is the implication that the Sun seems out of place in the midst of the planets with retrograde motion. Why does it occupy this position? Regiomontanus also wonders why an argument for arranging the planets by sex is overlooked – with the females (the Moon and Venus) lowest, the males highest, and the hermaphrodite in between.[77] The alternatives associated with Regiomontanus's questions lie at the interface among mathematical, astrological, and physical considerations. In keeping with his habit of raising fundamental questions, Regiomontanus treated the order of the planets as an unsolved problem illustrated by the contradictory stances of Ptolemy, Martianus Capella, Geber, Biṭrūjī, and others. Copernicus would work on precisely this problem and was thrilled to see that reordering the planets (and the Earth) around the mean Sun gave their spheres a necessary order.[78]

Although his years in Hungary advanced his work on the *Defensio*, Regiomontanus found himself in an untenable personal position before he could finish the project. Deep political and personal rifts developed between King Matthias Corvinus and his chancellor, Archbishop János Vitéz, whom the king accused of treason. Regiomontanus was thus caught between his two powerful feuding patrons. He had known Vitéz for more than a decade and had dedicated his *Tabulae directionum* to him. To the king, he had dedicated his *Tabula primi mobilis* and planned to do likewise for the *Defensio* (his draft of the dedication survives in the autograph). He evidently chose simply to step out of his impossible situation. His correspondence suggests that he seriously considered a return to the university, this time to Erfurt.[79]

THE NUREMBERG YEARS (1471–75)

In the end, he settled in Nuremberg (1471), where his career took a new turn. He set up the first press devoted primarily to the mathematical sciences, even as he planned to improve planetary predictions with better observations.[80] In 1472 George of Trebizond, Bessarion, and Vitéz all died. However, as his press got underway, he forgot none of them. His publication program developed into an effort systematically to print what he deemed the best and most useful works in the mathematical sciences, duly corrected of their textual errors. About 1472 the first two works he issued mixed classical and contemporary material: Marcus Manilius's long, didactic, first-century poem *Astronomica* (one of Vitéz's favorite works) and Georg Peurbach's *Theoricae novae planetarum.*[81] The next works were number-filled: his massive *Ephemerides* and his thirty-year *Calendar* (in Latin and German editions) in 1474.

With these two works in press, he also issued a trade list that publicized not only his four editions in print to date but also his ambition to produce forty-five others, as well as maps and instruments.[82] His long list included both standard texts that he valued without regard to our historical periodization and some twenty of his own works. Among the latter were two born of the feud between Bessarion and George of Trebizond: the *Defensio Theonis contra Georgium Trapezuntium* and the *Epitome of the Almagest* (which he called his *breviarium*). Indeed, his proposed output had an adversarial undercurrent: six works in his list contain the word *contra.*

His last edition, which he did not title, was equally antagonistic. Since it does not appear on his trade list, he presumably decided rather suddenly to print the work now known as *Disputationes contra deliramenta cremonensia* (ca. 1475), perhaps with his Italian trip in mind.[83] His most important change to the draft manuscript was to add a preface aimed at the critics of his contentious printing program, some of whom he perhaps expected to encounter in Italy. Widely reprinted in astronomical compendia of the fifteenth and sixteenth centuries, this dialogue marks a watershed in astronomical pedagogy. It undermined the credibility of the older, widely used *Theorica planetarum communis.* Readers and printers paid attention. Multiple editions of the *Disputationes* diffused widely Regiomontanus's criticisms of the old *Theorica planetarum communis,* which fell into disuse, and promoted his advocacy of Peurbach's *Theoricae novae planetarum* (issued in some fifty editions into the seventeenth century).[84] Beyond berating the old *Theorica planetarum communis* for not treating planetary models in three dimensions, Regiomontanus refuted its claim that the motion of any point on Mercury's deferent was uniform about the equant point. The uniform rotation of Mercury's epicycle *centre* about the equant point (i.e., equal angels in equal times) could *not* be generalized to the rest of the circumference of that deferent.[85]

What was the significance of this point? For those who knew the *Almagest* or Campanus of Novara's *Theorica planetarum* well, there was no news here. But for the many students whom the *Theorica planetarum communis* had introduced to astronomy, the error suggested that the equant point could function as the centre of uniform rotation of a rigid sphere or ring (Mercury's deferent in this case). It was *not* the case that every point on its circumference moved with uniform motion. The analysis that had driven Langenstein to do away with uniform motion as a sine qua non of astronomical theory was now highlighting much more broadly the unresolved tension between physical spheres and the equant point.

In addition to the content of his works and program, Regiomontanus proved versatile as a creative printer who pioneered designs and typographic techniques. He did not shy away from complications. He printed with a single pull of the press the text of his Manilius together with its decorative, white-vine, constructed initials. In his edition of Peurbach, he inaugurated the printing of astronomical and geometrical diagrams and raised their level of complexity in later editions. In the *Theoricae novae planetarum*, the planetary diagrams were hand-coloured as a group. Copy after copy displays the same shade of aquamarine for the diagram of the first solar *theorica*, the same shades of yellow and brown for the lunar eclipse diagrams, and so on. Whether or not the colouring was done in-house, Regiomontanus took steps to standardize the decoration of the entire edition.[86] In the Peurbach (ca. 1472), the diagrams run the width of the text (easier to lay out). Around 1474–75, in the *Disputationes*, Regiomontanus was imbedding the geometrical diagrams in the text, approximating his own practice in some of the astronomical manuscripts he had copied in Vienna.[87]

Thanks to Regiomontanus's Herculean efforts in computation and his pioneering designs and typography, finding daily planetary positions became even simpler. The astrologer with access to his massive *Ephemerides* of 1474 no longer needed to calculate daily planetary positions: he could simply read them off – for the next thirty years. The convenience of simply looking up a position or carrying out a small interpolation not only saved the average practitioner much laborious calculation but also lifted the burden of understanding the intricacies of the astronomical models themselves. To print the *Ephemerides* and *Calendars*, Regiomontanus introduced yet another typographical innovation. He printed his data in tabular form, constructing a dense numerical grid from long thin horizontal metal lines interspersed with type-sized vertical metal bars – to say nothing of his eclipse diagrams and use of red ink.

His Latin and German calendars (*Calendarium* and *Kalendar*) of 1474 used even more media. At the back, Regiomontanus appended a thick, cardboard-like sheet with instruments. On the recto were volvelles and thread markers for lunar motion, and on the verso was a universal sundial (*quadratum horarium generale*) equipped with small, riveted, articulated brass arms and a thread.[88] His fastidiousness extended to typographical errors. When he caught them during his print runs, he stopped the press to correct them. Those he found thereafter, he corrected by hand in the copies still in his possession.[89]

By comparison with his ambitions as a printer, Regiomontanus's production was modest. However, his program and his editions proved significant, both as artifacts and as inspiration. After Regiomontanus's death in 1476, Erhard Ratdolt in Venice reprinted some of his predecessor's output and made good on some of his plans (e.g., by publishing Euclid's *Elements*), as Johannes Schoener and Johann Petreius did later in Nuremberg.[90]

Among the works that Regiomontanus left unfinished was a piece of complex astronomical clockwork. Inspired by having seen Giovanni Dondi's fourteenth-century astrarium in Pavia before 1464, he was building one himself in Nuremberg. Like Dondi, he may have seen his device as the material embodiment of physical constraints and mathematical theory. His printing advertisement bragged about this "near-miracle" and implied that it was almost finished. The design and the fate of the instrument are unknown.[91]

After moving to Nuremberg, Regiomontanus began a cooperative observational venture with the goal of producing new tables. In one of his letters, he criticized the predictions of all known tables, noting his own observations in Vienna, Rome, Padua, Hungary, and Nuremberg. He also made overtures to potential collaborators, asking them for their observational data.[92] Regiomontanus had long been a careful observer. In his new quarters, however, he started a systematic observation program before the fateful 1475 trip to Rome, from which he never returned. After Regiomontanus's death, his associate Bernard Walther (d. 1504) pursued a thirty-year program that yielded several hundred high-quality observations in Nuremberg, for which he eventually used a large zodiacal armillary. It was once thought that Walther did not intend his observations to improve astronomical theory (even though others, including Copernicus, did later use them in this way). Walther's records of conjunctions and stationary points may, however, have served horoscopic purposes directly (e.g., in relation to actual births or deaths) since he knew that the *Alfonsine Tables* were inaccurate, as Regiomontanus had emphasized in a letter to Bianchini.[93]

COPERNICUS'S USE OF REGIOMONTANUS'S WORK

Like many of his Cracow predecessors who came to Bologna, Copernicus was adept in astronomy before arriving in Italy. He was familiar with Regiomontanus's reputation and some of his works, which were already studied at Cracow thanks to Martin Bylica of Ilkusch.[94] Copernicus owned Regiomontanus's *Tabulae directionum* (Ratdolt's 1490 Augsburg edition), which he had bound with his *Alfonsine Tables* (1492), probably early on (the binding is Cracovian).[95] In addition, he probably knew of Albert of Brudzewo's commentary on Peurbach's *Theoricae novae planetarum,* which also drew attention to Regiomontanus.[96]

After studying at Cracow, Copernicus travelled to Bologna to study law (1496–1500). Having briefly returned to Poland, he went back to Italy to study medicine in Padua (1501–03), eventually receiving his law degree from the University of Ferrara. He thus crossed the Alps at least four times. The shortest route went through Vienna, but since Copernicus was aware of Regiomontanus's importance, a Nuremberg detour may have appealed to one interested in cutting-edge work in astronomy.[97]

When Copernicus arrived in Bologna in 1496, the longstanding reputation of Cracow masters had preceded him: they had dominated the chair of astronomy/astrology at mid-century. While studying law, Copernicus became an associate of Domenico Maria Novara, who held Bologna's chair of astrology in the 1490s. The scanty evidence about their relationship points to a close cooperation between the two men, including occasional joint observations.

For reasons that remain obscure, Novara called Regiomontanus and Bianchini his teachers.[98] Whether personal or textual, this filiation makes it likely that Novara was aware of the *Epitome*'s existence before it appeared in print. Copied in Ferrara in 1493, one manuscript that has largely escaped notice shows that specialists tried to obtain the work. Its many scribal errors (often corrected, presumably by the owner) indicate that it was copied from a text different from Bessarion's clean and highly legible manuscripts (now in the Biblioteca Nazionale Marciana).[99] In short, there must have been yet another manuscript in circulation.

Thanks to the press, access to the *Epitome* improved quickly. In 1496, the year of Copernicus's arrival in Bologna, Johannes Hamann printed *Regiomontani Epytoma Almagesti* in Venice. Novara the Bolognese astrologer and Copernicus the astronomically competent law student from Cracow were among its earliest potential readers.[100]

We do not know when Copernicus first encountered the *Epitome*, but he would eventually use it extensively. Several generations ago, Ludvik Birkenmajer illustrated Copernicus's dependence on the *Epitome*, which the mathematical analyses of Noel Swerdlow have shown to be pervasive and fundamental.[101] Copernicus drew on Regiomontanus's work at crucial junctures. The earliest are the traces of the *Epitome*'s language in the computations of planetary spheres that preceded the conversion to heliocentrism before the *Commentariolus*, but they also pervade the detailed quantitative implementation of his new theory in *De revolutionibus orbium coelestium*. In between, Copernicus seems to have leaned on the *Epitome* for his understanding of Ptolemy's models and their parameters. There, he also learned about more recent astronomy, including Battānī's post-Ptolemaic revisions, computations, and measurements and perhaps Biṭrūjī's alternative order of the planets.[102] It is of considerable significance that the *Epitome of the Almagest* stressed some of the unfinished business of astronomy, such as the order of the Sun and the inferior planets, to which Regiomontanus explicitly ascribed "no certainty" (*nulla certitudine*) at the beginning of book 9.[103] Most dramatically, the *Epitome* stands behind Copernicus's move to his new astronomical system, which placed not the physical Sun but the mean Sun at the centre of the Earth's orb.[104] The latter case may be summarized as follows.

In the planetary theory of the *Almagest*, Ptolemy tried to account for two primary "anomalies" (or departures from uniform motion) in the motions of the planets. What he called the first anomaly (or "the anomaly with respect to the zodiac") concerned the nonuniform velocity of the planets, including the Sun and Moon, as they moved through the zodiacal constellations. In its twelve-year west-to-east circuit through the zodiac, for example, Jupiter's motion is sometimes faster and sometimes slower. The second anomaly (or "the anomaly with respect to the Sun") is most manifest in the planets' retrograde motion, a temporary east-to-west reversal of their paths against the stars. For the inferior planets, retrogradation occurs near inferior conjunction with the Sun, whereas for the superior planets, it occurs near opposition to (180 degrees from) the Sun – precisely why they are called "anomalies with respect to the Sun." At the beginning of book 12 of the *Almagest*, Ptolemy had stated that the "second anomaly" of each superior planet could be represented by two equivalent models, one epicyclic and the other eccentric.

For the inferior planets, however, he judged the eccentric model impossible (curiously and wrongly). In his *Commentary on the Almagest* (1451), George of Trebizond not surprisingly repeated Ptolemy's claim.

In book 12 of the *Epitome*, however, Regiomontanus says nothing explicit about Ptolemy's position. Instead, he proves without comment that the eccentric and epicycle-and-deferent models are equivalent for the inferior as well as the superior planets. Regiomontanus's lack of commentary deserves one. In the absence of a context, this equivalence proof looks like a silent correction of Ptolemy. F. Jamil Ragep could therefore take Regiomontanus's silence as a sign of borrowing from ʿAlī Qushjī, who had criticized Ptolemy's claim in book 12.[105] The *Defensio*, however, tells a different story. Here, Regiomontanus reiterates his equivalence proof and his diagram, but he also attacks George of Trebizond for claiming that Ptolemy denied the equivalence of the eccentric and epicyclic models for the second anomaly of the inferior planets. In this instance, George's reading of the *Almagest* was in fact correct (although the *Almagest* itself is wrong). Without explaining why the *Almagest* reads as it does at the beginning of book 12, however, Regiomontanus claims that his equivalence proof expresses Ptolemy's own intention, on the grounds that the latter had proven equivalences elsewhere in the *Almagest*.[106] Regiomontanus evidently could not bring himself to believe that Ptolemy considered the eccentric model for the inferior planets impossible. In an apparent fit of fault-finding gone overboard, Regiomontanus dismissed George's correct reading of the text as the latter's failure to understand Ptolemy.

With its proximate origins in this controversy, Regiomontanus's equivalence proof for the second anomaly effectively set up the framework for Copernicus's reordering of the heavens around the mean Sun, as Swerdlow's analysis has shown.[107] For the inferior planets, the eccentric model of the second anomaly – denied by Ptolemy and almost everyone else – transforms directly into Copernicus's planetary system.

Into this new system, glimpsed thanks to the inferior planets, Copernicus also brought the superior planets. Having done so, he had in effect replaced the second anomaly (or "the anomaly with respect to the Sun") of all the planets by the single annual motion of the Earth around a stationary point that Copernicus calls the "mean Sun."

This seemingly curious qualification deserves a comment because, contrary to popular belief, the "true Sun" (or the centre of the physical Sun) was not the centre of Copernicus's universe. The mean Sun is a mathematical point that coincides with the true Sun only twice a year. It is important to understand why the mean Sun mattered to Copernicus's predecessors and why it literally remained central to his own universe.

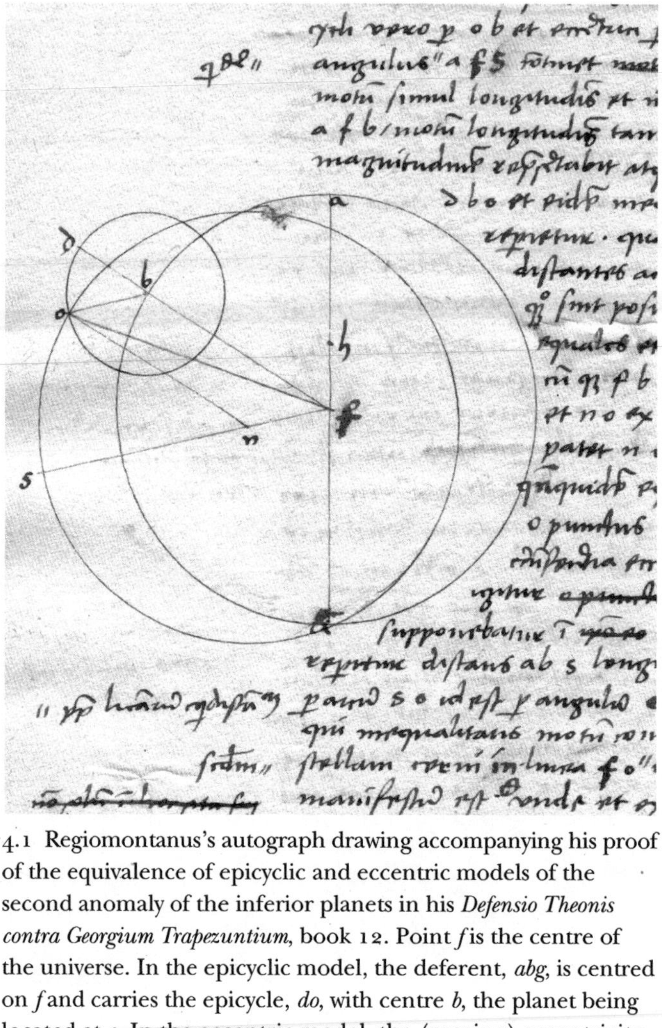

4.1 Regiomontanus's autograph drawing accompanying his proof of the equivalence of epicyclic and eccentric models of the second anomaly of the inferior planets in his *Defensio Theonis contra Georgium Trapezuntium*, book 12. Point *f* is the centre of the universe. In the epicyclic model, the deferent, *abg*, is centred on *f* and carries the epicycle, *do*, with centre *b*, the planet being located at *o*. In the eccentric model, the (moving) eccentricity is *nf*, the eccentric being centred at *n*.

Before Copernicus, the true Sun and the mean Sun each represented both a position and a velocity with respect to the reference frame of the fixed stars. As seen from the Earth, the motion of the (true) Sun against the stars is not uniform. It speeds up and slows down slightly over the course of the year. To predict where the true Sun will be, one can start, as a first approximation, with its mean speed. In short, treat the Sun as if it were moving at a uniform rate around the Earth (e.g., on a uniformly rotating sphere concentric with the Earth). The actual position of the

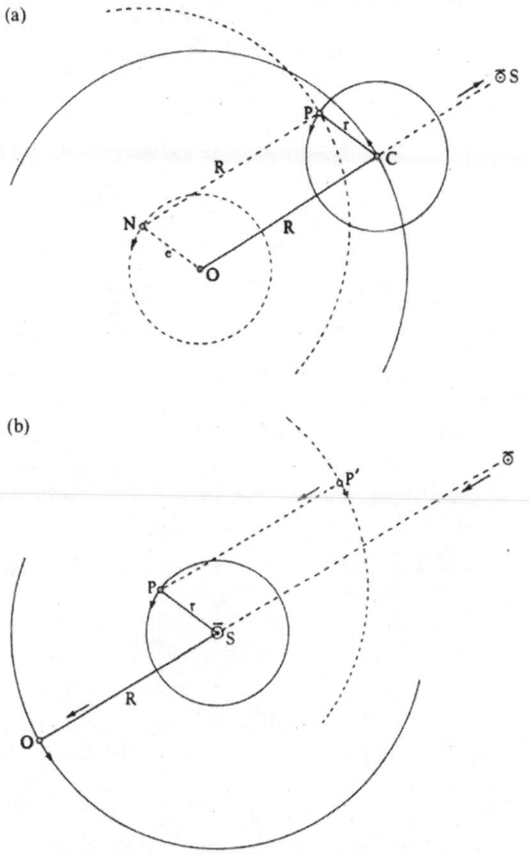

4.2 The epicyclic model, drawn in solid lines in diagram (a), for the second anomaly of an inferior planet, P. The deferent with radius R = OC is centred on the Earth, O, and carries the planet, P, located on the epicycle with radius r and centre C; OC is directed toward the mean Sun, $\bar{S}$. The dashed lines show the *eccentric model* for the second anomaly. The planet, P, is on the larger dashed eccentric circle with radius R = NP and eccentricity e = ON, whose centre, N, moves about the Earth at O at the same rate as P about C in the epicyclic model. The motion of P with respect to the Earth, O, is the same whether one uses e = r and NO (the eccentric model with the dashed sides of the parallelogram) or OC and r (the epicyclic model with the solid sides). Today, we would say that the vectors that define the two models are equivalent. Diagram (b) shows the heliocentric transformation in solid lines. The motion of P with respect to the Earth, O, is still preserved if the mean Sun is no longer merely a direction in space – beyond OC in (a) – but the centre of the planet's motion (SP = r = e). Preserving the geometrical equivalence with the previous models requires putting the Earth, O, in motion around the mean Sun at distance R.

Sun can then be predicted by adding (or subtracting) a correction to (or from) the mean. This is equivalent to modelling the Sun's behaviour by placing it on an epicycle, the centre of which is carried by a deferent. The motion of the epicycle centre, located on the rim of the deferent, is that of the mean Sun. The rotation of the epicycle will then make the true Sun vary on either side of the mean, identified with the epicycle centre.

The importance of the mean Sun goes well beyond its role in solar theory. In classical Ptolemaic theory, it is also a fundamental feature of each planetary model. The epicycles that carry the superior planets all rotate about their centres at the same rate as the mean Sun. That is, the line from the epicycle centre to the planet always points in the same direction as the mean Sun (see the superior planets in figure 4.3). In the models for the inferior planets, it is the deferents that move with the mean Sun: in each case, the line from the Earth to the epicycle centre is locked onto the mean Sun, and the motion of the planet on the epicycle accounts for the positions of Mercury and Venus now ahead of and now behind the Sun (see the centre of figure 4.3).

Copernicus's new system was therefore a geometrical inversion of planetary models in which the "mean Sun" was a foundational element.[108] In the *Almagest*-based epicyclic models for the planets, the epicycles of the superior planets and the deferents of the inferior planets all rotate in lockstep with the mean Sun: there is one 360-degree revolution in 365 and slightly less than one-quarter days. In the alternative eccentric configurations permitted by Regiomontanus's proofs in book 12, chapters 1–2, of the *Epitome*, the centres of the "eccentrics" carrying the superior planets revolve around the Earth in lockstep with the mean Sun.[109]

The clue to Copernicus's mathematical transformation lies in a page of handwritten notes in his bound volume containing the printed editions of Regiomontanus's *Tabulae directionum* and the *Alfonsine Tables*. Swerdlow's analysis of these "Uppsala Notes," as they are known, shows that the results of Copernicus's computations were based implicitly on explorations of Regiomontanus's alternative eccentric model. After assigning twenty-five parts to the "eccentricity" (effectively the distance between the Earth and the mean Sun in Regiomontanus's eccentric model), Copernicus used this standard as his measure to compute the radii of the planets' spheres. He thus discovered that these dimensions fell into a necessary sequence about the mean Sun (i.e., the radius of the orb of Mars was 38, that of Venus was 18, and so on). This nonarbitrary order seems to have convinced him that the Earth (with a radius of 25) could be placed between Mars and Venus and in motion around the mean Sun.[110] The

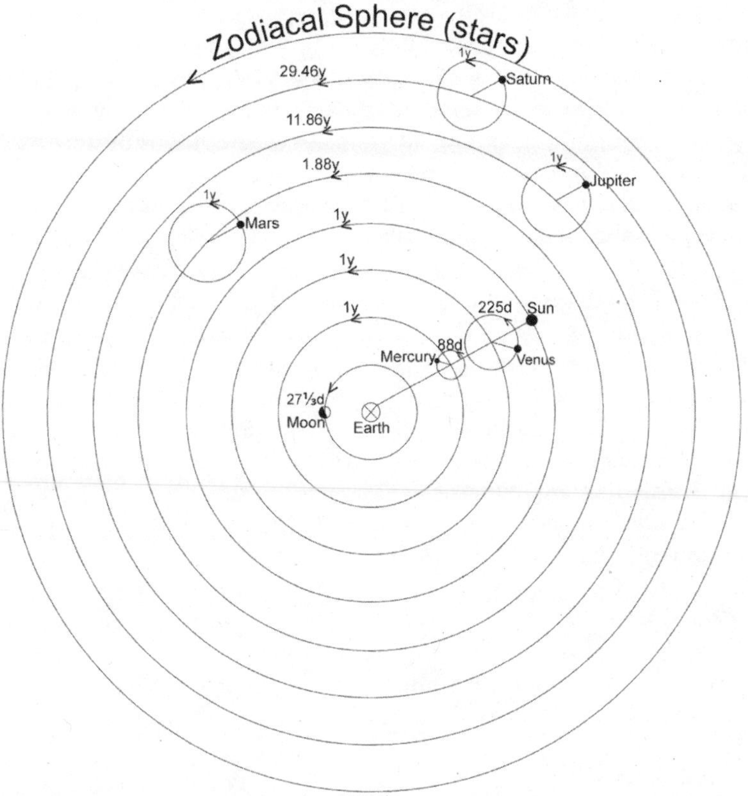

4.3 Ptolemaic schema illustrating the solar linkages of the planets (not to scale). The line from the Earth to the (mean) Sun "carries" the centres of the epicycles of the inferior planets (i.e., Mercury and Venus). Conversely, for each superior planet's epicycle, the line from the epicycle centre to the planet is parallel to the Earth-Sun axis.

motivation for the inquiry and the calculations, however, likely derived from the longstanding problem of the order of the planets.[111] Copernicus's clear dependence on Regiomontanus's *Epitome* for the equivalence of the two models reinforces the notion that, even in his early work, he preferred it to the *Almagest*.[112] Indeed, Copernicus may never have known that Ptolemy had proclaimed the eccentric model impossible for producing the second anomaly of Venus and Mercury. Instead, book 12, propositions 1–2, of the *Epitome* – in fact, a reversal of Ptolemy masquerading as an exposition – would have allowed Copernicus to start with a fully symmetrical "Ptolemaic" foundation: retrograde motion could be produced in all five planets using the geometrically equivalent (but physically different?) eccentric and epicyclic models.

The first known form of Copernicus's new theory appears in a short *theorica* now known as the *Commentariolus*, which he had drafted certainly by 1514 but perhaps as early as 1508. To reach the results recorded in the "Uppsala Notes" and the new system centred on the mean Sun, he used the *Epitome* as a ladder. After his climb, he did not throw it away. On the contrary, he continued to use the *Epitome* as both a source of data and a foil. In the *Commentariolus* itself, Copernicus seems to have set off the postulates behind his new system against the propositions of book 1 of the *Epitome*. In *De revolutionibus*, Copernicus mentioned the *Epitome* much more rarely than he used it, but Georg Joachim Rheticus's *Narratio prima* gives a sense of his teacher's reliance on Regiomontanus. Noel Swerdlow and Otto Neugebauer have documented Copernicus's explicit dependence on Regiomontanus in discussions of Ptolemy's procedures for deriving numerical parameters from observation – to say nothing of summaries of Battānī for his discussions of precession, the tropical year, and the sidereal year.[113]

Not least, Copernicus's faith in the *Epitome* extended to following Regiomontanus in *not* undertaking to derive his astronomical models themselves from observations. Both men believed that, whatever their problems from a physical point of view, Ptolemy's models were basically adequate to their task from the geometrical and predictive points of view.[114]

Thanks to the *Defensio Theonis contra Georgium Trapezuntium*, however, we can appreciate just how skeptical Regiomontanus was about the physical adequacy of Ptolemy's models. If one reads this controversy alongside the hints scattered throughout Regiomontanus's correspondence and other works, the composite picture is one of ongoing dissatisfaction with key aspects of Ptolemaic astronomy. Regiomontanus continued to search for some form of homocentric astronomy that eliminated eccentrics and epicycles, and he criticized the *Almagest*'s "circular," rather than spherical, astronomy. His correspondence stressed the Ptolemaic models' inconsistency with the observed sizes and brightnesses of some planets, suggesting that Regiomontanus would have welcomed models that improved upon Ptolemy's, both physically and predictively. Indeed, he was positively searching for alternatives. Copernicus would make a significant contribution to the first part of that program with his compelling solutions to the classic problem of retrograde motion and the more recent one of "contrary" motions in the heavens – the east-to-west daily motions versus the west-east motions of the seven wanderers (see

Biṭrūjī above). It was Johannes Kepler, however, who would succeed at the most extraordinary task of completely recasting the astronomical models from scratch thanks to his dogged persistence, exceptional insight, and the data from Tycho Brahe's systematic observations.

5

Framing the Appearances in the Fifteenth Century: Alberti, Cusa, Regiomontanus, and Copernicus

Raz Chen-Morris and Rivka Feldhay

INTRODUCTION

IN HIS *Commentariolus* (ca. 1510–14), while postulating the motion of the Earth as a necessary condition for explaining the system of the world, Nicholas Copernicus (1473–1543) denounced the philosophers' defence of the immobility of the Earth in the following words: "For indeed, the things by which the natural philosophers principally try to establish its [the Earth's] immobility, for the most part rest upon appearances; all these things fall down here especially because we are engaged [with something] beyond that very same appearance."[1] Thus it seems that Copernicus's subtle suggestion of a daring reversal of the obtaining hierarchical relation between the visible, regarded as more pertinent to knowledge, and the invisible, often considered by the tradition to be obscure, was central to his astronomical endeavour from its earliest stages. It turns out, he claims, that the immobility of the Earth, which is among our most basic sensual experiences, is just apparent, namely visible but not true. At the same time, he posits the mobility of the Earth – not experienced, hidden from our eyes, or invisible – as a real phenomenon that explains the apparent motions of the stars and the planets. In his recent *The Copernican Question*, Robert S. Westman discusses this passage as part of Copernicus's aspiration to establish ancient Pythagorean speculations as a better logical procedure for saving the appearances than the Ptolemaic Earth-static view.[2] Copernicus's understanding of the "appearances," however, seems to involve more than just questions of logic or the disciplinary relationship between mathematics and physics in the Renaissance. By investigating the relationship holding between observed phenomena and the enquiry of one's point of view, Copernicus testifies to the changing common sense, among practitioners such as Leon Battista Alberti and Paolo Toscanelli, concerning the meaning of the visible and its relationship to the invisible. We argue that the new sensibility, which is at the heart of our chapter, could transfer from mathematicians to a philosopher-theologian like Nicholas of Cusa, intrigue the attention – albeit negative – of Johannes Regiomontanus, and facilitate the acceptance of the bizarre idea of a moving Earth. Thus Copernicus's "assent" to the imperceptible motion of the Earth may be interpreted as a first step toward the emergence of new astronomical practices that showed how something "beyond" the appearances could be transformed into astronomical knowledge.

Recent research has traced the emergence of a new attitude to observational practices in the late Middle Ages and the Renaissance.[3] Katherine Park has argued that in antiquity and throughout the Middle Ages, "observation" (*observatio*) and "experience" (*experimentum*) were two distinct notions with different epistemological connotations. *Observatio* was related to the collection of facts about detached phenomena accumulated by anonymous observers over time. *Experimentum* was the specific and identified act of assessing specific situations, usually through intervention in the phenomena and their manipulation.[4] Park points out that these two divergent practices had merged only in the second half of the fifteenth century in the works of such "experiential" observers as Georg Peurbach, Johannes Regiomontanus, and Bernard Walther, whose "observations became more frequent, more systematic, and less tied to rare events."[5] These new practices, Gianna Pomata suggests, laid the foundations for a novel epistemic genre that replaced experience with an "emphasis on seriality, mathematization ... exact and calibrated measurement ... clear-cut distinctions between direct and indirect experience, and the separation of observation from theory."[6] In this context of emerging new practices of observation, Copernicus's remark, which draws attention to the epistemological status of appearances, acquires additional import and urgency, demanding to reposition the observer and to demarcate the borders between the visible world and the invisible realm anew. These concerns are clearly on Copernicus's mind when, years later, in his *De revolutionibus orbium coelestium* (1543), he returns to the problem of "appearances" while dealing with the way the motion of the Earth reflects upon the motions attributed to the stars and planets by observers on Earth. Neither the risings and settings of the zodiac signs and the fixed stars nor the stations of the planets and their retrogradations are movements in the heavens, he maintains, but a motion of the Earth, "which the planets borrow for their own *appearances*." He further asserts that "[a]ll these facts are disclosed to us by the principle governing the order in which the planets follow one another, and by the harmony of the entire universe, if only *we look at the matter*, as the saying goes, *with both our eyes*."[7] The term "appearances" – translated from the Greek "phenomena"[8] (φαινόμενον) – is rooted in the "visible," in what a spectator's eye is capable of seeing, in the way something is observed. At the same time, it evokes the most ancient and ever-tormenting questions, such as what is the relation between what one sees and what really happens, or to put it differently, are the appearances we see "merely apparent"? Is visibility the essence of knowledge, or is it a necessary but not sufficient condition for acquiring knowledge, or does it only lead to other things that may be invisible, such as mathematical entities, or divine entities, or God himself? And is the truthfulness of visibility confined to the natural realm of seeing, or could it be broadened or deepened by meditation, instruments, or other artificial means?[9] Such questions are echoed in Copernicus's recurring references to the "appearances" and to the acts of "looking at," seeing, and observing. As a humanist, Copernicus could not be blind to, unaware of, or indifferent to the historical resonances of the terms he used. His position calls for a historical examination of the conditions of possibility for the endorsement of the idea that a physical phenomenon – the motion of the Earth, not accessible to the senses – is the cause of the apparent but not true motion of the planets. It suggests that a possible transformation in the understanding of the "appearances" was taking place sometime during the fifteenth century, a conclusion that is much in the spirit of Park and Pomata.

Our chapter focuses on the epistemic hierarchy and shifting boundary between the domain of the visible and that of the invisible as seen from the perspective of fifteenth-century scholars. It was at that time that the demarcating line was being reconsidered through reflection on practices of seeing, observing, and calculating, especially reflection on the position of the observer. The process was involved with the discourse on natural and artificial manners of seeing and informed by a new understanding of human and divine vision. These issues were at the centre of a conversation among mathematicians-astronomers, artists, philosophers, and theologians, as shown below.

Our point of departure is Leon Battista Alberti's (1404–72) *De pictura* (1435–36), which we read as an ambitious project to broaden the scope of the visible that challenged the accepted boundaries between the natural and the artificial. Our analysis focuses on the way Alberti's sensible and mathematical, yet invisible, grid of perspective constitutes the spatial relationships on the surface of the painting and offers a new perception of beauty radiating from things represented to the observer's understanding. Nicholas of Cusa (1401–64) elaborated on Alberti's project with different means.[10] For him, mathematics was not merely a method but also a model used in the constitution of the world for human understanding. In stating his thesis that the intellect is to truth like the polygon is to the circle in which it is inscribed,[11] he revealed the motivation behind his investigations of the quadrature problem, namely to observe critically, from an imagined divine point of view, the limitations of the human intellect. Applying the results of his investigations to the theological realm, Cusa broadened Alberti's discourse on the visible-invisible relationship and provided new kinds of legitimization for naturalizing the invisible within the discourse on human knowledge.

Cusa's preoccupation with mathematical procedures came to the notice of Johannes Müller von Königsberg, known as Regiomontanus (1436–76),[12] probably via the Italian mathematician Paolo Toscanelli (1397–1482), a common friend to both of them. Their critique of Cusa's "speculations" was an attempt to establish the autonomy of mathematical practices (and practitioners) and to protect it from the meddling of philosopher-theologians. The work of Regiomontanus, however, constituted the background for Copernicus's exploration of the possibilities and limits of this autonomy. Unlike Regiomontanus, Copernicus aspired to free the practices of mathematicians not only from theological and philosophical considerations but also from their "passive" subservience to "appearances." This did not mean that appearances were just illusions. Rather, it meant that there were new ways of rigorously reassessing the appearances, not just by elaborating the models for celestial motions and by philosophizing about their ontological nature but also by re-examining the point of view of the observer on Earth. The social network that connected Alberti, Cusa, Regiomontanus, and Toscanelli testifies to the existence in Italy of a cultural field in which mathematicians – both practical like Alberti and theoretical like Regiomontanus – as well as philosopher-theologians like Cusa took a position and articulated their critique of each others' views. This was the field that Copernicus probably acquainted himself with when he came to Bologna in 1496, and this field may have inspired his daring to experiment with the idea of a moving Earth.[13]

In considering the age-old historiographical puzzle of what enabled the competent, cautious astronomer Nicholas Copernicus to embrace the idea of an invisibly moving Earth, one cannot be satisfied with a narrow explanation, as brilliant as it may be. The partiality of a "technical" explanation to the very core of an astronomical system is best exemplified in some of Noel Swerdlow's conclusions regarding Copernicus's road to the heliocentric universe.[14] Swerdlow's contention is that a necessary step on that road was Copernicus's investigation of the first two propositions of book 12 of Regiomontanus's *Epitome of the Almagest* (1496). These propositions proved the validity of an eccentric model of the second anomaly for the inferior planets, believed by Claudius Ptolemy to be impossible. Once this model is established, Swerdlow claims, it is easy to realize the possibility of a heliocentric conversion of such a model. By such a conversion, the eccentric is realized to represent the planet's motion around the Sun, the order of the planets is established, and the retrograding motions are explained by means of the differences of velocities between the planets and the Earth. The crucial point in Swerdlow's description of Copernicus's context of discovery for our argument is the following: "Regiomontanus either failed to recognize the immediate heliocentric conversion of the model – which I think unlikely in view of his profound understanding of planetary theory throughout the *Epitome* and his careful and original analysis of the eccentric model of the second anomaly – or recognized the conversion very clearly but *refused to believe that it was physically possible.*"[15]

If Regiomontanus was very likely aware of the possibility of a heliocentric conversion, as Swerdlow maintains, one may rightly assume that there was no mathematical-technical reason for him to reject it. Likewise, there was no mathematical-technical reason for Copernicus to adopt it and infer further the motion of the Earth. Why, then, did Copernicus find such a solution attractive, whereas Regiomontanus simply stopped short of all that?

Of course, there is no clear answer to such a question. However, Copernicus's claim that he engaged with something "beyond the appearance" of the Earth's immobility (*praeter apparentiam*) seems to encourage an investigation into the discourse – and the networks in and through which this discourse was embedded, communicated, and disseminated – about the relationship of appearances to their "beyond" and the way it was negotiated in fifteenth-century Europe. With such investigation, we hope to add a layer to the discussion of the fifteenth-century European background to the work of Copernicus.

BROADENING THE SCOPE OF THE VISIBLE: ALBERTI'S PROJECT AND ITS INTERPRETATION BY CUSA

We begin with two examples that seem to represent two different attitudes to the visible and the invisible. Both imprinted the long cosmological-astronomical traditions of East and West with everlasting impact. The first example comes from Aristotle, who associated epistemological possibilities with sensibility and presupposed a direct relation between the human capability to see and the ability to know: "Of substances constituted by nature some are ungenerated, imperishable, and eternal, while others are subject to generation and decay. The former are excellent and divine, but less accessible to knowledge. The evidence that might throw light on them, and on the problems which we long to solve respecting them, is furnished but scantily by sensation." Suprasensual and perfect objects, such as celestial entities, stand in contrast to terrestrial things such as animals and plants. The latter, "living as we do in their midst," because of their affinity and nearness, are known with certitude "in [their] completeness." In contrast, human knowledge of celestial things is scarce, almost beyond one's grasp. In their loftiness and divinity, celestial things are known only as far as "our conjectures could reach."[16]

The second example is from Ptolemy's preface to the *Almagest.* There, he evokes the issue of visibility while introducing the three parts of "theoretical philosophy," claiming to follow Aristotle but in fact articulating a very different position. With Aristotle, Ptolemy argues that the invisible and motionless deity is hardly given to human knowledge since it is "completely separated from perceptible reality." Hence its activity "can only be imagined." As an object of knowledge, therefore, the invisible (deity) – the cause of all motion – is judged by Ptolemy to be an object for "guesswork" rather than knowledge. In contradistinction to Aristotle, however, the heavenly bodies and their attributes enjoy a very high status as objects of knowledge. Ptolemy calls them "divine and heavenly things." They are "on the one hand perceptible, moving and being moved, but on the other hand eternal and unchanging." Ptolemy assigns to the subject matter of mathematics a middle position between the "completely invisible and ungraspable nature" of divine things and "the unstable and unclear nature of matter." Only mathematical entities can "provide sure and unshakable knowledge," if one deals with them in a rigorous manner, using the methods of arithmetic and geometry.[17] "Perceptibility" and "stability" are Ptolemy's two conditions of possibility for astronomical knowledge, the main difference from Aristotle being what counts as "perceptible" or "visible" to each.

Against the backdrop of these epistemological attitudes, one can assess Alberti's radical and spirited quest for a new standard of visibility in the opening sentences of his *De pictura,* the treatise that laid the foundations for the theory of artificial perspective. There, he contends that against the mathematicians' abstract entities, the painters "wish to set this thing up as visible."[18] This demand for visibility challenged the dichotomy between the divine, spiritual realm and the material world. Alberti insists upon blurring the dichotomy between the visible, material signs and the abstract, suprasensual mathematical entities. To accomplish his goal, he adopts several measures. Initially, he implements what seem at first glance to be merely crude definitions for geometrical entities: "A point is a sign which one might say is not divisible into parts."[19] It soon transpires, however, that this statement is not mere reiteration of the Euclidean definition. A sign, he stresses, is "anything which exists on a surface so that it is visible to the eye." A "point" turns out to be any very small mark on any surface because "things which are not visible do not concern the painter, for he strives to represent only the things that are seen."[20] This objective, however, does not mean the painter is constrained to describe only present phenomena. On the contrary, Alberti celebrates paintings' "truly divine power" to literally make the "absent" and invisible "present" and thus, for instance, to "represent the dead to the living many centuries later." The visible mark on a surface is therefore not a crude and dull material stain but, being part of the art of painting, has the magical ability to materialize invisible elements. This quality of the painted visible marks is embedded in the nature of the picture itself. Although pictures are material, they are "so *transparent* and like glass that the visual pyramid passes right through it ... Therefore, a painting will be the intersection of a visual pyramid ... represented artificially with lines and colors on a given surface."[21]

For the picture to correspond exactly to this particular and invisible intersection of a pyramid, Alberti devises his famous veil, disclosing that "among my friends I call [the veil] the intersection." By stretching threads crisscrossing each other over a wooden frame, Alberti does not merely fix the painter's eye on the painted object but also stabilizes the visual pyramid and thus materializes the otherwise transparent intersection.[22] In this way, the veil divulges the pyramid of vision, together with its special points, such as the apex, and its base with its outlines.

The technical procedure for the materialization of abstract concepts and ideas – in this case, of invisible mathematical entities – was also used for even more intangible and illusive entities such as beauty. According to Alberti, the artist does not imitate and represent nature itself but aims at the forms of beauty that are recondite within natural phenomena.[23]

Alberti forbids the painter to follow only his own mind but urges him to observe nature. However, by "nature," he means a form of beauty lurking beyond the phenomena and concealed behind them.[24] To attain it, one has to collect and reconstruct data from concrete, physical bodies and recreate the form of real nature.[25]

Painting over a two-dimensional surface makes salient those values crucial to Alberti's ideal of beauty, such as "symmetry" and "harmony" between the different parts of the painting. Albertian perspective constitutes the perfect relationship between the bodies that appear in an artificially organized space. Moreover, painterly representation is an instrument for the improvement of nature and especially for self-improvement: "The gifts of nature are to be cultivated and enriched by industry, study and especially practice, and hereafter nothing that pertains to glory should be seen to be negligently overlooked by us."[26] From the context of these words, it seems that by the "gifts of nature," Alberti means first and foremost those capabilities and skills that nature bestowed primarily on the human race and only secondarily on physical nature itself. In other words, painting makes it possible to concentrate and disclose the "beauty" hidden and dispersed throughout the natural world. Through the recognition that the true nature of the world is more beautiful and varied than the human eye is physically exposed to, one can improve one's intellectual and especially moral capacities.

Such a desire to see what is beyond appearances finds similar expression in other fifteenth-century writers and intellectuals. Most notably, the philosopher and man of the church, Nicholas of Cusa, accepted the challenge set by Alberti's system of perspective and in a bold attempt suggested a new theology based on this desire to see the invisible: "However, we want to see the indivisible beginning itself."[27] This demand underpins all his great works, from *De docta ignorantia* to *De possest.* In one of his late treatises, Cusa suggests an experiment with beryl, which he uses as a sort of lens: "Beryl stones are bright, white, and clear. To them are given both concave and convex forms. And someone who looks out through them apprehends that which previously was invisible. If an intellectual beryl (*intellectualis berillus*) that had both a maximum and a minimum form were fitted to our intellectual eyes, then through the mediation of this beryl the indivisible beginning of all things would be attained."[28] Using a lens made of a beryl stone allows one to view things that were invisible before. In fact, Cusa suggests that such an operation can supply the mind with a vivid image of the invisible unification of opposites (*oppositorum coincidentia*). The beryl's power of magnification and diminution has no other function but to serve as an example for how to apply a wondrous effect for attaining a glimpse into the unknowable.

The Cusan meditation goes beyond medieval notions of allegory. In the twelfth century, Abbot Suger of St Denis, for example, also meditated over precious stones. In his case, the stones ornamented the "wonderful cross of St. Eloy" and "that incomparable ornament commonly called 'the Crest'" set in the cathedral of St Denis. For him, the point of departure for this meditation was textual, as the stones reminded him of Ezekiel 28:13: "Every precious stone was thy covering." Suger contemplated the allegorical properties of "the many colored gems" and he described his experience: "[The] worthy meditation has induced me to reflect, transferring that which is material to that which is immaterial, on the diversity of the sacred virtues: then it seems to me that I see myself dwelling, as it were, in some strange region of the universe which neither exists entirely in the slime of the earth nor entirely in the purity of Heaven; and that, by the grace of God, I can be transported to that higher world in an anagogical manner."[29] For Suger, the stones and their textual attributes inaugurate only an interpretative process. The image that, in contrast, Cusa perceives through beryl has a different function. No textual pointers are supplied to the reader, and the refraction through the beryl lens is supposed to provide a direct picture of the most abstract idea: "I frequently endeavored to reach conclusions in accordance with our intellectual vision (*intellectualem visionem*), which transcends the power of our reason (*rationis vigorem*).[30] Hence, in order that I may now very clearly develop for the reader a concept, I will put forward a mirror and a riddle (*enigma*) by which each reader's frail intellect may be aided and guided at the outer limits of the knowable. [This work] furnishes sufficient practical instruction as to the manner in which it is possible from enigmas to arrive at the loftiest vision."[31]

Undoubtedly, these words directly allude to St Paul's dictum "videmus nunc per speculum in enigmate." Cusa, however, presents the enigma not as a sign of a human postlapsarian state but as a positive vehicle for salvation. The riddle or enigma provides a bridge for the intellect to perceive what is beyond the power of reason and measurement to apprehend. The principle that allows the beryl stone to fulfil such a function is a special relation of similitude between material, sensible things and abstract concepts. Cusa describes a hierarchy of likenesses based on perceptual relationships between an object and its image, such as heat in itself and heat as a perceptual likeness of itself. The grades of this hierarchy are accordingly the perceptual likeness (referring to senses), the intellectual likeness (referring to the human intellect), and the intelligential likeness (referring to celestial beings such as angels). The first beginning, called intellect, emanates a flow of light communicating its own intelligence. The human mind, for its part, in its ascension toward God, applies measure as a means to connect the different stages of this hierarchy of similitude: "Hence, man finds within himself, as in a measuring scale, *ratione mensurante*, all created things."[32]

In using his creative powers man is similar to God. Through his measurements and the application of mathematics (i.e., artificial forms), he participates in the hierarchy of similitude: "So man is the creator of conceptual beings and of artificial forms that are only likenesses of his intellect, [and] just as God's creatures are likenesses of the Divine Intellect, so man has an intellect that is a likeness of the Divine Intellect, in respect to creating. Hence, he creates likenesses of the likenesses of the Divine Intellect … Therefore, man measures his own intellect in terms of the power of his works; and thereby he measures the Divine Intellect, just as an original is measured by means of its image."[33]

The notion of man as a measure of creation shines throughout the treatise *De beryllo.* Measuring is the true intellectual tool that can communicate between visible reality and divine meaning. Cusa asserts that visible reality is a book that expresses divine intention: "For *sensibilia* are the senses' book; in these books the intention of the Divine Intellect is described in perceptible figures (*sensibilius figuris*). And the intention is the manifestation of God the Creator."[34] Cusa's mode of reading is radically different from the allegorical reading of his medieval predecessors. To decipher the divine meaning embedded in the visible world, the human mind has to measure and compare. This act of setting things one against the other, matching and weighing them relatively, will lead the human mind to appreciate the juxtaposition of contraries embedded in the material and visible world. Every human investigation of nature must be mathematically conceived, as Cusa had already asserted in *De docta ignorantia*: "However, all investigations judge by proportional comparison of the certain to the uncertain, [and] therefore every inquiry is comparative and uses the means of comparative … Therefore, every inquiry proceeds by means of a proportional comparison, whether an easy or a difficult one … However, since proportion indicates an agreement in some one respect and, at the same time, indicates an otherness, it cannot be understood independently of number. Accordingly, number encompasses all things proportionally compared."[35] In sensing the world, one measures it, and in measuring it, one accommodates it for the act of true knowledge. This accommodation is not an act of passively abstracting the geometrical shape from perceptible things but is an active and creative process of reading quantities and geometrical figures into nature.

Mathematical entities and numbers, which proceed from our mind and exist as we conceive them, are not substances or beginnings of perceptible things but are only the beginning of rational entities of which we are the creator.[36] In *De beryllo*, Cusa sums it up, adopting Protagoras's celebrated dictum: "Protagoras, then, rightly stated that man is the measure of things. Because man knows – by reference to the nature that by sensing he knows – sensed things to exist for the sake of such sensory knowledge, he measures the sensed things in order to be able to apprehend, through his senses, the glory of the divine intellect."[37] Cusa's interpretation of Protagoras's dictum testifies to his allusion to Alberti's *De pictura*, for such an act of accommodation of the visible world to the measure of man is notable especially in the act of painting: "As man is the best known of all things to man, perhaps Protagoras, in saying that man is the scale and measure of all things, meant that accidents in all things are duly compared and known by the accidents of man. All of which should persuade us that, however small you paint the objects in a painting, they will seem large or small according to the size of any man in the picture."[38] For Alberti, painting a picture by following the right measure and according to ideal proportions is a means of capturing beauty hidden within natural phenomena. Cusa, in contrast, espouses the manipulation of geometrical diagrams as the vehicle for exposing the point of the unification of opposites. In his writings, Cusa presents several kinds of diagrams and their manipulations. These diagrams are neither didactic tools nor rhetorical devices of persuasion but are a practical tool to "acquire knowledge about the beginning of opposites and about their difference and about all that is attainable concerning the beginning and the difference."[39]

Thus, for example, Cusa presents a geometrical line, *ab*, that is a "likeness of Truth." The line is then folded at some point, *c*. This movement of the folded line creates different angles and, according to Cusa, figures "the movement by which God calls forth non-being to be." The diagram is an enigma – to be contemplated through a mathematical manipulation: "Hence, when, in a similar way, the Creator-Intellect moves *cb*, He unfolds exemplars (which he has within Himself) in a likeness of Himself – just as when a mathematician folds a line into a triangle, he unfolds a means of a movement-of-enfolding the triangle that he has within himself, in his mind."[40] The enigmatic diagram (for Cusa, all geometrical diagrams can serve as such enigmas) allows human contemplation to bypass its own limitations and to see what is beyond sensory experience and common-sense conceptualization: "Because our intellect cannot conceive of what is simple (for the intellect makes a concept by way of the imagination, which takes from *sensibilia* the beginning, or the subject of its images, or the figures), it cannot conceive of the essence of things. Nevertheless, the intellect sees beyond the imagination and [its imaginary] concept ... the indivisible essence."[41]

The diagram is to serve, just like the beryl lens, as a means to look beyond mundane experience in order to grasp the elusive beginning of the world and thereby to provide a point of departure for a true speculation: "Now, if you apply eyeglasses (*oculare*) and see according to the maximal and minimal modes the beginning of every mode, in which all modes are enfolded and which no other mode can unfold, then you will be able to make a truer speculation regarding the divine mode."[42] Cusa's diagrammatic reasoning culminated in his long-term preoccupation with the mathematical problem of squaring the circle. As we shall see, his inventive isoperimetric methods of comparing and measuring produced a mathematical solution that was heavily criticized by contemporaneous mathematicians but not totally rejected by modern historians of science. Following Cusa's diagrammatical reasoning allows us to reconstruct his own logic in trying to find a "visible" geometrical point to represent the "invisible" coincidence of opposites, which for him meant an intellectual vision of God. Moreover, through his writings on the quadrature, Cusa engaged the best European mathematicians of the period – whom he personally knew – in a conversation about the quadrature across disciplinary and professional boundaries. The echoes of this conversation were likely to have reached Copernicus in Bologna and Ferara some decades after they took place among Cusa, Regiomontanus, Toscanelli, and perhaps even Alberti.

CUSA'S MATHEMATICAL MUSINGS

Cusa's exploration of the measurement of a circle by a polygonal figure should be understood as a further articulation of his practices described above. The quadrature of the circle is an "enigma" by means of which seeing, measuring, comparing, and conceptualizing are brought to the limit. A thorough mathematical investigation of the problem of the quadrature, he suggests, may allow the passage beyond reason. Thus a glimpse of God's point of view may be attained by a higher kind of intellectual faculty, namely by "intellectual vision." Cusa's oscillation between doing mathematics within the context of Alberti's project and conceptualizing his conclusions in theological terms is a chapter in the history of "invisibles" in the fifteenth century that may have made possible Copernicus's later leap into a cosmological invisible such as the motion of the Earth.

From early on, Cusa's philosophical ideas about the coincidence of opposites were closely tied to his thinking about mathematical entities and their "hidden" visibility, which he had first presented in ordinary speech:

> [T]he minimum [curved] coincides with the maximum [straight] – to such an extent that we can *visually* recognize that it is necessary for the maximum line to be maximally straight and minimally curved. Not even a scruple of doubt about this can remain when *we see* in the figure here at the side that arc CD of the larger circle is less curved than arc EF of the smaller circle, and that arc EF is less curved than arc GH of the still smaller circle. Hence, the straight line AB will be the arc of the maximum circle, which cannot be greater. And thus *we see* that a maximum, infinite line is, necessarily, the straightest; and to it no curvature is opposed. Indeed, in the maximum line curvature is straightness.[43]

The desire to see the invisible was lurking behind these words and echoed Platonic notions about the intermediate position of mathematical entities between sensible beings of the physical world and the abstract beings of divine science.[44] Cusa understood mathematical beings in terms of ideas in the mind of God, for He had "ordered all things in measure and number and weight."[45] Cusa surmised that mathematical entities had a similar creative role in the mind of man: "In creating the world, God used arithmetic, geometry, music, and likewise astronomy. We ourselves also use these arts when we investigate the comparative relationships of objects, of elements, and of motions."[46] But Cusa went beyond his predecessors in emphasizing the active role of geometry in moulding reality for human understanding and in preparing the mind for seeing beyond reason, as is clarified in this chapter.

The publication of Cusa's *De geometricis transmutationibus* (1445) seems to have signalled a turning point in the kind of interest he showed in mathematics and a turning point in his attempts to attract the attention of professionals to his solutions. In the next fourteen years, he wrote eleven mathematical treatises dedicated to the quadrature and corresponded with fellow mathematicians, philosophers, and theologians about them. Historians Marshall Clagett[47] and Joseph Hofmann[48] believed that he was truly intrigued by nonorthodox solutions to problems considered by Aristotle and many later Aristotelian philosophers to be insoluble. Both thought that receiving the new translation of Archimedes of Syracuse by Jacobus Cremonensis (ca. 1453) from his friend Pope Nicholas V had supplied Cusa with hope for a fresh formulation of his solutions in a more focused way. Both agreed that he showed some competence in the subject. For the argument of this chapter, however, the main thing to realize is the way that, by seriously dedicating himself to solidifying mathematically his results and methods with the view of reaching a new kind of theological speculation, Cusa's work subverted the boundaries between the practices of painters, mathematicians, and philosophers-theologians, while forging a specific new form of a human-observer position. A few key moments in this development will clarify his trajectory. "Knowing the equality of the straight and the curved" is defined by Cusa as his main goal in the introduction to the first mathematical treatise, *De geometricis transmutationibus*. In the introductory letter, he summarizes four different ways to establish, or four "kinds of knowledge" relevant to, the "unknown art" that he desires: (1) rectification by finding an exact number for the ratio of the circumference to the diameter of the circle, (2) finding a proportion between two curved lines corresponding to a proportion between two straight lines, (3) finding two continuous proportions between straight lines, and (4) finding a proportional fourth to three given straight lines.[49] Each of these four ways will be developed by him as a strategy for solving his problem. However, it seems that his attempt at rectification, declared in the first premise of *De geometricis transmutationibus*, attracted most attention among contemporaries as well as among historians and became a major source of critique. Cusa there states and attempts to prove that "[t]he semidiameter of an isoperimetric circle, when a triangle has been inscribed, stands in relation to a line led from the centre of the circle to which the triangle is inscribed to one quarter of a side [of the triangle] in a ratio of 5:4."[50]

Cusa's results did not substantially differ throughout his various attempts at rectification. On the basis of his own calculations, done using Cusa's methods, Hofmann arrived at the value of 3.1423 for the ratio, which he thought was a reasonable result.[51] Clagett was mostly critical, but his strategy was to let Cusa's contemporaries – especially Regiomontanus – speak for themselves. Both show respect for Cusa's inventive isoperimetric methods, mostly rooted in ancient and medieval traditions, not in his reading of Archimedes's texts, even though Archimedes's spirit is the main source of inspiration.[52] An idea of these methods emerges from the account Cusa gives of his main principles, which we briefly summarize:

1. Among regular isoperimetric figures, the triangle's area is the smallest.
2. As the number of the polygon's sides increases, its area grows.
3. Thus, of all regular isoperimetric figures, the circle's area is the greatest – the assumption being that a circle is a regular polygon of infinite sides.
4. No regular polygon has a ratio with the isoperimetric circle.
5. But the ratio between the areas of two isoperimetric polygons is like the ratio between the differences of the radii of their circumscribed and inscribed circles.

6 But the difference in the radii of these circles – the circumscribed and the inscribed – is greatest in the triangle and decreasing in the other polygons.[53]

On the basis of these assumptions, Cusa determines the proportion between the radii of the inscribed circles and the proportion between the radii of the circumscribed circles of isoperimetric polygons. Representing these proportions on two simple straight lines, he interprets the point of their intersection as the minimal difference between radii of circumscribed and inscribed circles, marking the radius of the isoperimetric circle. In his language, this is the point of the "coincidence of opposites," where the difference between the curve and the straight lines disappears.

Cusa's representation is obviously wrong, as the differences between the radii do not progress in relation to a straight line but in relation to an asymptote. However, given the kind of project he had in mind, it seems to us superfluous to restate his mistakes. This task, in fact, was done by his contemporaries, as shown below, and has already been undertaken by modern historians of mathematics.[54] More relevant would be to underline how his procedures display the constructive role that his notions of the gradual progression through a series of isoperimetric figures toward the infinite played in his mind while he was thinking through the transition from a human, limited perspective to the omni-observant perspective of God. His confidence that the tremendous efforts he invested had been crowned with success should be read against the goals he set himself:

After almost innumerable ways by which I strove to reach the established art (always, nevertheless, failing), nevertheless, while looking back at the principle which I used in the books on learned ignorance, a way appeared, opened up to me. The art I seek, beyond the things already related among geometers, allows the conversion of the curve to the straight, and of the straight to the curve. Since among these things there is no rational proportion, this secret must hide in some coincidence of the extremes. Since this [coincidence] is in the greatest (as is related elsewhere) and the greatest thing is a circle which is unknown: in the smallest thing which is a triangle it is demonstrated that it [the coincidence] must be sought.[55]

Realizing the coincidence of the maximum (i.e., the searched for isoperimetric circle, seen as a polygon with infinitely many sides) with the minimum (i.e., the given isoperimetric polygons) is the climax of a search whose aim is obviously complex – indeed, somewhat megalomaniacal. It concerns the constitution of a whole new field of studies – "unknown art" – that is "close" to geometry but not identified with it. Although "everybody admits that it is possible that a curved line which may be neither greater nor smaller than a given straight line"[56] exists, the way to draw it is not quite clear, in spite of the "laborious speculations of geometers." Cusa wishes and believes that his new art – the art of coincidences – can offer the simplest solution. Moreover, not only will such a solution enable the geometric transformation, but it will also prefigure "an introduction for rising to higher matters" (*introductio ad altiora ascendendi*).[57]

Living in a world where the profane sciences were highly relevant as a basis for theology, Cusa was extremely concerned with the reactions of mathematicians to his efforts. His exchanges with Toscanelli and with Regiomontanus expose traces of the complicated archaeology of the various ways by which the fifteenth-century discourse of mathematicians constituted itself in alliance both with and against the pretense of theologians to use it for their own needs.

Cusa's *De geometricis transmutationibus* was dedicated to Paolo Toscanelli. The two had met in Padua, where Cusa came as a young man of seventeen to study law and mathematics. Toscanelli began his mathematical studies at Florence and continued in Padua, where he moved from mathematics and philosophy to medicine. Although both left the university in 1424 to pursue their very different careers, they kept in touch thereafter. They had much in common. Toscanelli was a precursor of the great Renaissance engineer-scientists and manifested his peculiar mixture of practical and mathematical skills by such monumental enterprises as the gnomon in the Church of Santa Maria Novella. Cusa was interested in "knowledge by doing" and fascinated by the wisdom of the market place (*idiota*), while systematically criticizing "bookish learning." His intimate tone while seeking the help and approval of an old friend indicates both his boldness and his difficulty in diving into the deep water of mathematics: "Correctly I have decided to resort to the most experienced arbitrator and zealot of truth, and at once lay out my discovery to a most commendable friend so that it may be valued on the scales of the fairest judge. Do not, therefore, dearest friend, reject these things as crude and disordered, even if you are preoccupied with greater matters; for they are few to read and indeed very easy to understand."[58] The last sentence betrays some kind of anxiety or intimidation on the threshold of the field of mathematicians. Toscanelli did not fail to fulfil the role assigned to him as a critical mathematician. He requested clarification for the result declared in the first premise of *De geometricis transmutationibus*. In response, Cusa dedicated to him – "to

Paolo, physician, the most learned of all men" – an additional text, *De arithmeticis complementis* (1450), in which he purported to make all his claims concerning the first premise "accessible to calculation."[59] This response, "which I submit for your correction, you who never tires," does not contain a more accurate approximation to the value of the ratio requested, nor does it offer a new methodological perspective. However, it radically differs in purpose from Cusa's former presentation of his mathematical enterprise. Giving up the attempt to embed his results in the broad philosophical-theological framework in which they were conceived, Cusa here focuses on one issue: reducing the ratio of the curved to the straight line into a ratio – but not a numerical value – between two straight lines. Jean-Marie Nicolle remarks that Cusa attempted to achieve this task by strictly applying Euclidean rules, thus neglecting precise calculation and the discourse of the coincidence of opposites.[60] Nevertheless, one phrase seems to contain the seed for the later concept of *visio intellectualis*, as Cusa justifies his new strategy by saying, "the *intellect should see* the ignorance and the defect of numerical proportion,"[61] namely the failure or limit of human reason and the need for an additional perspective on things.

During 1450 Cusa wrote two additional texts on the quadrature of the circle, in which he continued to explore possible strategies for presenting his main results. The first, *Quadratura*, includes a sort of justification for his shunning of calculation of the exact number of the ratio between the curved and the straight line, for "the figure of polygons is not the same kind of quantity as the figure of the circle."[62] Obviously, this statement refers to the rule of homogenous quantities and to the way it had been understood by Aristotelian philosophers – although not by mathematicians – for centuries. The second text, *De quadratura circuli*, was written after he had received the main objection of Toscanelli, only later formulated as a letter. By the time Cusa received the letter, he was already finishing the first part of his longest and most serious presentation of his mathematical ideas in *De complementis mathematicis* (1453).

Cusa included Toscanelli's critique in the last sentences of the first part of *De complementis mathematicis*, where he admitted that his solution may be incomplete.[63] Under this pretext, he wrote the second part of the text without, however, offering any real answer to Toscanelli's objections. Later on, in 1457, he also added a kind of fictional dialogue between Paulus and Nicolaus, apparently based on a conversation he had had with Toscanelli. At a certain point in this dialogue, having to confront further doubts expressed by his friend, Nicolaus admits, "I still suppose/believe that the quadrature of the circle is possible, and as a consequence, that it is impossible to attain anything without this."[64] The passage signals a new turning point in Cusa's trajectory. Starting now from the presupposition that the quadrature of the circle is possible – on the basis of his previous years of studying its mathematics – he will now concentrate on making explicit the bridge between his mathematical work and his theological insights about intellectual vision.

In his last mathematical text, *De mathematica perfectione* (1458), which Cusa believed was the most significant of all his mathematical treatises, the notion of *visio intellectualis* emerges as a means of demonstrating that at the point of coincidence between the arc and the chord, one could see that the difference is "absolute minimum": "For attaining knowledge of their relation, I look back at an intellectual vision, and I say that I see where the equality of chord and arc is: i.e., in the minimum of each. From this seen equality I proceed to investigate the intent by means of a right-angled triangle."[65] Having now superseded the search for an exact numerical value for the ratio between the curve and the straight line, Cusa uses "intellectual vision" to "see" their equality at the limit. Hofmann thought that Cusa had been on the threshold of perceiving a new kind of quantity, close to the infinitesimal. Nicolle seems to be skeptical of this interpretation. Be that as it may, what should rather be emphasized is that the fundamentally approximate nature of the number representing the ratio of the chord and the line was perceived by him as the sort of precision attainable by human visual capabilities and no more. Yet it leads to an understanding of the indiscernible precision itself as God. A year after the completion of *De mathematica perfectione*, Cusa had written *De aequalitate*,[66] which was printed among his *Sermones* in the Paris edition. We believe it was written, among others, also as his response to the mathematicians, although Cusa presents it as a reading of John 1:4: "Vita erat lux hominum" (The life was the light of men). Essentially, the text is a discourse on being, knowledge, and the blessed life, mobilizing Trinitarian theology to ground the notion of "absolute equality," the kind of equality "found only in the domain of eternity," arrived at through "intellectual vision" and differentiated from imprecise equality holding between a plurality of things. Starting as a paraphrase of the first verses of John the Evangelist, the discourse develops the idea of God as unity, giving life (*vita*) and diffusing the light of reason (*lux*) among humans through his co-substantial Son (*verbum, logos*). Thus the means chosen by Cusa to throw light on the notion of equality turns out to be an exploration of the kind of equality co-existing with distinctiveness between the three personae of the trinity. He addressed the *sermo* to his secretary, Peter von Erkelenz, "for the exercise of your intellect, which is eager for truth and apt for comprehension … so that you might enter into theological discourses."[67] The point of departure is the concept of equality conceived by Cusa as a condition of the absolute unchangeable (*absolute inalterabili*) that precedes all "otherness" (*omnem alteritatem praecedens in esse et posse*). On the basis of such conceptual explication, it is then stated, "'Equality' is the most equal name of the First and Eternal Beginning,"[68] and "Absolute Equality is identical with the Creator of heaven and earth." Thus "Equality begets from itself its Word," and "[f]rom these two there proceeds Union that is Equality (we call this Union the Spirit of Love)."[69] But "only where [there is priority to otherness] can there be Equality. Hence, it is impossible that a plurality of things be altogether equal, since those things can be a plurality only if they are different from one another and are distinct in essence."[70] Cusa here strives to anchor a distinction between absolute equality prior to any otherness – "universalitas igitur in ipsis est aequalitas sine alteritate" – and equality between things that are by nature "a plurality," meaning that no one among them can be precisely equal to the other. The equality of the circle and the polygon with which he had been preoccupied over the previous fourteen years is the most natural example, although Cusa does not mention it explicitly. The nature of the intellect previously discussed in the writings of Cusa accounts for the very possibility of the distinction he now comes to attain: "You have read in my *De Beryllo*," he writes to Peter, "that intellect wishes to be known … the Teacher, who is the Word for God, has taught me that seeing and knowing are the same thing … But sight that sees the visible – sees it apart from otherness and in and of itself – sees that it itself is not something other than is the visible."[71] Cusa then implies that a well-trained intellect should strive to see beyond otherness, to see itself observing and then realize that at its limits, sight coincides with the "invisible," namely with absolute equality.

Some conclusions emerge from following Cusa's theological discourse on equality, which at this point he constitutes as the basic ground of all knowledge: "Every science and every art is founded on Equality."[72] Cusa's distinction between equality in the Trinity and equality among any other kind of entities that are not a unity but a plurality delineates a thread of thought that leads us toward his mathematical discourse. Through years of preoccupation with the mathematical problem of the quadrature, he came to "see" the fundamental approximate nature of the sort of precision attainable by human visual/intellectual (sight as cognition) capabilities. This means that when the geometricians state the equality of the circle and the polygon, there always remains an invisible difference, invisible in the sense of being nonexpressible by even the smallest rational fraction. Nevertheless, since "the truth that the soul sees in different things, it sees by means of itself,"[73] it is possible to see oneself really "seeing" the nature of the geometricians' equality and to distinguish it from the absolute equality existing in God alone and embodied in the Trinity. This is the observer's point of view offered by Cusa to those who aspire for "true knowledge," which culminates, for him, in the knowledge of God, always beyond human capacity, but still within human horizons. According to Cusa, the same applied to the mathematician-observer: while seeing himself observing his own limitations, he glimpses God's omnivoyant point of view without ever grasping it.

CUSA IN THE FIELD OF THE MATHEMATICIANS

In 1463–64 Regiomontanus, the first European with real competence in high-level Greek mathematics and astronomy, as well as in the Greek language, wrote a series of texts on the quadrature of the circle, inspired by the importance of the problem but also intrigued by what he deemed to be the inadequate treatment of it by Nicholas of Cusa. Among those texts, there are one in the form of a dialogue and two in the form of letters, which are addressed to Toscanelli,[74] thus signalling appreciation of Toscanelli's critical attitude toward Cusa's methods and conclusions.

The dialogue (1464) between Aristophilus (Cusa's imaginary defender) and himself (Critias) well represents Regiomontanus's ambivalent relationship to Cusa, combining reverence for the cardinal's seniority and highest ecclesiastical rank with more than a grain of contempt for his incompetent move into the field of mathematicians.[75] It opens with Aristophilus – Cusa's speaker – talking to himself in praise of "the man to whom Nature herself has brought forth from her treasury such an important gift to philosophy, even while he is hemmed in by arduous public affairs."[76] It soon transpires that Aristophilus, in fact, tends to blur the distinction between Archimedes – the "Sicilian flower among geometers" – and another person, an outstanding modern "who promises to give lucidly and briefly a straight line equal to a circular line: and hence to square the circle will hardly seem to be difficult."[77] It is clear, by now, that Aristophilus's words express the ironic tone assumed by the author of the dialogue from its inception. Aristophilus then proceeds to present Cusa's solution, encouraged by Critias's approval: "O conclusion worthy of mention! If it bespeaks the truth, the old longstanding incompatibility of the circle and the rectilinear figures will be completely abolished." At a certain point, however, Aristophilus is worried by a certain mocking tone in Critias's voice and decides to confront him with a direct question: "concerning the conclusion mentioned above which seems to be the source of the argument, what do you think of it? Do you accept it as true, or not?" Although Critias is reluctant to answer, he still emphasizes that “the authority of the man is great.” But Aristophilus, aware that his question remains unanswered, spits out, “Why do you twist about enigmatically?” The unavoidable response is that “[a] suitable judge of such a matter is to be sought elsewhere. But if you have any way of proving this conclusion speak out.” Pushed to his limits, then, the desperate Aristophilus responds, “I find no way at all. Still it seems that he had demonstrated it by a certain argument which I cannot now recall.”[78]

Regiomontanus, however, does not simply leave the reader aware that Cusa is simply not an interlocutor for mathematicians. Rather, he proceeds to show how Cusa’s conclusion can be checked for its truth or falsity by actually testing his hypothesis numerically: “Listen then. Our Archimedes in the booklet *On the Measurement of the Circle* demonstrated by means of numbers that the periphery or circumference of the circle exceeds three times its diameter by an amount that is less than its seventh part but more than 10/71 of the same diameter.”[79] The rest of the text is dedicated to demonstrating Regiomontanus’s interest in numerical approximations, along with his critique of Cusa’s failure to offer an approximation that would match even the Archimedean lower bound.[80] That was Regiomontanus’s main critique and, in fact, his only concern. Cusa’s interest in the epistemological status of approximation, his ability to point out how one may discover the limit of human knowledge and set that limit as a starting point for further inquiries, and his insistence on experimenting with alternative proofs all remained beyond the field of interest of Regiomontanus. He was clearly keen to defend the boundaries of positive mathematics against Cusa’s epistemological insights and theological speculations. Just as Toscanelli rejected Cusa’s interpretation of the quadrature problem and cast doubt on the uses he made of diagrams for representing “coincidentia oppositorum,” so did Regiomontanus reject Cusa’s distinction between mathematical equality and absolute equality, which inspired Cusa’s understanding of the limits of his mathematical ability to solve the quadrature problem and pushed him to emphasize that the straight and curved lines do not belong to the same genus. Regiomontanus well understood Cusa’s attempt to differentiate a realm of rational knowledge from that of intellectual vision. He did so by maintaining the mathematically “invisible” yet intellectually “visible” – the difference between the curve and the straight line – and by connecting it to the distinction between the curved and the straight line as two mathematical entities of a different genus. Hence Regiomontanus, in his own *De quadratura circuli,* addressed to Toscanelli, critically refused Cusa’s crucial distinction by saying, “Now in coming closer to the original [objectives] of our exercise, we assert as most certain that the circumference of a circle is of the same genus as any straight line, in fact that all lines whether straight or curved with any sort of curvature do not differ specifically."[81]

A very long passage follows, where Regiomontanus supports himself not only by citing *On the Measurement of the Circle* but also by relying on additional Archimedean texts, newly translated by Jacobus Cremonensis, namely *On the Sphere and the Cylinder, On Spiral Lines,* as well as on medieval Islamic sources such as the *Verba filiorum.* This use of mathematical sources to support his position may signal his awareness that by boldly erasing the boundary between two kinds of mathematical entities – lines and numbers – he was also defying the authority of philosophers to intervene in the discourse on mathematics, which he obviously experienced as a burden to practising mathematicians.

Regiomontanus's distance from Alberti's and Cusa's intellectual project to represent invisible and abstract entities in a visual form is also manifest in his views about the reform needed in astronomy and the place of observation within it. In his 1464 correspondence with Giovanni Bianchini of Ferara,[82] Regiomontanus presented the lamentable state of contemporary astronomy. Noel Swerdlow describes his answers to Bianchini's queries in terms of a catalogue of errors inherited from the ancients in a field that was essential for studying the motion of the fixed stars, Sun, and planets.[83] Often, Regiomontanus presented those errors while advocating his own observations, conducted together with his teacher and strengthened by evidence from his friends. Thus, for the declination of the ecliptic, he gave, for example, the numbers of Ptolemy, Albategnius (al-Battānī), Thābit ibn Qurra, Abū Isḥāq Ibrāhīm al-Zarqālī, and the *Alfonsine Tables,* and he concluded by saying, "Although we (my teacher [Peurbach] and I) found it about 23:28 with instruments, I have often heard Master Paolo [Toscanelli] of Florence and [Leon] Battista Alberti saying that they themselves had observed carefully and did not find it greater than 23:30; which fact also convinces us to correct our tables, that is, the table of declination and others that are based upon it."[84] These examples demonstrate how Regiomontanus was interested in presenting himself as setting new standards for the good astronomer and as having a reformative mind, compared with the majority of practitioners, whom he denounced: "I cannot but wonder at the indolence of the common astronomers of our age who, just as credulous women, receive as something divine and immutable whatever they come upon in books either of tables or their canons, for they believe in writers and make no effort to find the truth."[85] Swerdlow, however, draws attention to the scarcity of Regiomontanus's *actual* observations, as well as to their rather crude nature, describing them as "little more than conjunctions of alignments with stars or another planet," a conclusion Swerdlow reached after checking records of the observations in a publication from 1544.[86] He also demonstrates that Regiomontanus was always concerned with the model rather than with the theoretically complicated cluster of problems related to "testing" established parameters with a view to changing them if necessary.[87] Thus Swerdlow asserts that "Regiomontanus is really thinking more of a model than of the tables."[88] Nevertheless, throughout the years, Regiomontanus never stopped complaining of the erroneous observations of his predecessors and put his trust in those astronomers ready to make new observations and compare them with sound and good calculation. Truth in the mathematical sciences is immutable and eternal, he claimed in the "Oration on the Dignity and Utility of the Mathematical Science,"[89] given as an inaugural lecture in the course on Alfraganus (al-Farghānī), which he was probably teaching in Padua in 1464: "The theorems of Euclid have the same certainty today as a thousand years ago. The discoveries of Archimedes will instill no less admiration in men to come after a thousand centuries than the delight instilled by our own reading." But the only way to penetrate truth is through observations and their comparison with the figures of the ancients. The same message shines through his 1471 letter to Christian Roder,[90] the rector of the University of Erfurt, in which he expressed his critique of Cusa's quadrature as a "ridiculous" attempt to rival Archimedes. In the same letter, Regiomontanus also articulated his critique of a certain type of astronomers who tend to ignore the main task of the profession, namely observation. Thus he wrote, "Here are esteemed great astronomers, those who can produce any kind of calculation of celestial motions. But these are people who never make observations of their own."[91] A similar critique was repeated in his preface to his *Tabula primi mobilis* (1467–71).

The scarcity and crudeness of Regiomontanus's observations, joined to the fact that he never actually tried to re-establish the elements of astronomy, stand in sharp contrast to his rhetoric about observations. This gap points to a rather limited inclination to reflect on the problems involved with observation and on the vexing question of the point of view of the observer so emphasized in the work of both Alberti and Cusa. In his project of printing corrected mathematical texts, which he vigorously pursued during his last years in Nueremberg,[92] Regiomontanus shared with his fellow humanists an operational view of language, whether mathematical or natural. His Paduan "Oration" points to a practical, instrumental view of the origins of mathematics in Egyptian measuring of the fields. Here was a "positivist" mathematician "avant la lettre" shunning Alberti's project of visualizing the invisible – including the point of view of the observer – as well as Cusa's imaginative use (or abuse) of his isoperimetric methods for the sake of "envisioning" God's point of view.

A speculative note may be proper at this point. It seems that something deeper than just contingent coincidence is signalled by the juxtaposition of Regiomontanus's critique of Cusa's quadrature solutions with the kind of dissatisfaction that he expressed toward his fellow astronomers' "superficial" approach to the practices of observation and toward their results. It may provide an indication that "positive mathematics" was not enough for a breakthrough that could include the observer's point of view from the blind spot of a position on a moving earth not situated at the centre of the universe.

Despite the critique of mathematicians like Toscanelli and Regiomontanus, who found Cusa's work "obscure and lacking in positiveness," their common participation in an intellectual community was not insignificant. Cusa's musings with geometrical diagrams were not simply written as a guide to solitary meditation. His texts became part of a lively exchange of ideas based on personal encounters, mutual respect, and friendship. These exchanges and debates were an intellectual crossroad where some of the more important endeavours of fifteenth-century scientific thought were communicated and probed. Although Cusa and Toscanelli remained friends throughout their lifetime, Toscanelli found a kin spirit in the much younger Regiomontanus, who came to Italy in 1461 at the invitation of Cardinal Basilios Bessarion. In Bessarion's villa in Rome,[93] Regiomontanus met Toscanelli and Cusa (who had met Bessarion years before at the Council of Florence). Apparently, both visited Bessarion's academy of 1461. Moreover, Alberti had been a member of the Papal Curia since 1420, and although we know of no direct contact between him and Regiomontanus, Alberti surely frequented Bessarion's villa as well. Regiomontanus testified that he had studied Cusa's *Quadratura* while still in Vienna and had discussed it with his teacher Peurbach, who had been no less critical of Cusa than himself and Toscanelli. In his 1464 letter to Bianchini, Regiomontanus praised Toscanelli's and Alberti's observations, which exemplified, for him, good observations fitting his high standards of precision. Toscanelli testified that his observations were painstaking, and a surviving manuscript contains the observations and calculations that he performed respecting the orbits of the comets of 1433 and 1449–50, Halley's comet of 1456, and the comets of May 1457, June–August 1457, and 1472.[94] It thus seems true to say that he shared with Regiomontanus the kind of mathematics characterized by "positiveness," the lack of which in Cusa's work they both found lamentable.

In spite of harsh critique by the mathematicians, the continued praises bestowed by Regiomontanus on Cusa testify that an ongoing conversation, including face-to-face acquaintance, shaped the habits of life and thinking of the small number of fans of mathematics attracted to the Italian centre.[95] Thus they were tied into a kind of network characterized by a variety of positions but not deprived of a common ethos. Writing letters to each other, dedicating their works to respected friends highly appreciated for their talents and/or position in society, and inventing fictive dialogues in which distinguished mathematicians were represented, portrayed, and criticized all became common strategies of inclusion and exclusion that destabilized disciplinary boundaries and paved the way for reflection on the foundations of human perception, mathematics, cosmology, and the divine.

REGIOMONTANUS AND COPERNICUS

In traditional astronomy, the cognitive primacy of sensory perception, assumed by Aristotle,[96] was tied to the senses' inherent limitation. This limitation, based on the paradigmatic role of visual experience, had played a crucial role in shaping the foundational tenets of astronomical theorization since antiquity. Theory was wholly dependent on what the astronomer saw, and appearances were assumed to be valid and authentic regardless of the specific theory suggested.[97] All there was to be explained was in front of the astronomer's eyes, and these explanations were supplied under the assumption of order. Thus, according to Simplicius of Cilicia, "Astronomy … demonstrates the arrangement of the heavenly bodies on the basis of the declaration that the universe is really and truly a cosmos." Simplicius further stresses that it is the astronomer's role to find a way to order appearances according to this principle: "if we suppose that their orbits are eccentric circles or that [they] describe an epicycle, their apparent irregularity will be saved."[98]

Appearances are not illusions or visual deceptions. They are true and have to be explained in accordance with the assumption that the motions of the heavenly bodies are "by nature uniform and circular." In relation to a static observer situated at the centre of the universe, the planets truly move backward. It is the task of the astronomer to suggest a system of circles to explain why the planets move in such peculiar ways without damaging the cognitive value of the observer's ocular experience: "The apparent irregularity [anomaly] in their motions is the result of the position and order of those circles in the sphere of each by means of which they carry out their movements, and in reality there is in essence nothing alien to their eternal nature in the 'disorder' the phenomena are supposed to exhibit … It will be shown that either of these hypothesis [eccentric orbits or epicycles] will enable [the planets] to appear, to our eyes, to traverse unequal arcs of the ecliptic (which is eccentric to the universe) in equal times."[99] The deferent (or eccentric) and epicycle are both calculated in relationship to the point of view of an observer situated at the centre of the universe. This uncontested location of the observer is what allows the mathematician to construe these circles as credible explanations.[100] This dependence of the mathematical theory on visual experience is clearly revealed in Ptolemy's presentation of the equant as an explanation of the anomalies of the planets. The equant is a point that is not directly related to the observer but to a "point bisecting the line joining the centre of the ecliptic and the point about which the ecliptic has its uniform motion." Ptolemy readily admits that this procedure is not taken from any "apparent principle" and is "without" proof and that its only justification is that it is "in agreement with the phenomena." He further emphasizes that the coherence of the mathematical models (i.e., "the similarity of the hypotheses of the planets") is less significant than saving ocular experience and that "individual phenomena are demonstrated in accordance" with the more basic principle of preserving "uniform circular motion … without exception."[101]

Postulating the equant involves a knotty methodological move. It implies that the point from which planetary motions can be viewed as uniform is an imaginary point unrelated to the position of the observer. This fact is still clear in the second half of the fifteenth century when Peurbach asserts that in contrast to the eccentric spheres, which are physically real and are calculated in regard to the observer's central position, the equant is based on a "circle imagined about the point [i.e., the centre of the equant] that is the point on the line of the apogee as far from the centre of its orb as this centre is distant from the centre of the world."[102]

Astronomical knowledge, according to the tradition upheld by Regiomontanus, necessarily assumed the reality of celestial appearances. One may aspire to keener eyesight, but one is not to doubt what one sees. Astronomers apply invisible spheres and circles only to substantiate the authenticity of their observations. Probing the demarcation between the phenomenal realm and the realm of invisible structures initiated in the fifteenth century by people like Alberti and Cusa challenged this traditional conception of astronomy on several levels. First, the observer's standing is not predetermined and static, and appearances are relative to one's point of view. Second, one can peer beyond appearances to gauge invisible structures and entities through the use of different kinds of devices. Obliterating visual experience, the device allows the observer to perceive what is not (in Alberti's words, "it makes the absent present"),[103] indicating that things are not always what they seem.

These two notions may have shaped Copernicus's propensity to accept the invisible motion of the Earth as a basic principle of his system. Reiterating the opening citation of our chapter from Copernicus's early *Commentariolus*, we can follow Copernicus's problematizing of the appearances: "For indeed, the things by which the natural philosophers principally try to establish its [the Earth's] immobility, for the most part rest upon appearances; all these things fall down here especially because we are engaged [with something] beyond that very same appearance."[104]

That the heavens appear to move cannot serve as a proof of the immobility of the Earth, as appearances may sometime hide more than they reveal, depending mainly on the observer's position. This understanding leads Copernicus to assume that "[w]hatever motion appears in the sphere of the fixed stars," "whatever motions appear to us to belong to the sun," and "the retrograde and direct motion that appears in the planets" are all false and that "the motion … belongs not to them but to the [motion] of the Earth."[105] The Sun and the sphere of the fixed stars are immobile, and the planets move "uniformly, just as the principle of perfect motion requires."[106] The humanly constrained observer sees the heavens upside down, as nothing appears to be in its right place. Taking into account the proportion between the distance from the Sun to the Earth and the distance from the centre of the universe to the zodiac will reallocate the heavenly bodies: "Now, since it is assumed that the semidiameter of this sphere has an imperceptible quantity compared to the height of the sphere of the fixed stars, it follows that the Sun will appear to be carried around by this motion just as if the Earth were located in the centre of the universe. Since, however, this results from the motion not of the Sun but rather of the Earth, so for example, when the Earth is in Capricorn, the Sun is seen in a straight line along the diameter in Cancer, and so on in the same way."[107] Not only are the real places of the celestial bodies confused, but if one assigns the daily rotation of the Earth "turning swiftly on its poles in the order of the signs" to the eighth sphere, "the entire universe appears to be driven around in a headlong whirl." Throughout the *Commentariolus*, Copernicus does not tire of pointing out that appearances lead astronomers to erroneously ascribe the wrong motions to the celestial bodies and that one should adopt a critical attitude toward the testimony of the eyes. Visual experience has to be calculated away as resulting from the observer's motion: "the planet is seen sometimes to move retrograde and often to stand still, which does not result from the motion of the planet but rather from the motion of the Earth in the great sphere changing the position of the observer."[108]

The eyes are limited as the sources of factual information. The planets reveal their true positions only on specific and rare occasions. The naive astronomer, who is not considering that appearances are ascertained only relative to one's point of view, is led astray by the eyes in configuring the true places of the celestial bodies: "It is therefore clear that the true positions of Saturn, Jupiter, and Mars are only visible to us when they are 'midnight stars,' which generally happens in the middle of their background motion. For then they are exactly in a straight line with the mean position of the Sun, and cast off their parallax."[109]

Copernicus, however, is far from contending that appearances are mere illusions. The astronomer's observations are the result of heavenly bodies moving uniformly in the celestial realm, and any astronomical theory must take these appearances into consideration. In rejecting Ptolemy's lunar theory, Copernicus musters evidences based on observations and measurements claiming not only that Ptolemaic theory does not preserve uniform circular motion of the Moon but also that "it follows by a mathematical proportion that when the moon in quadratures rests at the lowest part of the epicycle, it should appear about four times larger (if only the whole moon were illuminated) than the new and full Moon." That is, the distance that the Ptolemaic theory assumes does not correspond to the observed variation of the apparent lunar diameter and parallax: "If, however, one investigates more carefully, he will find that both [the apparent size and parallax] differ only very slightly in quadratures from what they amount to at new and full Moon."[110] Assessing accurately the quantitative aspect of an apparent image is enough to discard a theory, but it does not supply the astronomer with enough data to assess what actually occurs in the heavens.

To go beyond appearances, the astronomers have to calculate their own actual viewpoint and its location in relation to the entire universe. Alberti's insistence that the painter must determine beforehand the position of the eye in order to correctly observe the painted object is retrieved in Copernicus's need to situate his observer in the universe before beginning to describe its motions. This need is the result of Copernicus's re-evaluation of celestial appearances. Reorganizing the hierarchical relationship between the mathematical model, appearances, and the invisible yet real motions of the Earth induces rethinking of the epistemological status of astronomical theory in general. Andreas Osiander's famous introduction states the new complexity as clearly as possible: "Nor is it necessary that these hypotheses should be true, nor indeed even probable." Osiander's phrasing brings to the fore the incongruence between the astronomer's task to provide "a record of the motions of the heavens," established with "diligent and skillful observations," and uncertain hypotheses made of invisible circles and epicycles to assist in calculating past and future motions.[111] Copernicus's questioning of the status of appearances made the demarcation line between the two sides of the ancient dichotomy between observational facts and the explanatory mathematical models (or hypotheses) poignantly clear.

However, accepting Osiander's position was unviable, and to establish the superiority of the mathematical model over appearances, Copernicus's initial move is to cast doubt on the stipulated position of the observer. The opening sections of his *De revolutionibus*, following the Ptolemaic order of exposition, present the reader with the basic spherical panoramic shape of the universe and of the Earth. After these introductory moves, Copernicus starts diverging from the Ptolemaic literary model. Ptolemy, at this point, proceeds to establish the Earth as an immobile viewpoint at the centre of the universe and then dedicates a technical chapter to how one can measure the different arcs and angles on the concave surface of the heavenly sphere. Only then does Ptolemy turn to the details of the planetary motions. Copernicus, in contrast, immediately after setting the astronomical scene, proceeds to argue that the motion of the heavenly bodies is perpetual, circular (or compounded of circular motions), and uniform. Making this assumption reveals immediately the inherent astronomical problem that some of the heavenly bodies present certain irregularities. This order of exposition allows Copernicus to raise the question of the observer's position as crucial for the determination of the planets' motions: "It must therefore be agreed that though their motions appear to us irregular they are regular, either because the axes of their circles are different, or perhaps because the Earth is not in the middle of the circles in which they revolve, and to us observing from the Earth the transits of these stars, it comes about that because of the different distances, they seem larger when they are nearer than when they are further away (as is shown in *Optics*)."[112]

Since, in contrast to Ptolemy's order of exposition, the question of the observer's position is not yet determined, Copernicus can inquire "what is the relationship of the Earth to the heavens?" while adding a sarcastic admonition "that in scrutinizing the loftiest heights we do not fall into ignorance of what is nearest, and by the same mistake attribute to the heavens what belongs to the Earth."[113] Copernicus then sets to putting the observer's point of view in motion, assuming that all appearances are relative to that point: "Now the Earth is the point from which the rotations of the heavens is observed and brought into our view. If therefore some motion is imputed to the Earth, the same motion will appear in all that is external to the Earth but in the opposite direction, as if it were passing by."[114]

Like Cusa in *De visione dei*, Copernicus asks readers to transcend their particular point of view in order to imagine a new point of view from which they may examine their own act of observation. Cusa had written this treatise as a response to Vincent of Aggsbach, a Carthusian monk, who attacked the followers of Cusa at the monastery at Tegernsee for their type of intellectual mysticism. In his answer, Cusa instructs the monks of Tegernsee to assemble and gaze at a special picture or an icon that conveys the illusion of an all-seeing eye. This illusion is no mere curiosity or wonder but is meant to transport the readers by "human fashion" to "things divine"[115] – an effect recorded since Ptolemy. The monk beholding the portrait feels as though he is being watched and followed by the portrait's gaze, and he at first receives the impression that the icon's gaze is turned upon him. The immediate metaphor suggested by Cusa is the analogy of the icon's gaze with divine love: "Lord, Thy glance is love. And just as thy gaze beholdeth me so attentively that it never turneth aside from me, even so is it with Thy love."[116] The icon does not have to represent God or Christ; on the contrary, one of the examples Cusa suggests of such an icon is the self-portrait of Rogier van der Weyden that was painted for the Brussels town hall in 1438.[117] Beholding one's own stare is what allows one to transcend one's particular point of view and to grasp its position relative to other points of view. Cusa aims to lead his readers to understand that their immediate and naive relationship with the portrait is partial and to become aware that the vision of God is shared by all. Thus, following Cusa's instructions, a picture is impressed on the imagination of the meditating monk of how the infinite number of relative points of view is subsumed by the absolute perspective of God. In reflecting on one's own point of view, a different and invisible (or divine) point of view is becoming available: "I behold in the face of the picture a figure of infinity, for its gaze is not limited to one object or place, and is thus infinite, seeing that it is not more turned to one than to another of them that looks upon it. Yet, albeit its gaze is infinite in itself, it seemeth to one regarding it to be limited, since it looketh so fixedly on any beholding it as if it looked on him alone and on naught else."[118] With a central image that is an optical illusion, Cusa moves the spectator from a traditional reminder of the unreliability of human perspectives to a more radical self-displacement, the recognition that one's status as an observer is itself illusory; it is the observer who is the observed. Cusa affirms both the centrality of the observers and their potential eccentricity.[119]

Going beyond one's local and immediate point of view entailed the realization that appearances are a function of the observer's location. The new forms of visibility proposed by Alberti's techniques of perspective and by Cusa's geometrical visualizations were part of a more general cultural re-assessment of the role of perception in the cognitive process leading to knowledge. This role had special relevance to the epistemological status of astronomy, the observational science par excellence. This status is exactly what is at stake when Copernicus criticizes those who think they can prove geometrically that the Earth is in the middle of the universe. The core of Copernicus's argument is the limits of sense perception and the need to surpass them. He points out that his readers assume that the viewer is situated at the centre of the universe or a very small distance from it. For such a viewer, the horizon seems to be a great circle that bisects the ecliptic, whose centre is again the observer's point of view, and such a "horizon … is the dividing line between the visible and invisible" (*quo definiuntur apparentia, a non apparentia*).[120] Observing the constellations of the zodiac rise and set, the viewer can then identify the straight line between the opposing stars as the diameter of the ecliptic running through its centre, which is "at least apparently" (*prout apparet*) the Earth. The second stage of the argument accounts for the fact that a line from the centre of the Earth and a line from its surface are "virtually parallel" (*similes parallelis*), yet they appear to be a single line (*apparent esse linea una*). This phenomenon proves that "the sky is immense in comparison with Earth," but that is "as far as the senses can judge," and nothing else seems to have been demonstrated (*nec aliud demonstrasse videtur*).[121] The mobility or immobility of the Earth cannot be derived from one's sense experience, as these phenomena presuppose the observer's point of view. Calculating the observer's position, Copernicus can transcend visual experience and gauge a new invisible point of view from where a new picture of the universe is revealed. These calculations incorporate novel mathematical techniques coming from the East, yet Copernicus mobilizes these techniques to answer the challenges that Alberti's artificial perspective and Cusa's theological speculations offered to visual experience in the preceding century.

PART THREE

The Multicultural Astronomical Background to the Copernican Revolution

6

Fifteenth-Century Astronomy in the Islamic World

Sally P. Ragep

INTRODUCTION

IN A PERSONAL LETTER HOME TO HIS FATHER in Kāshān, a province in Iran near Isfahan, the fifteenth-century Islamic astronomer and mathematician Jamshīd Ghiyāth al-Dīn al-Kāshī (d. 832/1429) describes scientific activities at the Samarqand Observatory and Madrasa under the auspices of the ruler of Samarqand, Ulugh Beg. Kāshī informs his father that there are 500 students studying mathematics in twelve places scattered throughout Samarqand out of more than 20,000 students, all "steadily engaged in learning and teaching."[1] In his intimate and detailed letter, one of many letters written in reply to his father's queries about life in Samarqand, Kāshī provides us with evidence of the existence of a thriving scientific community of scholars, who wrote, read, and disseminated scientific materials.[2] Moreover, Ulugh Beg's Madrasa was just one of many madrasas established in Samarqand and, indeed, throughout Islamic lands.[3]

The fact is that by the fifteenth century, Samarqand scholars had inherited a rather rich corpus of Islamic astronomical works and were actively engaged in study and debate with scientific treatises that had been many centuries in the making. That so many astronomical treatises survived through numerous tumultuous upheavals (including the Mongol invasions) is testimony to how entrenched the notion of a scientific education was within Islamic society. Many of these works would inspire original compositions, commentaries, supercommentaries, and glosses; and these writings would circulate widely and play critical roles in the development of Islamic theoretical astronomy well into the nineteenth century. It is significant that concerted efforts would be made to seek teaching approaches that could accommodate the older Islamic scientific traditions along with new scientific developments well after "European science" came on the scene.[4] So when considering the bold step taken by Nicholas Copernicus, an interesting question to keep in the back of our minds as we explore the astronomical traditions of Islamic societies is the extent to which a deeply rooted, "naturalized" tradition of science may, paradoxically, have inhibited the type of change that made a Copernicus possible.[5]

ISLAMIC ASTRONOMICAL WORKS IN THE FIFTEENTH CENTURY: DO NUMBERS COUNT?[6]

The numbers of students whom Kāshī reports were engaged in studying the mathematical science in Samarqand are staggering but still feasible, even taking into account Kāshī's propensity to exaggerate.[7] He depicts the existence of a well-established group of assembled scholars and students who were devoted to the study of the mathematical sciences at that time. So in an attempt to shed more light on what numbers may (or may not) indicate about what was being studied, table 6.1 presents the findings of a preliminary survey of fifteenth-century works on Islamic astronomy. It shows that roughly 120 authors wrote some 489 treatises during what we may call the long fifteenth century (taking into account authors who began writing in the late fourteenth century as well as those who continued into the early sixteenth); and their works are represented by several thousand extant manuscript witnesses located in repositories throughout the world.[8]

The subject matter can be classified broadly as practical and theoretical, and these topics include cosmology (both celestial and terrestrial realms), instruments, handbooks (*zīj*es) and tables,[9] calendars, timekeeping, and astrology.

The numbers and types of astronomical works indicate that the vast majority fit into the category of being practical. They deal with computational instruments, timekeeping, and various aspects of Islamic ritual, such as prayer times, determination of the *qibla* (sacred direction to Mecca), and crescent visibility – for determining the beginnings of the Islamic lunar months, particularly Ramaḍān, the month of fasting.

Not many works deal with astrology per se (11 out of 489, or a mere 2 per cent), but this is not surprising if we take into account that astrologers were most likely relying on earlier compositions. Moreover, if we include the *zīj*es (40 out of 489, or 8 per cent), which could also serve the needs of astrologers, it boosts the total to 10 per cent.[10] Ulugh Beg's *Zīj*, for example, devotes an entire section to astrology.[11]

Table 6.1 Types of Islamic astronomical works in the fifteenth century

Type	*Titles*	*Authors*
Instruments	209	57
Practical astronomy, including timekeeping, *qibla* determination, etc.	135	56
Astronomical handbooks (*zījes*) and tables	40	27
Theoretical works	61	28
Astrology	11	10
Lunar stations (*anwāʾ*)	3	3
Unknown	30	22
Totals	489	203*

* Based on 120 individual authors; the discrepancy in number of authors listed is due to those astronomers who wrote in multiple categories.

The findings for astrology underscore the need to be cautious about drawing definitive conclusions about demand based on interpreting numbers alone. The role of the astrologer was multifaceted within medieval Islamic society.[12] The practice of astrology – at least the general subject of astrology as a science that interprets celestial signs and makes predictions – was widespread and quite popular in certain circles given that few could deny the allure of a discipline that dangles "the promise of predictive power over a full scale of phenomena ranging from cosmic events to the outcome of a battle or the length of an individual's life."[13] Many Islamic astrologers (like Western ones) found a niche, especially in the courts, as consultants to rulers offering the enticing promise of guidance and prediction.[14] Some have speculated that Ulugh Beg's belief in astrology and horoscopes was an incentive for his building the observatory.[15] But opponents of the practice of astrology and alchemy – and there were many, ranging from Hellenized philosophers to religiously based opponents – found much fault among the practice itself and its practitioners, not the least of these being claims of special abilities for interpreting God's divine will.[16]

One of the works studied in Samarqand was Abū Rayḥān al-Bīrūnī's (d. ca. 442/1050) *Al-Qānūn al-Masʿūdī*, a book based primarily on Claudius Ptolemy's *Almagest* but also containing within it criticism of the astrologers.[17] The example of Bīrūnī highlights the ambiguous status of astrology in Islamic society since his "true" attitude toward astrology has been debated. Some uphold that Bīrūnī really believed that the basic tenets of astrology were spurious and that its practitioners were unscrupulous; others argue that this interpretation ignores his motivations for composing his important astrological primer *Kitāb al-Tafhīm li-awāʾil ṣināʿat al-tanjīm* (The Book of Instruction in the Elements of the Art of Astrology) as well as twenty-three or so other astrological treatises.[18] But irrespective of his motivations, the point to keep in mind is that

Bīrūnī's astrological-astronomical works continued to be studied in Samarqand (and throughout central and south Asia and elsewhere) centuries after their original compositions.

In comparison to practical works, the findings show that about 13 per cent (61 out of 489) of the titles of astronomical works are devoted to theoretical works on cosmology. To account for this discrepancy, one might expect practical astronomical works to be more prolific because of their utilitarian nature, the notion of science being used as a handmaiden for religious rituals and needs in the "service of Islam."[19] Another reason might be the presumption that practical works require less specialization or rely more on formulaic methods for execution. Notwithstanding these reasons, many practical treatises, like their theoretical counterparts, could contain a didactic component.[20] Many were also extremely sophisticated and highly specialized. As early as the ninth century, one sees honorifics appended to the names of astronomers and mathematicians, indicating rather well-defined fields of specialization, and the vast majority of these fields reflect practical occupations.[21]

Among these honorifics, one that deserves special attention is that related to the office of the *muwaqqit* (timekeeper), a position associated with mosques and madrasas. According to David King, the science of reckoning time (*ʿilm al-mīqāt*) constituted the "essence of Islamic science," and "for the history of science on the global scale, *ʿilm al-mīqāt* in Islam was as important as *ʿilm al-hayʾa* (theoretical astronomy), *ʿilm al-zījāt* (mathematical astronomy), or *ʿilm aḥkām al-nujūm* (astrology)."[22] Putting aside King's essentialist tone, many *muwaqqit*s undoubtedly came up with creative and ingenious solutions to extremely complex problems in spherical astronomy, which demanded scientific knowledge beyond constructing tables for timekeeping and the regulation of prayer times.[23] One renowned *muwaqqit* was ʿAlāʾ al-Dīn ibn al-Shāṭir, the fourteenth-century head timekeeper of the Umayyad Mosque in Damascus, whose innovations in non-Ptolemaic planetary theory have been linked to the work of Copernicus. The extent of the influence of Ibn al-Shāṭir's planetary models on Copernicus has been discussed and debated for well over fifty years,[24] with recent research addressing "when, where, and in what form"[25] Copernicus may have incorporated Ibn al-Shāṭir's models into the *Commentariolus*.[26]

Ibn al-Shāṭir is often portrayed as being unique for his stellar achievements in both theoretical and practical matters of astronomy. However, many Islamic scholars straddled multiple subjects, and there could also be considerable overlap between disciplines. So, for example, we find that it is in his fifteenth-century commentary to Naṣīr al-Dīn al-Ṭūsī's "theological" work, the *Tajrīd al-ʿaqāʾid*, that ʿAlī Qushjī discusses astronomy and puts forth the "case that astronomy should dispense with its dependence upon Aristotelian physics" and that "the Earth's rotation is a possibility."[27]

It is not surprising, then, that the survey findings show that of the 23 per cent (28 out of 120) of fifteenth-century authors who wrote on theoretical issues of astronomy, some 61 per cent (17 out of 28) also wrote on practical subjects. Nonetheless, Kāshī informs his father that there is a distinct demarcation among the Samarqand scientists: "none of them is such that he is acquainted with both the theoretical ('scientific') and the applied ('practical') sides of observations."[28] Kāshī is probably judging a scientist's ability according to mastery of recognized criteria for ranked proficiency levels. This explains why he informs his father that Qāḍīzāde al-Rūmī, the head teacher at Samarqand (who was also Ulugh Beg's tutor),[29] is someone "who possesses the theoretical knowledge contained in the *Almagest* but not its applied side. He has not done anything that pertains to the practical. He is the most learned among them, but even he is only a beginner in theoretical astronomy."[30] Qāḍīzāde was certainly not inept in the practical applications of astronomy; nor was he a novice in theoretical astronomy (as shown below).[31] Kāshī's assessment indicates that the bar was set rather high to demonstrate mastery of theoretical astronomy!

Some sixty-three astronomical works were said to have been produced at Samarqand.[32] Furthermore, Kāshī cites the following astronomical and mathematical works of his predecessors that were being studied: Claudius Ptolemy's *Almagest*, Euclid's *Elements*, Naṣīr al-Dīn al-Ṭūsī's (d. 672/1274) *Al-Tadhkira fī 'ilm al-hay'a* and the commentaries on it by Niẓām al-Dīn al-Nīsābūrī (d. 730/1329–30) and al-Sayyid al-Sharīf al-Jurjānī (d. 816/1413), Quṭb al-Dīn al-Shīrāzī's (d. 710/1311) *Nihāyat al-idrāk fī dirāyat al-aflāk* and *Al-Tuḥfa al-shāhiyya fī al-hay'a*, Maḥmūd ibn Muḥammad ibn 'Umar al-Jaghmīnī's *Al-Mulakhkhaṣ fī 'ilm al-hay'a al-basīṭa* (composed 602–03/1205–06), and Abū Rayḥān al-Bīrūnī's (d. ca. 442/1050) *Al-Qānūn al-Mas'ūdī*.[33]

Fortunately, we can confirm many of the titles Kāshī mentions and supplement his overall depiction of a Samarqand education with a personal account of Fatḥ Allāh al-Shīrwānī (d. 891/1486), contained in his commentary on Ṭūsī's *Tadhkira*. Shīrwānī reports that he travelled to Samarqand from his native Azerbaijan in pursuit of scientific studies after reading al-Sayyid al-Sharīf al-Jurjānī's *Tadhkira* commentary with the Shī'ī scholar Sayyid Abū Ṭālib at the Shrine of Imām 'Alī Riḍā in Mashhad.[34] He tells us that he then spent five years at the Samarqand Madrasa studying Niẓām al-Dīn al-Nīsābūrī's commentary on the *Tadhkira* (among other things), before receiving his diploma (*ijāza*) in 844/1440 with Qāḍīzāde al-Rūmī.[35]

Shīrwānī's detailed account of student life highlights several points: that a student would have sought out a prescribed program of study for a higher education and that to obtain a diploma, a student had to undergo a rather gruelling process demonstrating proficiency through oral testing, listening, and reading. Shīrwānī is quite specific in his descriptions of the way lectures were held at the Samarqand Madrasa and in his account of the slow and careful process involved in reading texts by examining the subjects in detail through explanations, discussions, and establishing connections between the texts and their sources.[36] Shīrwānī's text, corroborated with historical sources such as those of the fourteenth-century Egyptian encyclopaedist Muḥammad ibn Ibrāhīm ibn al-Akfānī and the sixteenth-century Ottoman scholar Aḥmad ibn Muṣṭafā Ṭāshkubrīzāde, reaffirms that students were required to progressively master a body of scientific teaching textbooks categorized as beginner (*mukhtaṣar*), intermediate (*mutawassiṭ*), and advanced (*mabsūṭ*). For the discipline of theoretical astronomy (*hay'a*), the sources inform us that the assigned reading might have consisted of Ṭūsī's *Tadhkira* and/or Jaghmīnī's *Mulakhkhaṣ* for beginners, a work by Mu'ayyad al-Dīn al-ʿUrḍī (d. ca. 664/1266) for intermediate students, and Shīrāzī's *Nihāya* and *Tuḥfa* for the most advanced students.[37]

SAMARQAND: DIACHRONIC AND SYNCHRONIC SCHOLARLY PIPELINES

The majority of theoretical astronomical works from the survey seem to stem from central Asia and Anatolia; and this finding supports what we know about the fifteenth century, namely that in these regions a significant number of scientists were actively composing texts, debating theoretical questions, and creatively re-examining and reformulating topics in astronomy. Transoxiana, Khurāsān, and Khwārizm were particularly renowned for being "the philosophical and scientific granaries of Islamic civilization,[38] but since Islamic scholars travelled extensively, there were many scholarly pipelines throughout Islamic lands,[39] a prominent one being between central Asia and Anatolia.[40]

Qāḍīzāde al-Rūmī hailed from Bursa (once the capital of the Ottoman state) and had studied the mathematical sciences in late-fourteenth-century Anatolia as a member of a renowned circle of scholars referred to as the Fanārī School.[41] Qāḍīzāde would become an important link in disseminating the fruits of their scientific activities to central Asia. Mullā Shams al-Dīn al-Fanārī (d. 834/1431), under the auspices of the Ottoman sultan Bāyazīd I (791–805/1389–1403), had been charged with inviting the best and brightest intellectuals to collect and standardize scientific textbooks for the curricula of the burgeoning Ottoman madrasas.[42] Many of these textbooks stemmed from the work produced in thirteenth-century Marāgha as well as from treatises composed in the centuries prior to the Mongol invasions that devastated the region. Among the scholars Fanārī attracted to Bursa was ʿAbd al-Wājid ibn Muḥammad (d. 838/1435), who travelled to Anatolia from his native Khurāsān, where he subsequently became one of Qāḍīzāde's teachers before teaching at the eponymous Wājidiyya Madrasa in Kütahya.[43] ʿAbd al-Wājid's commentary on Jaghmīnī's *Mulakhkhaṣ* became an important text that was studied within the Fanārī circle and presumably at the Wājidiyya Madrasa; and Ṭūsī's *Tadhkira*, Shīrāzī's *Nihāya* and *Tuḥfa*, and ʿAbd al-Jabbār al-Kharaqī's (477–553/1084–1158) *Al-Tabṣira fī ʿilm al-hayʾa* were known and cited. ʿAbd al-Wājid's commentary may have also played a subsequent role in Qāḍīzāde's decision to write his own *Mulakhkhaṣ* commentary, which he composed in 814/1412 in Samarqand and presented to Ulugh Beg (along with his commentary on Shams al-Dīn al-Samarqandī's *Ashkāl al-taʾsīs*, a geometrical tract).[44]

The scholarly connections between Anatolia and central Asia, along with Samarqand's high reputation in the mathematical sciences, undoubtedly were strong factors in Qāḍīzāde's decision to undertake the long journey eastward.[45] Qāḍīzāde's arrival meant that Samarqand scholars acquired the collection of the circle of Mullā Fanārī, which included scientific treatises from the pre-Mongol and Īlkhānid periods.[46]

Just as our Samarqand scholars were indebted to the members of the Fanārī circle, they in turn were indebted to the scientific activities of their predecessors, who included astronomers working in the thirteenth and fourteenth centuries in Iran and Syria, among whom were Ṭūsī, Shīrāzī, ʿUrḍī, and Muḥyī al-Dīn al-Maghribī. This collective group, which attracted an impressive array of scholars from regions that spanned from China to Spain, has frequently been referred to as the "Marāgha School," a term coined as a convenient reference to the Marāgha members of the important but short-lived Mongol-sponsored Marāgha Observatory (ca. 657–81/1259–83).[47] Oddly, the Marāgha School lived on into the next centuries. So the work of Ibn al-Shāṭir in Damascus in the mid-fourteenth century was also considered part of the school.[48] But in fact, it was never a "school" in the sense of being a madrasa; and furthermore, few of the innovations and major works associated with the Marāgha Observatory, such as the Ṭūsī-couple and alternatives to Ptolemaic astronomical models, originated during the time the observatory was in operation.[49]

The perception of a Marāgha School has tended to mask the great debt the thirteenth-century scholars owed to their predecessors in the preceding centuries, especially those who flourished in the mid-twelfth to early thirteenth centuries in the vicinity of Merv in central Asia, under the auspices of the Khwārizm-Shāhs (r. 470–628/1077–1231). These earlier scholars were formative in establishing many of the foundations of *hay'a* that have often been associated with the Marāgha School. Thus the Marāgha scholars should be highly credited for their own significant scientific achievements, not the least of which centred on the work associated with what is considered the first large-scale observatory ever built,[50] but in addition, they should be duly recognized for their concerted efforts to resuscitate the works of their immediate and earlier predecessors. For example, by the mid-1240s, Ṭūsī was engaged in a monumental project – which spanned almost twenty years – of providing a body of textbooks, with commentary, of the Greek classics and early Islamic scientific works.[51]

When Shīrāzī provided his summary list of important *hay'a* works (composed in both Arabic and Persian) up until his time in the explicit of his *Nihāya* (680/1281), he was paying homage both to his contemporaries and to his predecessors. So included among his list are Ṭūsī's *Tadhkira*, *Al-Zubda*, and Persian *Risālah-i Mu'īniyya*, alongside Jaghmīnī's *Mulakhkhaṣ* and Kharaqī's *Muntahā al-idrāk*, *Tabṣira*, and *Al-'Umda li-ūlī al-albāb*.[52]

THE DISCIPLINE OF *HAY'A*

So what exactly were *hay'a* works?[53] They were part of a genre of astronomical literature termed *'ilm al-hay'a*, a corpus that attempted to explain the configuration (*hay'a*) or *physical* structure of the universe as a coherent whole, meaning the physical, simple bodies that composed both the celestial region ("cosmo-graphy") and the lower bodies of the terrestrial realm ("geo-graphy").[54] According to Qāḍīzāde, this way of dealing with *all* the bodies was an innovation of Islamic astronomers that differentiated their science from that of the astronomy of the ancient Greeks. In other words, it brought together the unchanging realm of the celestial aether and the ever-changing realm of the four elements, the world of generation and corruption, into a single discipline.[55]

By the eleventh century, *'ilm al-hay'a* had replaced *'ilm al-nujūm* (the science of the stars), particularly in eastern Islam, as the general term for the discipline of astronomy, which did not include astrology.[56] For example, this is what we find in Ibn Sīnā's (Avicenna's) *Aqsām al-'ulūm al-'aqliyya* (Classification of the Rational Sciences) and in most accounts of the discipline after this time. And with *'ilm al-hay'a* no longer being a subdivision of astronomy but the field itself, one branch then becomes *hay'a basīṭa* (plain *hay'a*), being cosmographical works containing no geometrical proofs.[57]

The genre of *hay'a basīṭa* literature was highly influenced by Ptolemy's *Almagest* (omitting its mathematical proofs) and by his *Planetary Hypotheses*, usually including discussions on the sizes and distances of the stars and planets. *Hay'a* became recognized as a strictly mathematical discipline with an emphasis on transforming mathematical models of celestial motion into physical bodies as a way of providing a picture or configuration (*hay'a*) of the universe as a whole. Its focus was on addressing the external aspects of cosmology, or issues related to "how" the celestial and terrestrial realms operate the way they do, not on dealing with questions of "why." It is significant that *hay'a* works do not discuss subjects related to the "causes" of natural phenomena and matters of Aristotelian metaphysics; this was because these internal aspects of cosmology related to natural philosophy and metaphysics and were thus considered inappropriate for inclusion in *hay'a* works and were dealt with elsewhere.[58]

To make Ptolemy's works more comprehensible, Islamic scholars provided an assortment of accounts of various aspects of Ptolemaic spherical astronomy and planetary theory, in addition to those falling under the category of *hay'a basīṭa* works.[59] Some were reworkings of Ptolemy's *Almagest*, which could contain original material within them, such as we find in Ṭūsī's *Taḥrīr al-Majisṭī* (Recension of the *Almagest*), which exists in dozens of copies, perhaps hundreds if we include the commentaries and translations.[60] Others were treatises devoted to criticizing and reconciling inconsistences in Ptolemaic astronomy and reforming certain models. For example, Abū ʿAlī al-Ḥasan ibn al-Haytham (d. ca. 430/1040) criticizes the irregularities and violations of uniform circular motion within the *Almagest* and the *Planetary Hypotheses* in his *Al-Shukūk ʿalā Baṭlamyūs* (Doubts about Ptolemy). He further proposes alternative models to resolve the problems of Ptolemy's planetary latitude theory in a work called *Al-Maqāla fī ḥarakat al-iltifāf*.[61] Some have included al-Farghānī's (Alfraganus's) ninth-century compendium of the *Almagest*, entitled *Jawāmiʿ ʿilm al-nujūm*, within the *hay'a basīṭa* tradition.[62] However, despite its popularity and wide circulation, especially in Latin translation,[63] it was more of a descriptive compilation of selected parts of the *Almagest*, lacking illustrations and at times coherence due to Farghānī's style of conflating topics with a piecemeal approach.[64]

Although Farghānī has a section on sizes and distances of the orbs, his presentation, like that of many early Islamic writers on astronomy, tended to follow the *Almagest*, in which the planetary models were presented geometrically rather than physically. These mathematical models were certainly compatible with a physical representation of the universe, as Ptolemy himself had shown in his *Planetary Hypotheses*; nevertheless, there was, and continues to be, some ambiguity about whether Ptolemy's models were "real" or were only meant to "save the phenomena."[65] What concerns us here are the perceptions of Islamic authors themselves, who traced the beginnings of the physical models of *hay'a* not to Farghānī but to Ibn al-Haytham.

As Ibn al-Haytham asserted, his fifteen-chapter *Al-Maqāla fī hay'at al-'ālam* (Treatise on the Configuration of the World)[66] was an attempt to explain to the reader how the various components of the Ptolemaic models operated and ultimately fit together.[67] It is not surprising that such a work could influence generations of scholars throughout Islamic lands[68] and have a major impact on astronomical planetary theory in the Latin West.[69] Based on available evidence, it does seem that theoretical astronomical works prior to Ibn al-Haytham tended to consist of general overviews, summaries, and technical (often selective) discussions. In comparison, *On the Configuration of the World* attempts to match the mathematical models of the *Almagest* with physical structures to account for the various motions of the celestial bodies. In so doing, Ibn al-Haytham does not provide parameters, proofs, a discussion of sizes and distances, or even illustrations, although there are indications that he wanted to include them;[70] and since his focus is more on explaining the hows of the celestial components, he keeps terrestrial topics to a minimum and omits philosophical and astrological topics altogether.

Ibn al-Haytham should be duly recognized as a "pioneering inspiration rather than a prototype to be emulated."[71] Nevertheless, to claim that he "single-handedly established physical cosmography in Islam" overshadows the fact that he was often making explicit what was already implicit in previous Islamic theoretical works.[72]

Unlike Ibn al-Haytham's *On the Configuration of the World*, beginning in the early twelfth century, most subsequent *hay'a* works, irrespective of their level of difficulty, include an introduction and a two-part division, one section dealing with the configuration of the celestial region and another section devoted to topics related to the configuration of the Earth. Some of these later *hay'a* works might also devote a chapter or section to the subject of sizes and distances and/or a discussion of chronology. This signature structure seems to have been codified with Kharaqī's *hay'a* works, especially his influential *Muntahā al-idrāk fī taqāsīm al-aflāk*[73] and the shorter, more popular *Al-Tabṣira fī 'ilm al-hay'a* (both composed ca. 526–27/1132–33).[74] However, Kharaqī's *Muntahā*, being an intermediate-level work, was extremely technical, replete with long explanations and extensive diagrams; and his *Tabṣira*, although abridged, was also fairly technical, especially for an elementary treatise. Ultimately, a growing need for a far more "user-friendly" elementary textbook on theoretical astronomy resulted in the composition of Jaghmīnī's *Mulakhkhaṣ*.[75]

Jaghmīnī's *Al-Mulakhkhaṣ fī ʿilm al-hayʾa al-basīṭa* stands out as the most widely circulated Arabic treatise on Ptolemaic astronomy ever written. The base Arabic text served as the starting point for at least sixty-one commentaries, supercommentaries, glosses, and translations (into Persian, Turkish, and Hebrew),[76] and these writings span at least seven centuries beyond its original composition date of 602–03/1205–06 in the region of Khwārizm in central Asia. The topics include basic astronomical definitions and concepts, parameters of the motions of the planets and the Earth's inhabited zone, and above all, a cosmography (*hayʾa*) of the universe that offered a scientific account of God's creation.

The impact and longevity of the influence of the *Mulakhkhaṣ* are not in question, as evidenced by thousands of extant copies of the original and its various derivatives contained in repositories worldwide. Therefore, it is not at all surprising that the survey found that Jaghmīnī's *Mulakhkhaṣ* was the most commented on and studied *hayʾa* treatise in the fifteenth century, followed by Ṭūsī's *Tadhkira*, completed just over half a century later.[77]

Both the *Mulakhkhaṣ* and the *Tadhkira* undoubtedly played key roles in disseminating the teaching of theoretical astronomy throughout Islamic lands for generations of scholars.[78] But in addition to these and works from the classical period of Islamic science, fifteenth-century Samarqand also inherited scientific works from twelfth-century Merv, thirteenth-century Marāgha, late-thirteenth- and early-fourteenth-century Tabriz (which inherited the Marāgha scientific tradition and observatory),[79] fourteenth-century Bursa, and other major centres of intense scientific activity, such as the Mamluk cities of Cairo and Damascus.[80] Many of these works inspired original compositions, commentaries, supercommentaries, and translations. And the scientific material amassed in fifteenth-century Samarqand would become the basis for textbooks of Islamic astronomical learning that would be disseminated widely from North Africa to India into the twentieth century.[81]

This was a period of significant innovations in theoretical as well as observational astronomy, the latter showcased due to Ulugh Beg's collaborative *Zīj* project, which provided important improved tables of astronomical parameters based on new observations. Ulugh Beg's Samarqand Observatory, built several years after he founded the Samarqand Madrasa, is undoubtedly one of the most acclaimed observatories of Islam from the standpoint of work and longevity (spanning some thirty years).[82] ʿAlī Qushjī – who ostensibly succeeded Qāḍīzāde as director of the observatory – was an active member of the program, but this did not deter him from criticizing some of its deficiencies in his Persian commentary on the *Zīj*, entitled *Sharḥ-i Zīj-i Ulugh Beg.*[83] Qushjī's bold stance applied to issues related to theoretical astronomy as well. Within his various works, he put forth the position that the astronomer had no need for Aristotelian physics and should establish his own physical principles independently of the natural philosophers, he raised the possibility of the Earth's motion, and he also proposed eccentric models as an alternative to epicyclic ones for the inner planets.[84]

Several *hayʾa* treatises were composed in the fifteenth century that were quite significant for the teaching of theoretical astronomy: Qāḍīzāde's *Sharḥ al-Mulakhkhaṣ* (composed in Samarqand in 814/1412 and presented to Ulugh Beg),[85] Qushjī's Persian *Risāla dar ʿilm-i hayʾa* (composed in Samarqand in 862/1458), and the longer Arabic version of it entitled *Al-Risāla al-Fatḥiyya fī ʿilm al-hayʾa* (presented to Mehmed II in 878/1473).[86] The study of *hayʾa*, by providing a picture of God's entire creation (both the unchanging celestial realm and the ever-changing sublunar one), offered another approach to serve God, namely a rational and noble approach for attaining a better understanding of God through His creation.[87] In so doing, *hayʾa* works became staple additions to the curricula of madrasas, the main institutions of Islamic societies.[88] Qushjī's *Risāla al-Fatḥiyya* became the second most favoured intermediate-level book in Ottoman madrasas after Qāḍīzāde's commentary.[89]

SAMARQAND SCIENTIFIC ACTIVITIES: A DISCRETE EPISODE?

A prevalent narrative regarding Islamic science tends to promote its history as being composed of discrete episodes and dependent in the main on courtly patronage or individual initiatives, with the religious institutions playing a limited role.[90] Samarqand scientific activities are thus depicted primarily as an outcome of the courtly patronage of "one Timurid prince, i.e. Uluġ Beg,"[91] who had a personal bent for mathematics and the means to see his vision implemented.[92] But without denying that funding and individual scientific achievements matter, can this adequately account for the hundreds of matriculating students, who presumably had some degree of prior scientific knowledge and were eager to study something as complicated as Ptolemy's *Almagest,* or Euclid's *Elements,* or Shīrāzī's *Nihāya*?

The view that a scientific education (or a scientific achievement) was "largely an individual affair"[93] is in agreement with George Makdisi's often-cited position put forth in his seminal work *The Rise of Colleges*, namely that a religiously endowed (*waqf*) institution, such as the madrasa, was *legally* bound to adhere to the stipulations of its endowment and therefore "was devoted primarily to law, the other sciences being studied as ancillaries."[94] Setting aside the veracity of this stance, the perceived problem of selective exclusion has been seized upon by those who tend to view Islamic society as a whole as having turned its back on rational discourse, especially on science, from at least the twelfth century onward. This sweeping generalization has been challenged by numerous intellectual historians in the fields of Islamic science, philosophy, and theology, especially as mounting evidence has indicated that some science (and most certainly philosophy and logic) continued to be studied and taught within a religious context.[95] Nonetheless, this hasn't halted Eurocentric scholars such as Edward Grant[96] or Toby Huff[97] from contrasting Islamic learning – unfavourably – with the institutionalization of scientific and philosophical learning at universities in Europe.

Some scholars have maintained that it is irrelevant whether or not the mathematical sciences entered the madrasas with some form of legitimacy. Sonja Brentjes asserts that "[t]he locus of teaching simply did not matter nearly as much as the teaching itself" and that "it is almost irrelevant to ask whether these ['secular'] sciences were taught in a teaching institute, a private house, or a garden."[98] The key to explaining Islamic scientific education and achievements over time and distance rested with individuals, personal relationships, and scholarly networks.[99] And for another cultural context, Nancy Bisaha notes the "interconnectedness" of seemingly "discrete histories" via travelling individuals as an important factor in defining what it meant to be a European.[100]

One certainly can understand the appeal of focusing on individuals, specific time periods, and locales: it can highlight the important role of minorities and migrating scholars in disseminating knowledge,[101] it can account for stellar scientific achievements of individual scholars,[102] and it allows us to spotlight less mainstream astronomical theories that emerged, such as the homocentric cosmology.[103]

Although it has its positive side, such an emphasis on the local can result in missing the big picture. Traditions are important; how else can we understand scholars, separated by time and place that could amount to centuries and continents, confront common sets of astronomical problems from a common conceptual and methodological standpoint? It is this point that helps us to understand that Islamic astronomy in the fifteenth century – whether in Samarqand or elsewhere – was not an isolated event or episode. It built upon centuries of work and stood on the shoulders of countless scientists (not necessarily giants). Moreover, it was this astronomy that most likely provided the immediate context of transmission to a bourgeoning European astronomy.

ON CONNECTING FIFTEENTH-CENTURY ASTRONOMY IN ISLAM TO COPERNICUS

Within Islamic societies, great scientific achievements have been due both to the efforts of individual scientists and to assemblies of scientists working together, such as we find in Marāgha or Samarqand; one assumes such collective activities make a difference. A danger of putting forth explanations based on the heroic individual scientist in search of knowledge is that it runs the risk of perpetuating the notion of innovations occurring only via the exceptional scholar, perhaps seeking out other exceptional scholars, whether contemporaries or predecessors. This portrayal relies more on individual geniuses and less on the role of group dynamics and on the importance of the infrastructure of knowledge, formed through either textual cultures or institutions.

It therefore seems to me important to emphasize the role of the various Islamic institutions – madrasas, mosques, courts, and khanqahs – in fostering knowledge, including scientific ideas, especially once the Ottomans came on the scene. Recall that Huff attributes the Copernican revolution not to the man himself but to an educational system, although he obviously assumes that this was unique to Europeans.[104]

Indeed, the Ottoman Empire – certainly by the time of the reign of Mehmed II in the mid-fifteenth century, if not much before – had evolved from a small frontier principality in the Balkans into a quite formidable political, economic, and military force that stretched from the Danube to the Euphrates. Moreover, replete with a centralized administration, it implemented, among other things, standardized educational reforms within its institutions.[105]

It is through these Ottoman institutions that one finds the connection between Islamic astronomy and Copernicus, as well as his immediate Latin predecessors. As has been detailed by Tzvi Langermann and Robert Morrison, a certain Moses ben Judah Galeano (Mūsā Jālīnūs) crossed a number of boundaries in the eastern Mediterranean, travelling, among other places, between the Ottoman court and Italy.[106] In his travels, Galeano brought knowledge of a number of devices and models of Islamic and Jewish astronomy, including a version of the Ṭūsī-couple and Ibn al-Shāṭir's models, that were to be so instrumental to the planetary systems presented in Copernicus's *Commentariolus*. We also have the case of the critical proposition of Qushjī, by which he shows how to transform the epicyclic models of Mercury and Venus into eccentric models, a possible linchpin in Copernicus's transformation of a geocentric system into a heliocentric one. That this proposition also appears in the 1496 printing of Regiomontanus's *Epitome of the Almagest*, with a diagram quite close to that of the extant Turkish manuscripts, is strongly suggestive of close connections between Istanbul and Vienna circles.[107]

But singular cases of transmission should be seen within a much wider context of intercultural interaction, as several authors in this volume have emphasized. Works like Ṭūsī's *Tadhkira* and his *Recension of the Almagest* were not marginal texts; they were extensively disseminated, as indicated by the extant manuscripts, and as we have seen, they were studied in madrasas and even at religious shrines, as indicated by both manuscript witnesses and secondary accounts. Therefore, it should not surprise us that such works could find echoes in European academic centres, which, after all, were literally at the doorsteps of the Ottoman Empire, if not contained within it.[108] It should equally not surprise us that an extant manuscript of Jaghmīnī's *Mulakhkhaṣ* resides in a Sarajevo repository, brought there from Edirne in the seventeenth century, for the *Mulakhkhaṣ*, along with other Islamic scientific texts – including many by fifteenth-century scholars such as Qāḍīzāde (in Bursa and Samarqand) and Sibṭ al-Māridīnī (in Cairo and Damascus) – became a staple astronomical teaching textbook in the territories of Bosnia and Herzegovina well into the twentieth century.[109] This history of transmission should, at the least, give us pause when making claims about the barriers separating Europe from Islam; at least geographically, the Islamic world extended into Europe well into the modern period.[110]

Let us look a bit more closely at the example of Jaghmīnī's elementary textbook on theoretical astronomy, the *Mulakhkhaṣ*. Claims have been made that the physicalization of the celestial orbs in Latin *theorica* works would have been a key stimulus for Copernicus in proposing his new cosmology.[111] My focus here is not on the existence of prior *theorica* works dealing with physical concerns but on the multicultural context in which a work such as Georg Peurbach's *Theoricae novae planetarum* might appear and subsequently inspire a cosmological leap. It is well known that Islamic scholars had been engaged in resolving issues concerned with a coherent physical cosmology since the eleventh century, if not earlier. Downplayed, however, is that many of these early works, such as those of Farghānī and Ibn al-Haytham, did not contain diagrams. However, later texts did; and works such as Jaghmīnī's *Mulakhkhaṣ* were composed for teaching within an institution for pedagogical reasons. As already mentioned, there exist thousands of Islamic astronomical manuscripts filled with diagrams containing presentations of an orb model for the motion of each celestial object. E.J. Aiton states that "Peurbach evidently drew upon Ibn al-Haytham's (Alhazen's) *On the Configuration of the World* or some later work based on this."[112] But Willy Hartner, writing some thirty years earlier than Aiton, not only compared Peurbach's Mercury model with that of Ibn al-Haytham's but also recognized Jaghmīnī's interest in the physical reality of the orbs and asserted, "The dependency of early Renaissance astronomers on ALHAZEN and AL-JAGHMĪNĪ is beyond doubt. Yet I am unable to tell at the moment from which of the two (possibly from both), and through which channels, they drew their information."[113] Jerzy Dobrzycki and Richard Kremer have speculated that although there is no textual evidence to date, Peurbach employed mathematical techniques from Islamic astronomers in thirteenth-century Marāgha for his models.[114] However, if we are to believe that cosmological revolutions can be inspired by visual evidence, should we not consider the source of visual, in addition to textual, evidence? Ibn al-Haytham is just one of many Islamic scholars who wrote such cosmological works, and his treatise inspired only a handful of commentaries, whereas the figures and commentaries of Jaghmīnī, Ṭūsī, and others fill repositories throughout the world and number in the thousands, and they had become a staple of Ottoman scientific research and pedagogy.

For those who ask why Islam did not give rise to a Copernicus, the answer can hardly be that the scientific traditions in Islam became extinct after the twelfth century. Or that there was a steady decline due to religious antagonism. The thriving scientific traditions in the Islamic world outlined above will, one hopes, dispel those misguided, though prevalent, narratives. Indeed, one could claim that the most thriving scientific traditions in the fifteenth and even early sixteenth centuries were in the Islamic world. Yet this long history of Islamic scientific inquiry may be the point. Scientific change may be far more difficult when the traditions (and paradigms, to coin a term) are so entrenched. So, for example, "sophisticated" Islamic scholars, well trained in the intricacies and arguments for a Ptolemaic system with its epicycles and eccentrics, would have dismissed homocentric astronomy. Thus, paradoxically, the strength of a scientific tradition, such as that in Samarqand, may have been a hindrance to adopting new, revolutionary ideas. Perhaps the lesson we then take from this cross-cultural comparison is that proposing revolutionary ideas may be easier for someone, such as Copernicus, whose scientific context was less rigid and was, in many ways, a work in progress.[115]

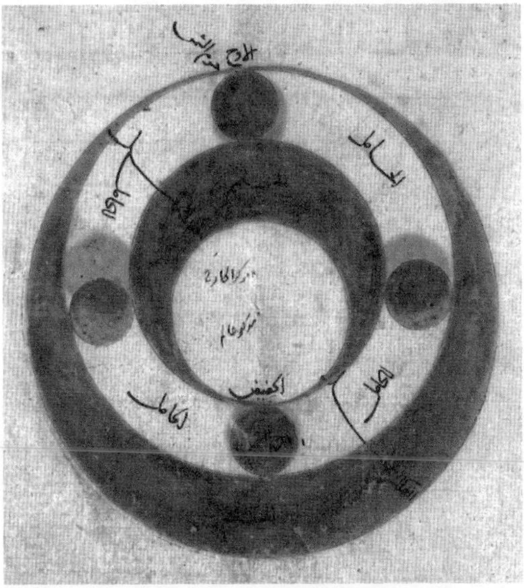

6.1 The Sun's orbs in Jaghmīnī's *Mulakhkhaṣ* (composed 602–03/1205–06).

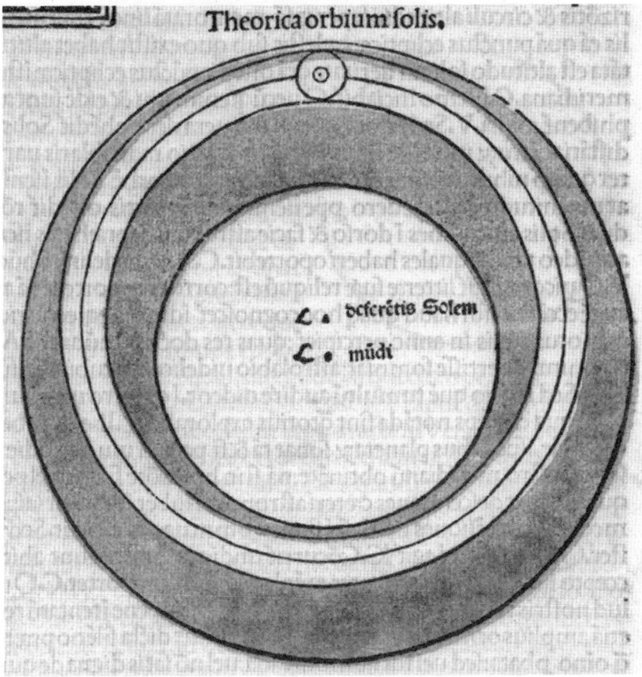

6.2 The Sun's orbs in Peurbach's *Theoricae novae planetarum*.

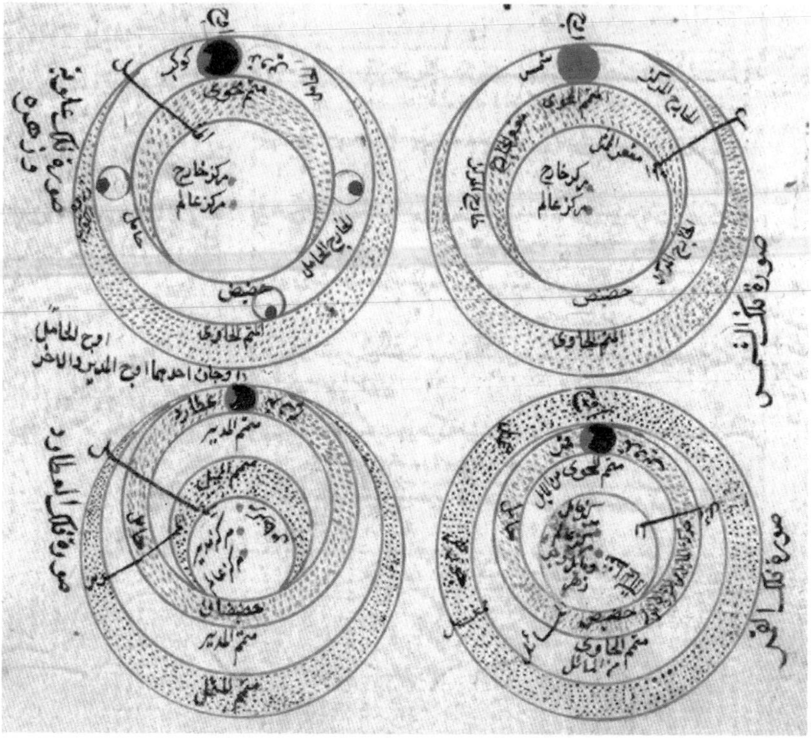

6.3 Orbs (clockwise from upper left) of the upper planets and Venus, the Sun, the Moon, and Mercury in Qāḍīzāde's commentary (composed 814/1412) on the *Mulakhkhaṣ*.

7

From Tūn to Toruń:
The Twists and Turns of the Ṭūsī-Couple

F. Jamil Ragep

IN DISCUSSIONS OF THE POSSIBLE CONNECTIONS between Nicholas Copernicus and his Islamic predecessors, the Ṭūsī-couple has often been invoked by both supporters and detractors of the actuality of this transmission. But, as I have stated in an earlier article, the Ṭūsī-couple, as well as other mathematical devices invented by Islamic astronomers to deal with irregular celestial motions in Ptolemaic astronomy, may be of secondary importance when considering the overall significance of Islamic astronomy and natural philosophy in the bringing forth of Copernican heliocentrism.[1] Nevertheless, the development and use of Naṣīr al-Dīn al-Ṭūsī's (597–672/1201–74) astronomical devices does provide us with important evidence regarding the transmission of astronomical models and with lessons about intercultural scientific transmission. So in this chapter, I attempt to summarize what we know about that transmission, beginning with the first diffusion from Azerbaijan in Iran to Byzantium and continuing to the sixteenth century. Although there are still many gaps in our knowledge, I maintain, based on the evidence, that intercultural transmission is more compelling as an explanation than an assumption of independent and parallel discovery.

THE MULTIPLE VERSIONS OF THE ṬŪSĪ-COUPLE

It will be helpful if we first analyze what exactly is meant by the "Ṭūsī-couple." The first thing to notice is that the term "Ṭūsī-couple" does not refer to a single device or model but actually encompasses several different mathematical devices that were used for different purposes (see table 7.1). Because this understanding is not always upheld in the modern literature, there has been considerable divergence, often leading to

Table 7.1 Versions of the Ṭūsī-couple

Name of device (Ragep)	*Description*	*Intended use*	*Other names*	*First appearance*
Mathematical rectilinear version (fig. 7.1)	Two circles uniformly rotating in opposite directions – the smaller internally tangent to the larger with half its radius – that produces rectilinear oscillation of a given point	Replacing the equant (and its like) in planetary models	Plane version (Saliba and Kennedy) Spherical version with parallel axes and radii in the ratio of 1:2 (Di Bono, 136) Device (*aṣl*) of the large and the small (circles)* Two-unequal-circle version	643/1245
Physicalized rectilinear version (fig. 7.2)	Three solid spheres based on the mathematical version that produces rectilinear oscillation of a given point	Replacing the equant (and its like) in planetary models	Physicalized two-circle version with maintaining sphere	643/1245
Two-equal-circle version (fig. 7.5)	Mathematically equivalent to the rectilinear version but using two circles of equal radius, each circle's circumference going through the centre of the other, one circle rotating twice as fast as the other to produce rectilinear oscillation of a given point	To account for Ptolemaic motions needing curvilinear oscillation on a great circle arc (but actually produces oscillation on a chord)	Plane version with equal radii (Di Bono, 137–8) Pseudo-curvilinear version	644/1247
Three-sphere curvilinear version (figs 7.6 and 7.7)	Three concentric spheres with different axes, one inside the other, rotating uniformly	Intended to account for Ptolemaic motions needing curvilinear oscillation on a great circle arc (mostly works as intended but with a minor distortion)	Spherical version (Saliba and Kennedy) Spherical version with oblique axes and equal radii (Di Bono, 136)	659/1261
Two-sphere curvilinear version (fig. 8.4)	Truncated version of the full three-sphere curvilinear version	For certain astronomical models (used by Ibn Naḥmias and later Copernicus)		ca. 1400

* It was often referred to as such in astronomical texts after Ṭūsī. He himself does not explicitly use this term to refer to the device although it is implied in the terminology he uses in the *Tadhkira*, as distinct from the *Ḥall* (see below).

Sources: Mario Di Bono, "Copernicus, Amico, Fracastoro, and Ṭūsī's Device: Observations on the Use and Transmission of a Model," *Journal for the History of Astronomy* 26, no. 2 (1995): 133-54; F. Jamil Ragep, *Naṣīr al-Dīn al-Ṭūsī's Memoir on Astronomy (al-Tadhkira fī ʿilm al-hayʾa)* (New York: Springer-Verlag, 1993), vol. 2, 427-56; George Saliba and Edward S. Kennedy, "The Spherical Case of the Tūsī Couple," *Arabic Sciences and Philosophy* 1, no. 2 (1991): 285-91.

confusion, about what exactly the Ṭūsī-couple is. This confusion, in turn, has made it difficult to trace transmission. So a quick historical overview is in order.[2]

MATHEMATICAL RECTILINEAR VERSION

The first version of the Ṭūsī-couple was announced by Naṣīr al-Dīn al-Ṭūsī in a Persian astronomical treatise entitled *Risālah-i Muʿīniyya* (*Muʿīniyya* Treatise), the first version of which was completed on Thursday, 2 Rajab 632 (22 March 1235).[3] Dedicated to the son of the Ismāʿīlī governor of Qūhistān, in the eastern part of modern Iran, the treatise is a typical *hayʾa* (cosmographical) work, one that provides a scientifically based cosmology covering both the celestial and terrestrial regions. But in presenting the Ptolemaic configuration of the Moon's orbs and their motions, Ṭūsī notes that the motion of the epicycle centre on the deferent is variable, which is inadmissible according to an accepted rule of celestial physics, namely that all individual motions of orbs in the celestial realm should be uniform. He goes on to say, "This is a serious doubt with regard to this account [of the model], and as yet no practitioner of the science has ventured anything. Or, if anyone has, it has not reached us." But "there is an elegant way to solve this doubt but it would be inappropriate to introduce it into this short treatise." He then teasingly turns to his patron: "If at some other time the blessed temper of the Prince of Iran, may God multiply his glory, would be so pleased to pursue this problem, concerning that matter a treatment will be forthcoming." In the chapter on the upper planets and Venus, as well as the one on Mercury, he makes a similar claim, namely that he has a solution that will be presented later. In addition to the problem of the irregular motion of the deferent (sometimes referred to as the "equant problem," although it is somewhat different for the Moon), Ṭūsī brings up another "doubt" or difficulty, namely that pertaining to motion in latitude – that is, north or south of the ecliptic. Claudius Ptolemy had rather complex models in his *Almagest* and *Planetary Hypotheses* that generated quite a bit of discussion among Islamic astronomers. One of these was Abū ʿAlī al-Ḥasan ibn al-Haytham (d. ca. 430/1040), who objected to the lack of physical movers for these models and provided his own in a treatise that is currently not extant. However, Ṭūsī refers to it in the *Muʿīniyya* and also notes that it is not entirely satisfactory; but as with his purported models for longitude, he eschews any details.[4]

Since Ṭūsī claims to have an elegant solution, one assumes that he would have presented it to his patron in short order. But, as we shall see, he waited almost ten years to present his new models. One clue to the delay could well be overoptimism on the part of the young Naṣīr al-Dīn; he claimed in the *Muʿīniyya* that he had solutions for all the planets, but as it turned out he was never able to solve the complexities of Mercury. Indeed, as an older man many years later, he was to admit this setback in his *Al-Tadhkira fī ʿilm al-hayʾa* (Memoir on the Science of Astronomy): "As for Mercury, it has not yet been possible for me to conceive how it should be done."[5]

The partial solution occurs in a short treatise that was again dedicated to his patron's son, Muʿīn al-Dīn. This work has come to us with a variety of names: *Dhayl-i Muʿīniyya* (Appendix to the *Muʿīniyya* [Treatise]), *Ḥall-i mushkilāt-i Muʿīniyya* (Solution to the Difficulties of the *Muʿīniyya*), *Sharḥ-i Muʿīniyya* (Commentary on the *Muʿīniyya*), and so on.[6] In all cases of which I know, the work is explicitly tied to *Risālah-i Muʿīniyya*, leading one to assume that it must have been written a short time after the treatise to which it is appended. This assumption, however, turns out not to be correct. Thanks to the recent discovery in Tashkent of a manuscript witness of the *Dhayl-i Muʿīniyya* with a dated colophon, we can now date this treatise, as well as the first appearance of the Ṭūsī-couple, to 643/1245: "The treatise is completed. The author, may God elevate his stature on the ascents to the Divine, completed its composition during the first part of Jamādā II, 643 of the Hijra [i.e., late October 1245], within the town of Tūn in the garden known as Bāgh Barakah."[7] As we can infer from the colophon, Ṭūsī was still in the employ of the Ismāʿīlī rulers of Qūhistān in southern Khurāsān. Tūn, present-day Firdaws, lay some eighty kilometres (or fifty miles) west-north-west of the main town of the region, Qāʾin, which was the primary regional capital of the Ismāʿīlīs.[8]

It clearly took Naṣīr al-Dīn longer than he anticipated to reach a solution, and even then it was not complete by any means. This "first version" of the Ṭūsī-couple consisted of a device composed of two uniformly rotating circles that could produce oscillating straight-line motion in a plane between two points. One of these two circles was twice as large as the second, the smaller one being inside the larger one and tangent at a point (see figure 7.1). The rotation of the smaller circle was twice that of the larger one. Although mathematically speaking the production of an oscillating point on a straight line could also be produced by the small circle "rolling" inside the larger, Ṭūsī is explicit that the larger circle "carries" (*mī bard*) the smaller one. The reason for this is that Ṭūsī will transform these circles into the equators of solid orbs rotating in the celestial realm, where any penetration of one solid body by another is expressly forbidden.[9] The transformation into solid orbs, the "physicalized rectilinear version," is shown in figure 7.2. Note that one needs a

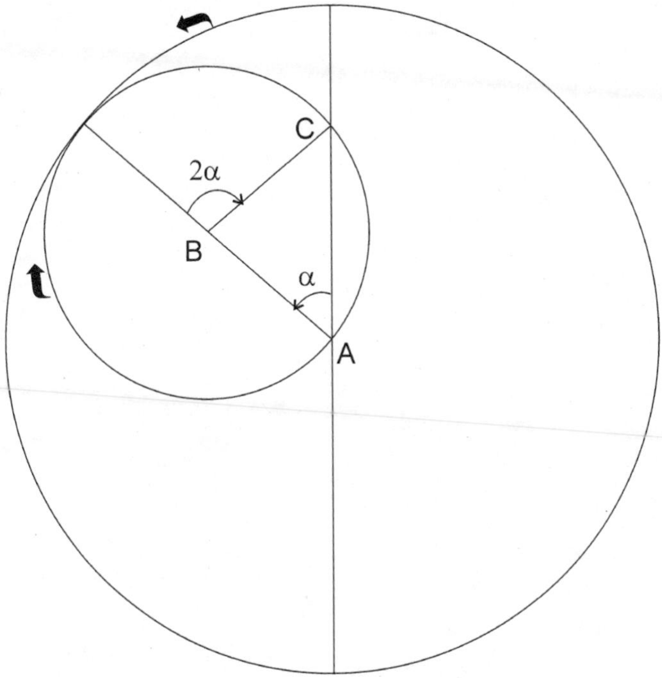

7.1 Mathematical rectilinear version of the Ṭūsī-couple.

third orb, what he calls the "enclosing sphere [*muḥṭīa*] for the epicycle," in order not to disrupt the epicycle; this third orb keeps D aligned with C and A. More on this later when I discuss Nicole Oresme.

Ṭūsī then proceeds to use the device to construct his alternative to Ptolemy's lunar model. It will be instructive, and important for tracing transmission, to compare this model from the *Ḥall* with the model Ṭūsī would present in *Al-Tadhkira fī ʿilm al-hayʾa*, which, unlike the *Muʿīniyya* and the *Ḥall*, was written in Arabic rather than Persian. The first version of the *Tadhkira* was completed in 659/1261 when Ṭūsī was in the employ of his new patrons, the Mongol Īlkhānid conquerors of Iran. Table. 7.2 provides a summary.

In the *Tadhkira*, Ṭūsī has made a number of changes in the lunar model that he first presented in the *Ḥall*. The most obvious is the change in terminology: "the dirigent orb" (*mudīr*) has now become the "large sphere," and the "epicycle's deferent orb" (*ḥāmil*) has been renamed the "small sphere." This change is most likely due to the confusion resulting from using the terms "dirigent" and "deferent," which are employed for other parts of the planetary models, to also designate the two outer spheres

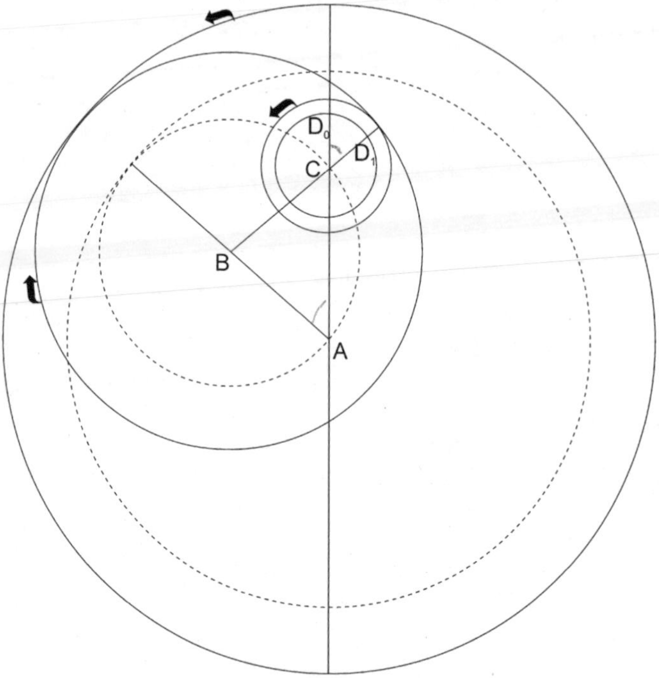

7.2 Physicalized rectilinear version of the Ṭūsī-couple.

making up the Ṭūsī-couple. Another more significant change is dividing the inclined orb of the *Ḥall* into two orbs in the *Tadhkira*, namely a different inclined orb (actually the inclined orb of the Ptolemaic model) and a different deferent. The resultant motion of these two orbs is 13;14°/day in the sequence of the signs, which is different from the 13;11°/day of the *Ḥall*'s inclined orb. In fact, this difference corrects the mistake in the *Ḥall*, where Ṭūsī made the inclined orb move at the rate of the mean motion of the Moon (*wasaṭ-i qamar*), apparently forgetting that this rate would result in the parecliptic motion being counted twice.

From this overview, we can conclude that the rectilinear Ṭūsī-couple and its applications to various planetary models emerged in stages and rather slowly. After Ṭūsī came up with the idea, apparently when writing the *Muʿīniyya*, it took many years before he felt comfortable enough presenting it in the *Ḥall*. But even then, the model still had a number of problems in both terminology and substance, which weren't solved until the writing of the *Tadhkira* some fifteen years later. But as we shall see, these differences help us in tracing the transmission of the device and models. They also help us to make the case, almost a

Table 7.2 Ṭūsī's lunar models from the *Ḥall* and the *Tadhkira*

Ḥall		*Tadhkira*	
Orbs	*Parameters*	*Orbs*	*Parameters*
Parecliptic orb (*mumaththal*)	0;3°/day (cs)	Parecliptic orb (*mumaththal*)	0;3°+/day (cs)
Inclined orb (*māʾil*)	13;11°/day (s)	Inclined orb (*māʾil*) Deferent orb (*ḥāmil*)	11;9°/day (cs) 24;23°/day (s) Net: 13;14°/day (s)
Dirigent orb (*mudīr*)	24;23°/day (s) or (cs)	Large sphere (*al-kabīra*)	24;23°/day (s)
Epicycle's deferent orb (*ḥāmil-i tadwīr*)	48;46°/day (opposite direction of dirigent)	Small sphere (*al-ṣaghīra*)	48;46°/day (cs)
Epicycle's enclosing orb (*muḥīṭ bi-tadwīr*)	24;23°/day (same direction as dirigent)	Enclosing orb (*al-muḥīṭa*)	24;23°/day (s)
Epicycle (*tadwīr*)	13;4°/day (cs)	Epicycle (*al-tadwīr*)	13;4°/day (cs)

Note: Motion in the sequence (s) or countersequence (cs) of the signs is determined by the orb's apogee point.

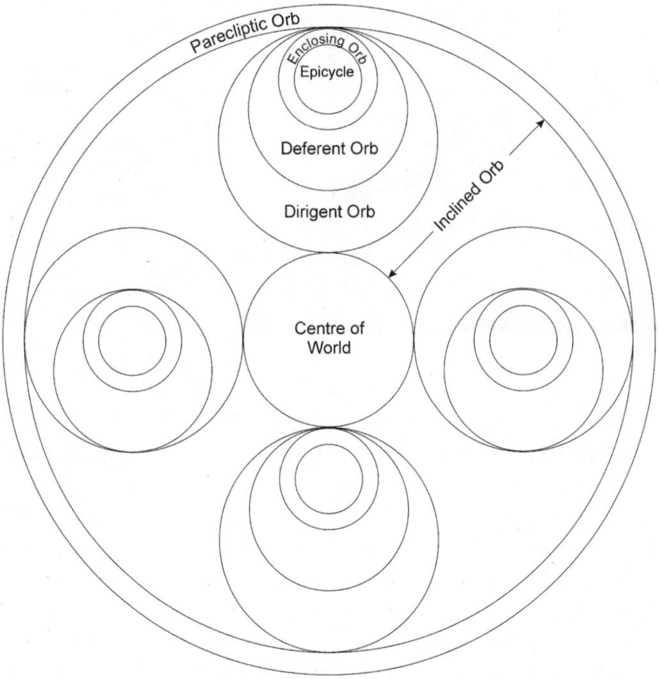

7.3 Lunar model from the *Ḥall*, showing six orbs in four different positions.

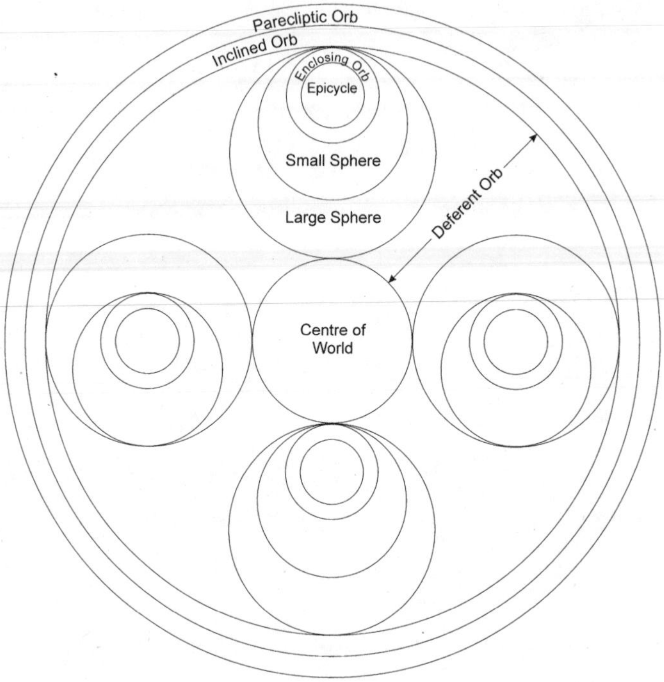

7.4 Lunar model from the *Tadhkira*, showing seven orbs in four different positions.

truism in the history of science, that such devices and models take time to evolve and be perfected. A sudden appearance of a complete and perfected theory or model should make us wary of claims of no transmission or influence.

TWO-EQUAL-CIRCLE VERSION

In addition to the rectilinear version of the Ṭūsī-couple, Ṭūsī also developed a curvilinear version that was meant to produce a linear oscillation on a great circle arc. This version was used to rectify a number of difficulties in Ptolemy's latitude theory, as well as a curvilinear oscillation caused by the prosneusis point in the latter's lunar model. In fact, as Ṭūsī mentions, it could be used wherever a curvilinear oscillation was needed, such as for motions of the celestial poles and vernal equinox, if observation showed such phenomena to be real.[10]

But before the final curvilinear version was introduced in the *Tadhkira* in 1261, it evolved slowly over a considerable period of Ṭūsī's lifetime. In the *Muʿīniyya*, when discussing the models for latitude, Ṭūsī notes that Ibn al-Haytham had dealt with latitude in a treatise and gives a brief sketch of his theory. But he finds this solution lacking, and criticizes it without going into details since "this [i.e., the *Muʿīniyya*] is not the place to discuss it." Despite this criticism, Ṭūsī does not claim to have a solution for the problem of latitude, unlike the case of the longitudinal motions of the Moon and planets.[11] In the *Ḥall*, Ṭūsī refrains from the earlier criticism of Ibn al-Haytham and instead presents the latter's model for latitude. Basically, this is an adaptation of the Eudoxan system of homocentric orbs, described in Aristotle's *Metaphysics*, applied to Ptolemy's latitude models, which used motion on small circles to produce latitudinal variation.[12] It is curious that Ṭūsī offers no model of his own, nor does he note, as he does later in the *Tadhkira*, that motions in circles will produce not only latitudinal variations but also unwanted longitudinal changes.

But a little over a year later, on 5 Shawwāl 644 (13 February 1247), to be exact, Ṭūsī published a sketch of another version of his couple that was meant to resolve some of the difficulties of Ptolemy's latitude models.[13] This version was presented in the context of his discussion of these models in book 13 of his *Taḥrīr al-Majisṭī* (Recension of the *Almagest*). After presenting a summary of Ptolemy's latitude model for the planets, and his special pleading regarding the complicated nature of these models, which include the endpoints of the epicycle diameters rotating on small circles to produce latitude in a northerly or southerly direction,[14] Ṭūsī provides the following comment:

> I say: this discussion is external to the discipline (*ṣināʿa*) [201b] and is not persuasive for this matter. For it is necessary for a practitioner of this discipline to establish circles and bodies having uniform motions according to an order and arrangement [such that] from all of them [circles and bodies] these various perceived motions will be constituted. For then these motions being on the circumferences of the mentioned small circles, just as they result in the epicycle diameters departing from the planes of the eccentrics in latitude northward and southward, so too will they result in their departing from alignment with the centre of the ecliptic, or from being parallel with diameters in the plane of the ecliptic with the exact same longitude, through accession and recession in the exact same amount of that latitude. And this is contrary to reality. And it is not possible to say that that difference is perceptible in latitude but not perceptible in longitude since they are equal in size and distance from the centre of the ecliptic.

> Now, if the diameter of the small circle were made in the amount of the total latitude in either direction, and one imagines that its centre moves on the circumference of another circle equal to it whose centre is in the plane of the eccentric in the amount of half the motion of the endpoint of the diameter of the epicycle on the circumference of the first circle and opposite its direction, there will occur a shift to the north and south in the amount of the latitude without there occurring a forward or backward [motion] in longitude.
>
> To show this, let AB be a section of the eccentric and GD be from the latitude circle that passes through the endpoint of the diameter of the epicycle. And they intersect at E. EZ EM are the total latitude in the two directions. And EH is half of it in one of them. We draw about H with a distance EH a circle EZ and about E with a distance HE a circle HTKL. We imagine the endpoint of the diameter of the epicycle at point Z to move on circle EZ in direction G to B and the center H to move on circle HTKL in the direction G to A with half that motion. Then it is clear that when H traverses a quarter and reaches T, Z will traverse a half and reach E. Then when H traverses another quarter and reaches K, Z will traverse another half and reach M. And when H traverses a third quarter and reaches L, Z will traverse another half and will reach E once again. And when H completes a rotation, Z will return to its original place so that it will always oscillate in what is between ZM on the line GD without inclining from it in directions AB. This is the explanation of this method. However, it requires that the time the diameter is in the north be equal to the time it is in the south; in reality, it is different from that. As for what is said regarding its motion on the circumference of a circle about a point that is not its centre, as stated by Ptolemy, this needs consideration to verify it according to what has preceded. We now return to the book [i.e., the *Almagest*].[15]

There are several things we can say about this device. First of all, as Ṭūsī notes, it does not accurately model Ptolemy's latitude theory since it results in equal times in the north and in the south.[16] Second, the motion of the epicycle endpoint is uniform with respect to the epicycle's mean apex, which again is contrary to what Ptolemy's model requires. Third, and more significant for our purposes, this model is actually a slightly modified version of the rectilinear Ṭūsī-couple that was first presented in the *Ḥall.* The problem, however, is that the motion of the endpoint of the epicycle's diameter is on a straight line, ZM, whereas the necessary motion should be on a great circle arc. This problem is curious. Surely, Ṭūsī is aware that the motion in latitude should occur on the surface of a sphere; why, then, does he have this rather stripped-down version of his couple that can result only in rectilinear oscillation? The answer, it seems, is that at this point he does not have a curvilinear version. He is dissatisfied with Ptolemy's small circles and also realizes that

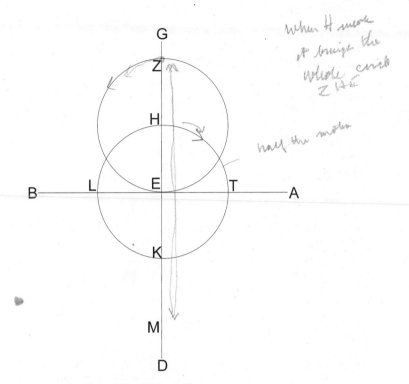

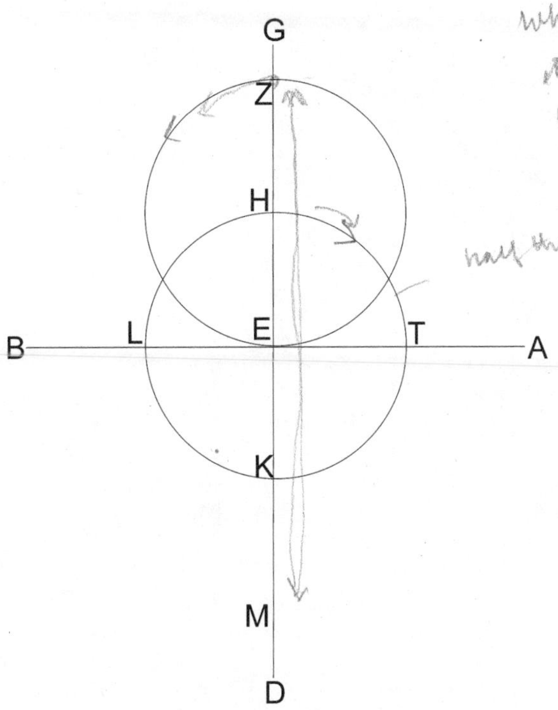

7.5 Two-equal-circle version of the Ṭūsī-couple.

Ibn al-Haytham's model does little more than provide a solid-sphere basis for the inadequate small circles, but all he has to offer is a kind of vague notion that his couple might be modified to create the necessary motion in latitude. He clearly is still in the thinking stage.

THREE-SPHERE CURVILINEAR VERSION

Ṭūsī does not in fact offer a true curvilinear version until almost fifteen years later, during the first part of Dhū al-qaʿda 659 (September or October 1261), at which time he publishes the first version of his *Al-Tadhkira fī ʿilm al-hayʾa.* There, he puts forth a model consisting of three additional orbs enclosing the epicycle that are meant to produce a curvilinear oscillation that results in the motion in latitude (see figures 7.6 and 7.7).[17] It is interesting that Ṭūsī presents this new model as a modification of Ibn al-Haytham's earlier attempt,[18] which, as we have seen, simply provides a physical basis for Ptolemy's small circles using

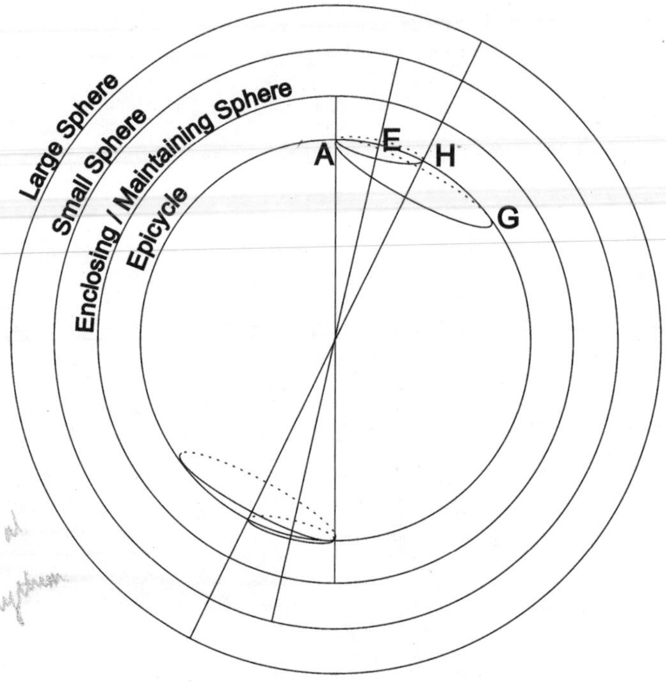

7.6 Complete curvilinear version of the Ṭūsī-couple, showing three embedded solid orbs (or hollowed-out spheres) with different axes enclosing the epicycle.

homocentric orbs, which we may call the Eudoxan-couple (see figures 7.8 and 7.9).[19] In addition to using the curvilinear version to resolve the difficulties related to the motion of the planetary epicycles in latitude, Ṭūsī notes that it may also be used for moving the inclined orb of the two lower planets in latitude and for resolving the irregular motion brought about by the Moon's so-called prosneusis point. Finally, he states that this version could also be used to model the variable motion of precession ("trepidation") and the variability of the obliquity if these two motions were found to be real.[20] As we will see, these suggestions for extended usage of the couple turn out to be significant.

USE OF THE COUPLE FOR *QUIES MEDIA*

There is another issue related to the rectilinear couple that may have a bearing on tracing transmission. Quṭb al-Dīn al-Shīrāzī, one of Ṭūsī's associates in Marāgha and subsequently one of the eminent philosophers and scientists at Mongol courts in Tabrīz, remarks in his *Al-Tuḥfa al-shāhiyya fī al-hay'a,* written after Ṭūsī's death in 684/1285, that "it is

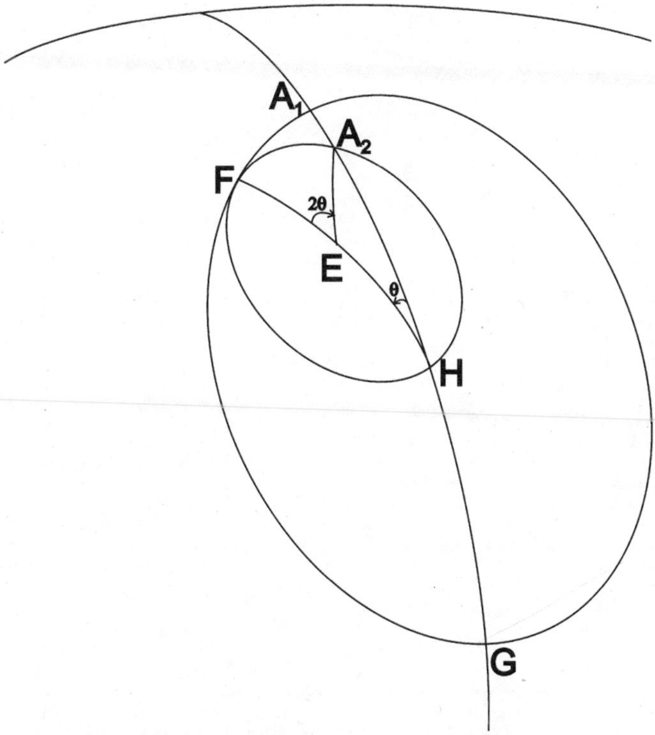

7.7 Polar view of the complete curvilinear version, showing the motion of the endpoint of the diameter of the epicycle along a great circle arc.

possible to use this [lemma] to show the impossibility (*imtināʿ*) of rest between a rising and falling motion on the line (*samt*) of a terrestrial diameter."[21] The idea here is that the Ṭūsī-couple, by showing that oscillating straight-line motion can be continuous, counters Aristotle's contention that there would be a "moment of rest" (*quies media*) between rising and falling.[22] This view was contested, and in fact Shams al-Dīn al-Khafrī (fl. 932/1525), in his commentary on the *Tadhkira*, disputes Shīrāzī on this point. As we shall see, there are echoes in Latin Europe of this debate, which could well be due to transmission.

SIGHTINGS OF THE ṬŪSĪ-COUPLE IN NON-ISLAMIC CULTURAL CONTEXTS BEFORE 1543[23]

We should note here that the development of the different versions of the Ṭūsī-couple, and the models based upon them, took place over a twenty-five-year period. The use, further development, and discussion of

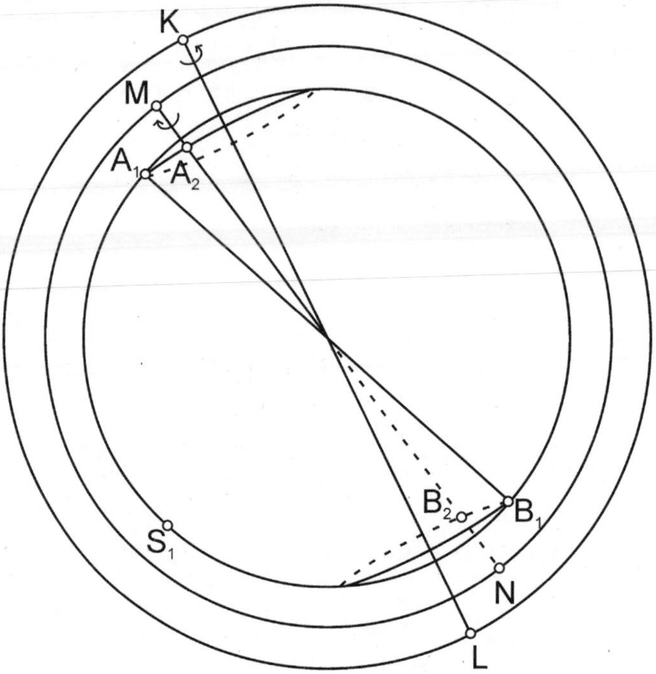

7.8 Ibn al-Haytham's Eudoxan-couple, showing two spheres.

the various versions of the couple in an Islamic context, such as I have noted above in the case of the *quies media* debate, can be traced over many centuries; the couple, which became known as the "large and small model [or *hypothesis*]"[24] (*aṣl al-kabīra wa-l-ṣaghīra*), was incorporated into other theories and systems, as well as explained in a number of commentaries and independent works. There can be no question that these later developments and discussions in an Islamic context, in whatever language, can be traced back to one or more of Ṭūsī's works. However, when we cross cultural boundaries, the situation becomes less clear-cut, and here one is faced with a variety of opinions about the origin of "Ṭūsī-couple sightings" in these other cultural contexts. With the exception of one example, and possibly a second, there are no cases of translations of Ṭūsī's writings on the couple into non-Islamicate languages. So in order to advocate that the appearance, or "sightings," of the couple in other contexts is due to intercultural transmission, we will be faced in most cases with the need to postulate either nonextant texts or nontextual transmission. Such arguments will thus need to be based on plausibility rather than direct evidence; but many arguments of

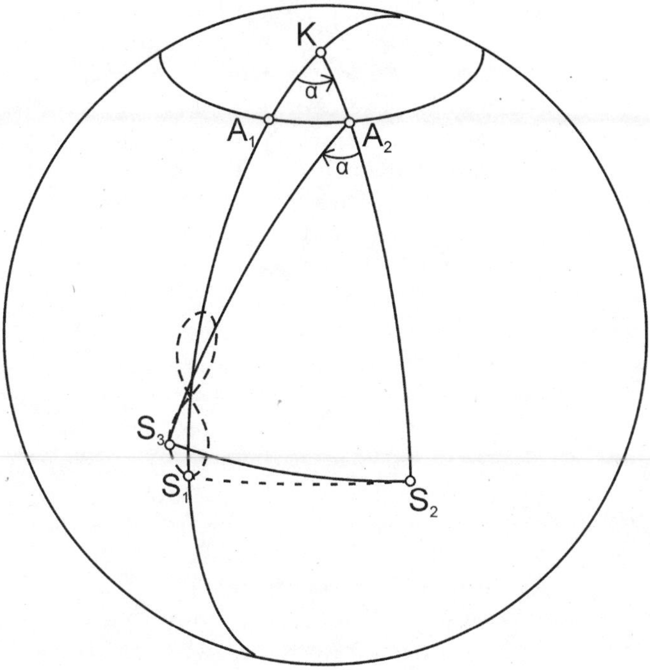

7.9 Motion of the endpoint of the diameter of the epicycle on a circular path rather than a great circle arc.

transmission in the history of science are based upon such plausibility arguments and often become virtually irrefutable, especially when precise numeration is involved. The case for the transmission of the Ṭūsī-couple is not quite so iron-clad, but given the various types of evidence that can be brought to bear, I argue that independent rediscovery, especially multiple times, becomes much less compelling.

But before presenting that evidence, I shall list and discuss the various sightings. Because of the problematic nature of some of the material, especially in the case of Oresme, I will devote considerably more space to some examples than to others.

Transmission to Byzantium

The first known appearance of the Ṭūsī-couple outside Islamic societies occurred around 1300, most likely through the efforts of a certain Gregory Chioniades of Constantinople, who is known for translating a number of astronomical treatises from Persian (or perhaps Arabic) into

Greek.[25] Included in these works is a short theoretical treatise that has been dubbed *The Schemata of the Stars.*[26] The lunar model in the *Schemata* uses the Ṭūsī-couple, and there are diagrams in one of the codices that greatly resemble diagrams in Ṭūsī's works.[27]

As I argue in a recent paper, the *Schemata* is mostly a translation of certain parts of Ṭūsī's *Muʿīniyya,* with the Ṭūsī-couple and lunar model coming from the *Ḥall*;[28] thus what we are dealing with is a case of the abridgement into Greek of a Persian original that we can confidently identify. It seems that Chioniades was tutored by a certain Shams al-Dīn al-Bukhārī (almost certainly Shams al-Dīn Muḥammad ibn ʿAlī Khwāja al-Wābkanawī al-Munajjim), who chose to teach his tutee using Ṭūsī's earlier Persian works rather than his revised and up-to-date *Tadhkira.*[29] It is not known whether this was for linguistic reasons (Chioniades perhaps knowing Persian but not Arabic) or because of a reluctance to give a Byzantine access to cutting-edge astronomical knowledge.[30] In any event, we can safely say that the version of the Ṭūsī-couple and lunar model found in the *Schemata* came from the *Ḥall* since both have six orbs for the lunar model and the same mistake in the inclined orb, namely 13;11°/day (s) rather than the correct 13;14°/day (s).[31]

The surprising conclusion is that the first known transmission of Ṭūsī's models came from his earlier Persian works, which contained a significant error. Furthermore, the only planetary model transmitted was the lunar model, and there is no hint in the *Schemata* of the models for latitude, either from the *Taḥrīr* or from the *Tadhkira.* Nevertheless, there can be no question that some of Ṭūsī's innovations had made their way into Greek by the early fourteenth century, and the existence in Italy of the only three known manuscript witnesses strongly suggests that the transmission of this knowledge had made it into the Latin world by the fifteenth century.[32]

I should also mention here that since Chioniades read the *Ḥall,* he would no doubt have been exposed to Ibn al-Haytham's latitude theory, which made up chapter 5 of that work.[33] This influence may well have relevance to the question of how that rather obscure theory might have reached scholars in Latin Europe.

The Ṭūsī-Couple and the Eudoxan-Couple in Latin Europe

Historians have identified multiple sightings of the Ṭūsī-couple and the Eudoxan-couple (i.e., Ibn al-Haytham's) in Latin Europe, starting in the fourteenth century. What follows is a chronological list, although certainly not exhaustive, of the figures associated with these sightings.

AVNER DE BURGOS

The Jewish philosopher and polemicist Avner de Burgos (ca. 1270–1340), a convert to Christianity who became known as Alfonso de Valladolid, proved a theorem in a Hebrew work identical to a rectilinear Ṭūsī-couple. Tzvi Langermann has noted that Avner "adduces his theorem in a mathematical context, the stated purpose of which is 'to construct (*li-ṣayyer*) a continuous and unending rectilinear motion, back and forth along a finite straight line, without resting when reversing direction [literally: "between going and returning"].'"[34] What is interesting here is that this use of the couple, as part of the *quies media* debate, is not something one finds in Ṭūsī but is to be found in the work of his associate and student Shīrāzī. As we will see, this may well have implications for the transmission of the couple to Europe.

NICOLE ORESME

Nicole Oresme (ca. 1320–82), in his *Questiones de spera*, which treats Johannes de Sacrobosco's *On the Sphere of the World*, describes some sort of model that will produce reciprocating rectilinear motion from three circular motions. Both Garrett Droppers and Claudia Kren raised the possibility that Oresme was somehow influenced by "Ṭūsī's device."[35] Recently, André Goddu has challenged this possibility and has raised another one, namely that Oresme hit upon a solution similar to Ṭūsī's for producing rectilinear motion from circular motions – although still leaving open the (weak?) alternative that Oresme may have come across some description of it.[36] Because Goddu's speculations, discussed below, depend upon several misinterpretations of both Ṭūsī and Oresme, we need to carefully consider what Oresme is proposing. Here is Kren's translation of the relevant passage with my suggested revisions:[37]

Concerning this problem [i.e., whether celestial bodies move in circular motion], I propose three interesting conclusions. First, it is possible for some planet to be moved perpetually according to its own nature in a rectilinear motion composed of several circular motions. This motion can be brought about by several intelligences, any one of which may endeavor to move in a circular motion, *nor would this purpose be in vain* [rev: and (the intelligence) is not frustrated in this endeavor].

Proof: Let us propose, conceptually, as do the astrologers, that A is the *deferent* [rev: deferent circle] of some planet, or its center; B is the *epicycle* [rev: epicycle circle] of the same planet; and C is the body of the planet, or its center; I take these [latter two?] as equivalent. Let us also imagine line BC from the center of the epicycle to the center of the planet, and CD, a line in the planet on which BC

> falls perpendicularly. Let circle A move on its center toward the east, and B toward the west. The planet, C, revolves on its own center toward the east. Moreover, since line BC is of constant length, because it is a radius, let us propose that the *distance* [rev: amount] B descends *in* [rev: according to] the motion of the deferent is *the distance which* [rev: as much as] point C *may ascend* [rev: ascends] with the motion of the epicycle. From this one can obviously observe that point C in some definite time will be moved in a straight line. *Let us then further assume that point B would ascend by its own motion on just the circumference on which it may descend with the motion of the planet* [rev: Let us then further assume that the circuit on which B would ascend by its own motion is as much as the motion of the planet descends]. It is further clear that point D will move continually on the same line; thus the entire body of the planet will be moved to some terminus in a rectilinear motion and will return again with a similar motion.[38]

To analyze this passage, and to understand Oresme's intention, we should note from the last sentence that the *body* of the planet is meant to move rectilinearly. Furthermore, not only does the centre of the planet (C) move in a straight line but a certain point (D), which is the endpoint of a planetary radius (CD), does as well.

Droppers, and Goddu who follows him, do not take the rectilinear motion of D into account; inexplicably, both have D at the end of a planetary radius whose starting point is C, the centre of the planet (see figure 7.10).[39]

In contrast, Kren does follow Oresme's text and provides a plausible reconstruction based upon a more or less correct interpretation of Ṭūsī's *Tadhkira* as she found it in Carra de Vaux's flawed 1893 French translation. Oresme provides no diagram, and Kren must admit that "as it appears in Oresme's *Questiones de spera*, the passage makes no sense whatsoever."[40] Nevertheless, following Kren's lead and making a few modifications, I believe we can reconstruct both Oresme's model and his intention.[41] In essence, what Kren proposes is that Oresme is not discussing the simple two-circle Ṭūsī-couple, which results in the rectilinear oscillation of a point between two extrema, but rather Ṭūsī's physicalized rectilinear version, which we have already encountered above.[42]

With reference to figure 7.2 and using Oresme's description, let us take A to be the centre of the deferent, B the centre of the epicycle, and C the centre of the planet. The solid lines indicate the outer surfaces of solid bodies, whereas the dotted lines indicate "inner equators" of these solid bodies. Note that the solid orbs are the actual moving bodies; they "accidentally" produce the mathematical Ṭūsī-couple indicated by the broken lines. So for this model to work, the epicycle (B) needs to move with twice the angular speed as the deferent (A) and in the opposite

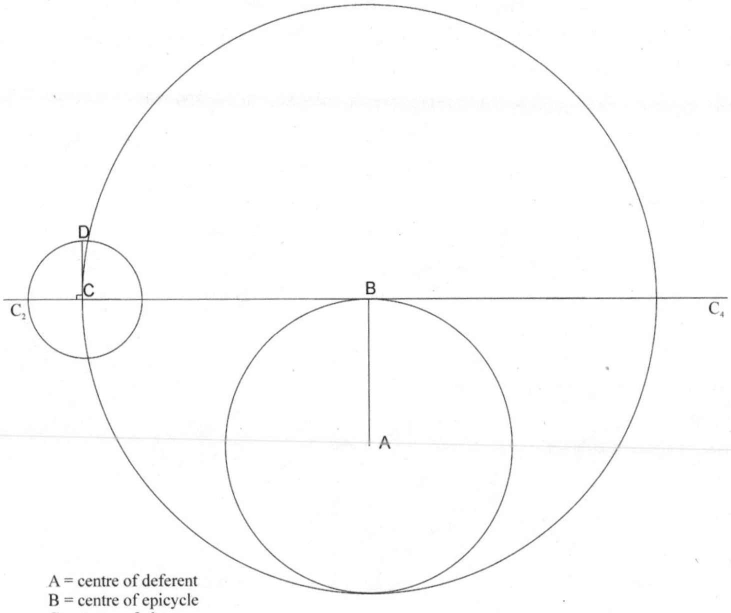

A = centre of deferent
B = centre of epicycle
C = centre of planet
C_2 and C_4 = extremes of straight-line path of centre C
CD = line in planet ⊥ to CB

7.10 Oresme's construction as proposed by Droppers.

direction. These movements will then result in the planet's centre (C) oscillating on a straight line. They will not, however, result in the apex of the planet (D) moving rectilinearly. As shown in the diagram, when the deferent and epicycle have rotated from an initial position (where A, B, C, and D were on the same line), D will move from D_0 to D_1. To deal with this issue, Ṭūsī introduces what he calls an enclosing sphere (*kura muḥīṭa*), which is shown in the diagram as an orb enclosing and concentric with the planet (C). This orb would then have the job of moving D from D_1 back to its initial position of D_0. Since $\angle BAC = \angle D_0CD_1$, the enclosing sphere needs to move with the same speed and direction of the deferent (A) in order to keep D oscillating on the straight line.

Kren has assumed that Oresme is simply copying Ṭūsī's physicalized rectilinear version, and she has some tortured readings that would introduce this fourth, enclosing orb into Oresme's account. But Oresme clearly says he only needs three circular motions, and in fact Ṭūsī's commentators indicate that one could replace orb C and the enclosing orb by combining their motions into a single orb. Ṭūsī does not do so, probably because for him orb C is an epicycle, not an otherwise stationary planet, and he does not want to lose its parameters, which are critical for Ptolemaic planetary theory, by combining it with another orb. But Oresme has no such constraints since for him the construction does not represent an actual planetary model. So the planet (C) can move as needed – in this case, with just the rotational direction and speed of the deferent (A) that will keep line CD aligned with the line of oscillation.

How well does this interpretation fit with the existing text? Actually, rather well, all things considered. Turning to figure 7.11, let us go through the various features as presented by Oresme:

1 A is the deferent, which "carries" (*deferre*) the epicycle (B); the planet (C) is moved by the epicycle. According to most standard medieval accounts, and presumably this idea is what Oresme intends by referring to the conceptualization of the astrologers, the epicycle is embedded in the deferent and the planet is embedded in the epicycle, as shown.
2 A radius (CD) of the planet would in general not be perpendicular to line BC in this construction; however, it would be perpendicular at the quadratures, as noted by Kren. As mentioned above, the alternative given by Droppers and followed by Goddu (see figure 7.10) does not fit the stipulation that D remain on the line of oscillation.
3 The directions of the motions (A eastward, B westward, and C eastward) is consistent with Ṭūsī's model.
4 Oresme emphasizes that BC is a radius of constant length, which probably indicates that he is aware that this stipulation is part of the proof for the Ṭūsī-couple. For this model to work so that point C remains on a straight line, Oresme needs to make B rotate twice as fast as A (or in his terms, point B will descend due to A, while C will ascend with twice the speed due to B). However, he seems to imply that the deferent and epicycle rotate at the same speed (or descend and ascend in equal amounts). Unless he has some other sense for "ascend" and "descend," Oresme does not seem to be in control of this rather critical part of the model.
5 If one accepts my emended translation, Oresme does understand that the planet will need to rotate in the direction opposite that of the epicycle. Again, we are not provided with any amounts, but it seems that Oresme is conceiving of D_0 being displaced to D_1 by the "ascending" motion of B, which would then need to be countered by the descending motion of the planet (see figure 7.2). The flow of the argument is then clear: he begins by "proving" that C will oscillate on a straight line and follows with his "proof" that D will follow

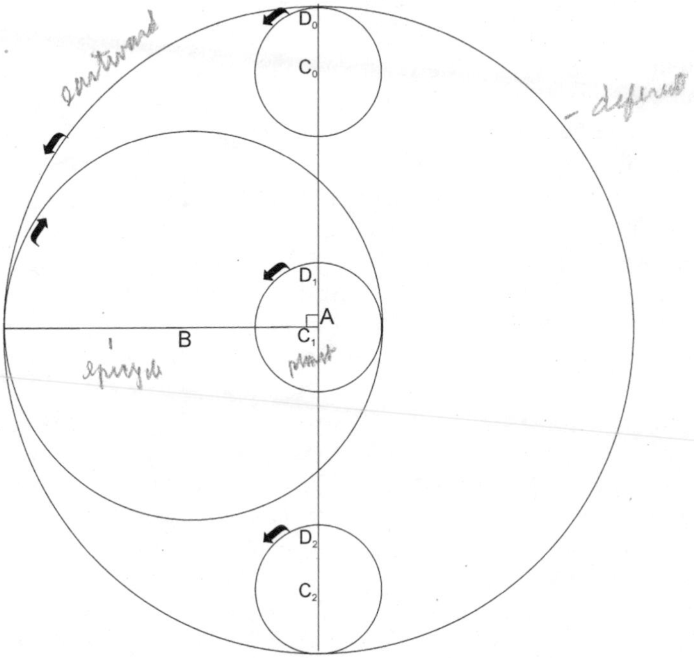

7.11 Oresme's physicalized rectilinear version of the Ṭūsī-couple.

suit and stay on the straight line by means of the additional motion of the planet.

What conclusions can we reach? On the one hand, Oresme is evidently aware of what we may call Naṣīr al-Dīn's physicalized Ṭūsī-couple as presented in the *Tadhkira*. But Oresme makes no claim to have invented this model on his own; and given his apparent lack of understanding of the necessity of having the epicycle move at twice the speed of the deferent, it would be implausible in the extreme to assume that he reinvented this model. On the other hand, the three-sphere version that Oresme presents, as a deferent-epicycle-planet construction, is not to be found explicitly in Ṭūsī or other Islamic sources of which I am aware; thus it seems likely that Oresme or an intermediary had adapted the model for this philosophical discourse. Finally, we should note that there is an echo of the use of the Ṭūsī-couple for the *quies media* debate that we first encountered with Shīrāzī. Oresme states, "By the imagination, it is possible that rectilinear motion be eternal, with the exception that in the point of reflection the movable would not be said to be moved nor at rest."[43]

JOSEPH IBN NAḤMIAS

In his *The Light of the World*, Joseph ibn Naḥmias, a Spanish Jew living in Toledo around 1400, used a double-circle device in his astronomical models that is mathematically equivalent to Ṭūsī's curvilinear version from his *Tadhkira* but in its truncated, two-sphere version. He also incorporates it into his recension of *Light of the World*. Note that despite living in the Christian part of the Iberian Peninsula, Ibn Naḥmias wrote *Light of the World* in Judeo-Arabic (Arabic in Hebrew script), although the recension is in Hebrew. In chapter 8 of the present volume, Robert Morrison details Ibn Naḥmias's use of the Ṭūsī-couple and also discusses the vexed question of its possible transmission to Ibn Naḥmias and other Jewish scholars.[44] I shall return to this question below.

GEORG PEURBACH

From an extensive mathematical analysis of the 1510 and 1512 annual ephemerides of Johannes Angelus, Jerzy Dobrzycki and Richard Kremer have concluded that they were based upon modifications of the *Alfonsine Tables*, these modifications consisting of mechanisms meant to produce harmonic motion that were somehow added to the standard Ptolemaic models.[45] Because Angelus seems to indicate that these were based upon a new table of planetary equations due to Georg Peurbach (d. 1461), Dobrzycki and Kremer speculate that the underlying models used by Peurbach incorporated one of the Marāgha models, perhaps the Ṭūsī-couple or the mathematically equivalent epicycle/epicyclet of ʿAlāʾ al-Dīn ibn al-Shāṭir. Aiton has also raised the possibility that Peurbach in his *Theoricae novae planetarum* may be referring to Ibn al-Haytham's Eudoxan-couple when he states, "On account of these inclinations and slants of the epicycles, some assume that small orbs have the epicycles within them, and that the same things happen to their motion."[46] Although speculative, these authors' conclusions do point to the possibility that European astronomers in the late fifteenth and early sixteenth centuries, other than Copernicus, used and adapted devices that we normally associate with Islamic astronomy. This is an important point that we will revisit when we discuss some of the objections that have been raised to astronomical transmission from Islam to Latin Europe.

JOHANN WERNER

In his *De motu octavae sphaerae*, Johann Werner (1468–1522) uses a two-equal-circle device to deal with the issue of variable precession, or trepidation. According to Dobrzycki and Kremer, "Werner allotted the trepidational motion of 'Thabit's' [Thābit ibn Qurra's] and Peurbach's models to the solstitial points of two concentric spheres. Two circles of trepidation, of equal radii and centred on the solstitial points of the next higher sphere, rotate in opposite directions so that trepidational variations in longitude do not introduce shifts in the obliquity of the ecliptic. Werner thus managed to generate linear harmonic motion by the uniform motions of two circles."[47] This model sounds a lot like the two-equal-circle version of the Ṭūsī-couple, but we need to be cautious. Werner does not use a 2:1 ratio for the motions of the two circles, and in his earlier analysis, Dobrzycki specifically states that this is not the Ṭūsī-couple as used, for example, by Copernicus.[48] However, since Werner's intention is to generate a linear oscillation to avoid shifts in the obliquity, one can indeed see a connection. However, further research would be needed to establish a relationship between Werner's use and earlier uses of the Ṭūsī-couple.[49]

GIOVANNI BATTISTA AMICO

Giovanni Battista Amico (d. 1538) used the three-sphere curvilinear version as described in the *Tadhkira* in his *De motibus corporum coelestium*, published in 1536;[50] in other words, he used the version with three spheres, two producing the curvilinear oscillation on the surface of a sphere and the third functioning as a counteracting sphere so that only the curvilinear oscillation of its pole is transmitted to the next lower sphere.[51] According to Mario Di Bono, "It is of particular interest that in the 1537 [revised] edition of his work Amico is aware that on the surface of a sphere the demonstration does not function as it should; but since the inclination of the axes is not great, he considers the error negligible."[52]

GIROLAMO FRACASTORO

Girolamo Fracastoro in his *Homocentrica*, published in 1538, refers to a device for producing rectilinear motion but does not incorporate it into his astronomy. The description and diagram make it clear that he is referring to the two-equal-circle version.[53]

NICHOLAS COPERNICUS

Noel Swerdlow and Otto Neugebauer succinctly summarize Copernicus's use of the various devices invented by Ṭūsī: "In *De revolutionibus* he uses the form of Ṭūsī's device with inclined axes for the inequality of the precession and the variation of the obliquity of the ecliptic, and in both the *Commentariolus* and *De revolutionibus* he uses it for the oscillation of the orbital planes in the latitude theory. In the *Commentariolus* he uses the form with parallel axes for the variation of the radius of Mercury's orbit, and by implication does the same in *De revolutionibus* although without giving a description of the mechanism."[54]

However, we will need to examine the situation a bit more closely.[55] Let us take *De revolutionibus orbium coelestium* first. In fact, the device put forth and the proof given in book 3, chapter 4, for variable precession and the variation of the obliquity are, *pace* Swerdlow and Neugebauer, for the two-equal-circle version, not for the two- or three-sphere curvilinear version (i.e., "Ṭūsī's device with inclined axes"). And in all other cases in which he uses it in *De revolutionibus* (for Mercury's longitude model in book 5, chapter 25, and for the latitude theory in book 6, chapter 2), Copernicus refers the reader back to book 3, chapter 4. We may then conclude that Copernicus wishes to use the two-equal-circle version exclusively in *De revolutionibus.* As Swerdlow and Neugebauer note, Copernicus's statement that he will be using chords rather than arcs (as necessitated by the use of the rectilinear rather than curvilinear version) is reasonable since the deviation from a curvilinear version is relatively minor.[56] But it does raise questions about the kind of modelling Copernicus uses in *De revolutionibus,* in contrast to the *Commentariolus.* In the *Commentariolus,* it is the truncated two-sphere curvilinear version that is used for the latitude models,[57] and it is the physicalized rectilinear version that is used to vary the radius of Mercury's orbit but in a truncated, two-sphere version without the enclosing/maintaining sphere.[58] The conclusion seems to be that Copernicus was attempting to provide actual spherical models for the two versions of the Ṭūsī-couple he uses in the *Commentariolus* but that he cut a corner or two by not dealing with the disruption of the contained orb, which, after all, is why Ṭūsī (and Amico) have their maintaining (or withstanding) spheres. In *De revolutionibus,* Copernicus abandons any pretense of full physical models for his Ṭūsī-couples and instead relies only on the two-equal-circle version, which, as we have seen, is a mathematical, not a physical, model.[59]

THE TRANSMISSION SKEPTICS[60]

Although difficult to gauge in a precise way, impressionistically it seems that a majority of historians of early astronomy have accepted, to a lesser or greater degree, the influence of late-Islamic astronomy on early modern astronomers, particularly Copernicus. This acceptance is perhaps most explicitly set forth by Swerdlow and Neugebauer: "The question therefore is not whether, but when, where, and in what form he [Copernicus] learned of Marāgha theory."[61]

Nevertheless, there have been a number of skeptics who have raised various issues that are worth exploring. In 1973, for example, Ivan Nikolayevich Veselovsky called attention to what is the converse of the Ṭūsī-couple, namely a device for producing a circular motion from straight-line motions, which was set forth by Proclus in his commentary on book 1 of Euclid's *Elements*.[62] Copernicus refers to just this passage in Proclus when he uses the Ṭūsī-couple for his Mercury model.[63] But there are numerous problems with attributing Copernicus's source to Proclus rather than Ṭūsī. In the first place, Proclus, as mentioned, is setting forth a way to produce circular motion from linear motions, which is the opposite of what the Ṭūsī-couple does.[64] Second, as noted by Swerdlow, Edward Rosen, and originally Leopold Prowe, Copernicus only received a copy of Proclus's book in 1539 as a gift from Georg Joachim Rheticus, which is many years after first using the couple in the *Commentariolus*.[65] Di Bono proposes, as a way to save Veselovsky's suggestion, the possibility that Copernicus may have seen a copy of the original Greek while in Italy, this idea gaining some plausibility because it was part of the library that Cardinal Basilios Bessarion had bequeathed to the Venetian Senate.[66] But again this suggestion raises numerous other problems, namely that Copernicus is then required to have read, or to have had read to him, a Greek manuscript and that he was then inspired by an obscure passage in it talking about something only vaguely related to a device that, as we have seen, was certainly available from other sources. And Copernicus himself does not even get the reference to Proclus correct; he has Proclus claiming that "a straight line can also be produced by multiple motions,"[67] but as we have seen, Proclus refers to the production of a circle, not a straight line. And in any event, Copernicus himself mentions "some people" who refer to the Ṭūsī device as producing "motion along the width of a circle,"[68] which indicates that the device is used by others (and almost certainly is not of his own making) and that Proclus is not one of these people since Proclus does not, and could not, refer to the motion as such.

Di Bono is certainly the most thoughtful skeptic, and his skepticism is nuanced and tempered. As an alternative to an Islamic connection, which he does not reject out of hand, he proposes that Copernicus, with the same aim of resolving the issues of irregular motion in Ptolemy's models, basically came up with the same set of devices and planetary models.[69] "As to Amico and Fracastoro, there is no need to imagine a source or a specific author from whom both authors derived the same device, nor to imagine a strict interdependence between them."[70] What is ironic here is that Di Bono begins his article insisting on examining the differences between the various models and their uses among the different astronomers he examines. As he puts it, "Moreover, as in this case even marginal similarities or differences may be of relevance, it is of the utmost importance not to cause such differences to disappear in the reduction to the mathematical formalism in use today."[71] But in the conclusion of the article, where he needs to reduce these differences in order to argue against transmission and for multiple rediscovery (or parallel development), he falls back upon Neugebauer's point that "[t]he mathematical logic of these methods is such that the purely historical problem of the contact or transmission, as opposed to independent discovery, becomes a rather minor one."[72] But the problem with this position is that the differences on which Di Bono is so insistent earlier in his article here fade to irrelevance since the "internal logic" supersedes any attempt to understand the historical developments involved; each actor is foreordained to come up with the "same" solution, even when these solutions are not the same. Yet another problem with Di Bono's position is that none of his European actors has left any hint that they developed the basic devices on their own. And where we do have a discussion of sources, namely in *De revolutionibus*, Copernicus on the one hand makes a somewhat irrelevant gesture toward Proclus – which has all the hallmarks of a humanist need to pad his text with a classical reference – and on the other hand, as we have seen, refers to others who have used the device. So Di Bono's contention that "the reciprocation device … could equally well have derived from an independent reflection [by Copernicus] on these same problems" seems to be undermined by what evidence is at hand.

A more recent skeptic is André Goddu, who agrees with Di Bono's skepticism about an Islamic influence but is equally skeptical about Di Bono's suggestion of a Paduan source. Instead, he proposes Oresme as the ultimate source of the reciprocating device in Europe, someone Di Bono does not mention in his own, wide-ranging article. As we have seen, Oresme does indeed describe a reciprocation device, but it is rather different from the one Goddu envisions.[73] Be that as it may, Goddu proposes the following: "The path to Copernicus would have proceeded from Oresme to Hesse, Julmann, and Sandivogius, and from them to Peurbach, Brudzewo, and Regiomontanus." But in making such a proposal, Goddu has confused, or conflated, two totally different models. Henry of Hesse (ca. 1325–97), a certain magister Julmann (alive in 1377), Albert of Brudzewo (1445–95), and perhaps Peurbach are not describing ("using" would be misleading here) some version or other of the Ṭūsī-couple but rather something like Ibn al-Haytham's Eudoxan-couple (see above). As for Sandivogius of Czechel (fl. 1430), what is being put forth is an additional epicycle for the Moon that would counter the original epicycle's motion; without this additional epicycle, we should be able to see both faces of the Moon, something that is not observed.[74] Goddu seems to be depending mainly on José Luis Mancha for his information on Hesse, Julmann, Peurbach, and Brudzewo, but

Mancha makes it very clear that what they are dealing with is Ibn al-Haytham's Eudoxan-couple, not the Ṭūsī-couple.[75] Thus when Goddu seeks to make Oresme the source for Hesse and subsequent writers, he is making a fundamental mistake, namely having something that is likely to have been some sort of Ṭūsī device be the source for a totally different type of model. Oresme was seeking to produce rectilinear motion from circular motion, whereas most of the other authors Goddu deals with (excepting Copernicus, of course) are simply reporting a way to physicalize the small circle motion of Ptolemy's latitude theory or are using the same device for the oscillation of the lunar apogee due to the Moon's prosneusis point.[76] That Goddu further claims that an adaptation by Copernicus of the Eudoxan model that Brudzewo describes is equivalent to the wholesale incorporation of Ibn al-Shāṭir's models into the *Commentariolus* is, to say the least, bizarre in the extreme.[77]

EMPIRICAL EVIDENCE FOR TRANSMISSION

Both Di Bono and Goddu ask for more evidence for transmission before passing judgment. This is a fair comment, and in what follows I present some of the evidence that has been discovered over the past twenty-five years or so.[78] I divide this evidence up into different pathways that transmission did take or could have taken.

The Byzantine Route

As mentioned above, it is now clear that the Ṭūsī-couple first made its way into another cultural context through Byzantine intermediaries, first and foremost Gregory Chioniades, who travelled to Tabrīz around 1295 and studied with a certain Shams Bukharos, whom we can now identify as Shams al-Din al-Wābkanawī.[79] That this transmission occurred through an adapted translation from Persian into Greek raises some interesting issues of intercultural exchange. Was this translation a result of the fact that the language of trade between Byzantium and Iran was mainly in Persian? If so, Chioniades may have had an easier time finding someone to teach him Persian than Arabic. And indeed, most of the Islamic astronomical works that found their way into Greek seem to have been from Persian sources.[80] This Persian bias may help us to understand why an ostensibly out-of-date treatise, such as Ṭūsī's Persian *Mu'īniyya* and its appendix, the *Ḥall*, which, as we have seen, contained the first versions of Ṭūsī's rectilinear couple and lunar model, were provided and taught to Chioniades rather than the mature versions found in Ṭūsī's later *Tadhkira*, which was in Arabic. But there could be other reasons. One of

Chioniades's successors, George Chrysococces (fl. 1350), relates the following story, which was told to him by his teacher Manuel:

in a short while he [i.e., Chioniades] was taught by the Persians, having both consorted with the King, and met with consideration from him. Then he desired to study astronomical matters, but found that they were not taught. For it was the rule with the Persians that all subjects were available to those who wished to study, except astronomy, which was for Persians only. He searched for the cause, which was that a certain ancient opinion prevailed among them, concerning the mathematical sciences, namely, that their king will be overthrown by the Romans, after consulting the practice of astronomy, whose foundation would first be taken from the Persians. He was at a loss as to how he might come to share this wonderful thing. In spite of being wearied, and having much served the Persian king, he had scarcely achieved his objective; when, by Royal command, the teachers were gathered. Soon Chioniades shone in Persia, and was thought worthy of the King's honor. Having gathered many treasures, and organized many subordinates, he again reached Trebizond, with his many books on the subject of astronomy. He translated these by his own lights, making a noteworthy effort.[81]

This passage of course reminds us, if we need reminding, that intercultural transmission at the time did take considerable effort and was not always a straightforward process. But it also teaches us that transmission was indeed possible. In this case, the transmission of the couple and models based on it is clear since they occur in Chioniades's *Schemata*. Less clear are the circumstances under which the *Schemata* itself was further transmitted. And did other knowledge contained in the *Muʿīniyya* and the *Ḥall*, but not contained in the *Schemata*, also get transmitted? An example of this latter case would be Ibn al-Haytham's Eudoxan-couple, which, as mentioned, was presented in a separate chapter in the *Ḥall* by Ṭūsī. Ibn al-Haytham's work itself is not extant, and the presentation in the *Tadhkira* is much more succinct than what is in the *Ḥall*. So a transmission of the Eudoxan-couple via Chioniades would provide an important link taking us to Henry of Hesse and beyond.

The *Schemata* is currently witnessed by three manuscripts: two in the Vatican (Vat. Gr. 211, fols 106v–115r [text], fols 115r–121r [diagrams]; and Vat. Gr. 1058, fols 316r–321r) and one at the Biblioteca Medicea Laurenziana in Florence (Laur. 28, 17, fols 169r–178r).[82] The Vatican manuscripts have diagrams, whereas the Florence one does not.[83] In Vaticanus Graecus 211, one diagram represents the mathematical rectilinear version of the Ṭūsī-couple (fol. 116r), and another represents al-Ṭūsī's lunar model from the *Ḥall* (fol. 117r), the one with six rather than seven orbs. The Florence manuscript was copied in 1323 according to the colophon on folio 222v, but it is not clear when the manuscript arrived in Italy. Vaticanus Graecus 211 was copied in the early fourteenth century and was recorded in the Vatican inventory of 1475; Vaticanus Graecus 1058 was copied in the middle of the fifteenth century and was perhaps in the Vatican inventory of 1475 but certainly, according to David Pingree, in the inventory made around 1510.[84] These sources provide us with evidence that the work, with diagrams, was available in Italy as early as 1475; on this basis, Swerdlow and Neugebauer favour this Italian transmission route for the Ṭūsī-couple to Copernicus, who studied and travelled in Italy between 1496 and 1503 (mainly Bologna, Padua, and Rome).[85] It may be significant that Copernicus spent part of the Jubilee year 1500 in Rome, perhaps to do an apprenticeship at the Papal Curia, which would have given him access to the *Schemata*.

The Spanish Connection

Relations between the two main branches of Christendom were fraught, and it seems likely that one of the reasons the twelfth-century translation movement brought Greek classics into Latin via Arabic translations, rather than directly from the Greek, was that it was easier to obtain Arabic versions of Greek texts in Spain than it was to obtain Greek manuscripts from Byzantium. Thus we must be cautious before assuming that Byzantine astronomy would have made its way westward before the fifteenth century. But there is another route that could have brought the new astronomy of thirteenth-century Iran to the Latin West. There is considerable historical evidence of ongoing diplomatic activity between the Spanish court of Alfonso X of Castile and the Mongol Īlkhānid rulers of Iran. The late Mercè Comes wrote an important article on the subject and noted a number of cases of similar astronomical theories and instruments appearing in both Christian Spain and Iran during the thirteenth century.[86] But perhaps the most striking example of a scientific theory from Īlkhānid Iran appearing in Europe is the attempted proof of Euclid's parallels postulate, produced in the important Tabrīz scientific milieu of the 1290s, which pops up in the work of Levi ben Gerson (Gersonides) in southern France, probably shortly after 1328, according to Tony Lévy, who made this important identification.[87] This is the proof found in the *Commentary on Euclid's Elements* published at the Medici Press in Rome in 1594 and incorrectly attributed to Ṭūsī; the proof was later discussed by the Italian mathematician Giovanni Saccheri.[88] If something as complicated as this proof of the parallels postulate could travel from Iran to Avignon in twenty-five years or so, the

Ṭūsī-couple, already translated into Greek, could presumably make it to France as well and be available for Nicole Oresme. As mentioned above, Ibn al-Haytham's Eudoxan-couple is a bit more difficult to trace, but the fact that Chioniades would have no doubt encountered it in his studies of the *Ḥall* provides another plausible vehicle of transmission, as does whatever means brought pseudo-Ṭūsī's parallels proof westward.

The Jewish Link

As we see with Gersonides, perhaps the most important agents of transmission from Islam to Christendom were Jewish scientists and mathematicians. Recent work by Tzvi Langermann and Robert Morrison has been ground-breaking in shedding light on a host of characters involved in this transmission. In addition to bringing Avner de Burgos's proof of the Ṭūsī-couple to our attention, Langermann has shown that in fifteenth-century Italy, Mordecai Finzi knew the *Meyashsher ʿaqov* of Avner de Burgos, in which, as we have seen, Avner proved that one could produce continuous straight-line oscillation by means of a Ṭūsī-couple. According to Langermann, Finzi clearly knew of the *Meyashsher ʿaqov*, as indicated by his copying of the interesting conchoid construction found in Avner's text.[89] It seems reasonable to assume, as Langermann does, that Finzi knew the other parts of the *Meyashsher ʿaqov*, including the Ṭūsī-couple proof. Furthermore, Finzi had extensive contacts with Christian scholars, as he notes in several places in his works and translations.[90] Thus here we have a Jewish scholar who most likely knew of the Ṭūsī-couple in contact with north Italian mathematicians a generation or so before Copernicus would be in the neighbourhood.

In chapter 8 of the present volume, Robert Morrison discusses another avenue through which the Ṭūsī-couple may have become known to Italian scholars via Jewish intermediaries. In addition to summarizing recent work on Ibn Naḥmias, Morrison traces the interesting career of a certain Moses ben Judah Galeano (Mūsā Jālīnūs). Galeano had ties to Crete and the Ottoman court of Sultan Bāyazīd II (r. 1481–1512) and also travelled to the Veneto region around 1500. Most interesting is that Galeano knew of the work of Ibn al-Shāṭir, whose models are so instrumental in the *Commentariolus*. Galeano also knew the writings of Ibn Naḥmias, whose models incorporated the Ṭūsī-couple and are quite similar to ones we find in Johannes Regiomontanus and Giovanni Battista Amico. Thus we have another route by which the Ṭūsī-couple may well have found its way to Italy in the late fifteenth century.

Manuscripts Galore

Something often overlooked in discussions of the transmission of devices like the Ṭūsī-couple (both within Islamic realms and interculturally) is that we are not dealing with a limited number of texts and manuscript witnesses. If we confine ourselves to Ṭūsī's works that present one or more versions of his couple and to works derived from them (i.e., commentaries, supercommentaries, and closely related works) that were composed before 1543 CE, we find at least fourteen texts represented by hundreds of witnesses (see table 7.3).[91] This table does not include philosophical, theological, and encyclopaedic works, or Quran commentaries, in which the couple is mentioned or discussed.[92]

I do not claim that the almost 400 manuscript witnesses enumerated in table 7.3 would have somehow been available to early modern European astronomers. Indeed, some of these manuscript witnesses were copied well after the sixteenth century. Nevertheless, a fair number of them currently reside in Istanbul and other former Ottoman lands, including those in eastern Europe. Although most of the Islamic manuscripts currently in European libraries were collected after 1500,[93] there were presumably Islamic scientific manuscripts that were available in various parts of Europe previous to that date.[94]

The last bit of empirical evidence for transmission is indirect but highly suggestive. Recently, it has come to light that the critical proposition that Swerdlow has claimed was used by Copernicus to transform the epicyclic models of Mercury and Venus into eccentric models, which is found in Regiomontanus's *Epitome of the Almagest*, was put forth earlier in the fifteenth century by ʿAlī Qushjī of Samarqand.[95] Although it is not known how Qushjī's treatise came to be known by Regiomontanus – which seems much more likely to me than independent rediscovery of the proposition[96] – a likely candidate is Cardinal Basilios Bessarion (d. 1472), the Greek prelate who almost became the Roman pope. Bessarion travelled to Vienna in 1460, where he met both Peurbach and Regiomontanus. That Qushjī's proposition occurs in the *Epitome*, which was completed around 1462, suggests that Bessarion is the intermediary. This idea gains further plausibility since he was originally from Trebizond and spent considerable time in Constantinople before its fall to the Ottomans in 1453. Consequently, he could have easily been in contact with Islamic scholars, who were in various centres in Anatolia, including Bursa, the home of Qāḍīzāde al-Rūmī, one of Qushjī's teachers and associates in Samarqand. Qushjī himself later came to Constantinople, in 1472, probably at the behest of Sultan

Table 7.3 Manuscript witnesses to the Ṭūsī-couple

Author	*Title*	*Date of composition*	*Manuscript witnesses*
Naṣīr al-Dīn al-Ṭūsī	*Ḥall-i mushkilāt-i Muʿīniyya* (Persian)	1245 CE	19
Naṣīr al-Dīn al-Ṭūsī	*Taḥrīr al-Majisṭī* (Arabic)	1247 CE	93
Naṣīr al-Dīn al-Ṭūsī	*Al-Tadhkira fī ʿilm al-hayʾa* (Arabic)	1261 CE	72
Quṭb al-Dīn al-Shīrāzī	*Nihāyat al-idrāk fī dirāyat al-aflāk* (Arabic)	1281 CE	37
Quṭb al-Dīn al-Shīrāzī	*Ikhtiyārāt-i Muẓaffarī* (Persian)	1282 CE	10
Quṭb al-Dīn al-Shīrāzī	*Al-Tuḥfa al-shāhiyya fī al-hayʾa* (Arabic)	1285 CE	49
Quṭb al-Dīn al-Shīrāzī	*Faʿalta fa-lā talum* (supercommentary on the *Tadhkira*; Arabic)	ca. 1300 CE	3
Ḥasan ibn Muḥammad ibn al-Ḥusayn Niẓām al-Dīn al-Aʿraj al-Nīsābūrī	*Tawḍīḥ al-Tadhkira* (Arabic)	1311 CE	53
ʿUmar b. Daʾūd al-Fārisī	*Takmīl al-Tadhkira* (commentary on the *Tadhkira*; Arabic)	1312 CE	1
Jalāl al-Dīn Faḍl Allāh al-ʿUbaydī	*Bayān al-Tadhkira wa-tibyān al-tabṣira* (commentary on the *Tadhkira*; Arabic)	1328 CE	1
al-Sayyid al-Sharīf ʿAlī ibn Muḥammad ibn ʿAlī al-Ḥusaynī al-Jurjānī	*Sharḥ al-Tadhkira al-Naṣīriyya* (commentary on the *Tadhkira*; Arabic)	1409 CE	51
Fatḥ Allāh al-Shīrwānī	*Sharḥ al-Tadhkira* (commentary on the *Tadhkira*; Arabic)	1475 CE	2
ʿAbd al-ʿAlī ibn Muḥammad ibn al-Ḥusayn al-Bīrjandī	*Sharḥ al-Tadhkira* (commentary on the *Tadhkira*; Arabic)	1507 CE	1
Shams al-Dīn Muḥammad ibn Aḥmad al-Khafrī	*Al-Takmila fī sharḥ al-Tadhkira* (supercommentary on the *Tadhkira*; Arabic)	1525 CE	2

Mehmed II. Admittedly, Bessarion was hardly the person to acknowledge the scientific achievements of Muslims; after all, he came to Vienna as a legate of Pope Pius II (Aeneas Silvius Piccolomini) in order to seek support for a crusade against the Turks that would recapture Constantinople.[97] But his intense interest in reviving the Greek scientific heritage in Europe would have overcome any hesitancy he may have had about bringing cutting-edge Islamic scientific thought to his young acolytes.

CONCLUSION

The possible transmission of the Ṭūsī-couple to Europe confronts us with a number of both practical and theoretical considerations. On a practical level, we need to trace the origins and development of the device and its appearance afterward over several centuries. As we have seen, it is critical that we be clear which version of the couple we are talking about and how it is being used. We also have needed to chart the various pathways by which the couple was, or could have been, transmitted.

On a theoretical level, we need to deal with several implicit issues in what has gone before by way of conclusion. The first we can call the issue of the hermetically sealed civilization. Many comments on intercultural transmission have somehow assumed that after the twelfth-century translation movement from Arabic into Latin, the gates of transmission became closed, and European Christendom and Islam were sealed off from one another until the colonial period brought them back into contact, this time with the relative civilizational – but more importantly, military – superiority reversed. This assumption has had a number of historiographical consequences. Much of premodern European history, both medieval and early modern, is written from a Eurocentric point of view. In many cases, this bias may be justified since, like politics, much of history is local.[98] However, this is not the case with all history. And here the insistence on an exclusively European-focused narrative can cause considerable distortion of the historical record. For example, discussing the development of trigonometry without bringing in the Indian introduction of the sine and, based on this innovation, the subsequent development of the other trigonometric functions and identities in Islamic mathematics leaves out an essential part of the story.[99] In the case of much postclassical (i.e., post-1200 CE) Islamic science, the assumption is made that Europeans would have had little contact because of cultural and linguistic differences. But this assumption by European intellectual historians is belied by the extensive evidence of political, economic, and cultural exchanges between various late-Islamic regimes and European realms.[100] European travellers did go to various regions of the Islamic world before the modern period, and there are certainly examples of Islamicate travellers in Europe.[101] But more to the point, it is also clear that Islamic scientific theories and objects did travel to Europe, as we have seen, through contacts such as those between Spain and Īlkhānid Iran, through Jewish intermediaries, and through Byzantine scholars and émigrés.

The above-mentioned research by Langermann and Morrison, as well as by İhsan Fazlıoğlu and other historians of the Ottoman period, points to something often overlooked, namely the important role of the Ottoman courts of Mehmed II, who was the conqueror of Constantinople, and of his son and successor Bāyazīd II in promoting scientific and philosophical study, which included providing patronage for Christian and Jewish, as well as Muslim, scholars. Many of these Christian and Jewish scholars travelled readily between the Ottoman and Christian realms.[102] And it should not be forgotten that, at the time, the Ottomans were a European power, with vast domains in eastern and central Europe, and had been such since the fourteenth century.

But there may have been more direct contact. Here, one needs to confront the myth of a linguistically impoverished Europe; even scholars sympathetic to transmission such as Swerdlow and Neugebauer feel compelled to remark that "[a] direct transmission of the Arabic [texts containing the non-Ptolemaic models used by Copernicus] is *of course* extremely unlikely."[103] But why "of course"? Some Europeans did know Arabic (how else could the twelfth-century translation movement have taken place?), and there is research showing that knowledge of Arabic was not unknown during the Renaissance.[104] At this point in our knowledge, we can only speculate that European astronomers either learned Arabic or worked with translators who did know enough to explain the non-Ptolemaic models of Ṭūsī, Ibn al-Shāṭir, and others. But it seems to me equally speculative to assume they did not. After all, Arabic is not all that esoteric – it is closely related to Hebrew, which was certainly studied by numerous European Christian scholars – and there were dictionaries and grammars available. And perhaps most importantly, why would someone seek to start from scratch when it was certainly known in the fifteenth and sixteenth centuries that Islamic astronomers still had much to teach their European counterparts?[105] But more generally from a historiographical point of view, it seems odd that so many European historians of the medieval and early modern periods have written histories that make their subjects seem isolated, devoid of curiosity, and impervious to outside influences.[106]

The next theoretical point to pursue is the question of "how much evidence is enough." It is a commonplace in the history of science to trace intercultural transmission through the reappearance of numbers, objects, models, propositions, and even ideas that we can locate in an earlier source. In fact, one might consider it our most precise way to document intercultural transmission. The gold standard in our field is arguably Hipparchus of Nicaea's value for the mean synodic month (reported by Ptolemy), namely 29;31,50,8,20 days (sexagesimal). Once Franz Kugler demonstrated in the 1890s that this value came from what is now known as Babylonian System B, the argument for Greek knowledge and use of Babylonian astronomy (at least its parameters) became incontestable. The same is also true of the fact that Hipparchus, despite what is reported by Ptolemy, did not make a recalculation using new observations. But why can we reach these conclusions? The answer is obvious. Would anyone seriously argue that two identical values to the fourth sexagesimal place is a coincidence? According to Di Bono and Goddu, the appearance of Ṭūsī's couple, Muʾayyad al-Dīn al-ʿUrḍī's lemma, Ibn al-Shāṭir's models, and so on in the work of Copernicus is not sufficient to prove transmission. But what makes this case different from the case of Hipparchus's value for the mean synodic month? The case made by Di Bono, and echoed by Goddu, is that somehow the "internal logic" is such that anyone confronting the problem of Ptolemy's irregular motions would come up with the same solutions.[107] But Di Bono makes it clear that his criteria for accepting transmission are so high that even a "high number of coincidences between Copernican and Arab models" is insufficient since it then "becomes very difficult to explain how such a quantity of models and information, which Copernicus would derive from Arab sources, has left no trace – apart from Ṭūsī's device – in the works of the other western astronomers of the time."[108] This argument is a curious one; given the tenuous nature of transmission, an insistence on multiple examples would render many cases moot, even one as strong as the transmission of the Babylonian synodic month.

Let us now turn to the issue of "internal logic" and parallel development. In fact, what we have in Islam and in the Latin West represent two very different historical developments. The criticism of Ptolemy on various fronts, including observational ones, begins quite early in Islam;[109] and certainly by the time of Ibn al-Haytham (d. ca. 1040), we have sustained criticisms of the irregularities in Ptolemy's planetary models.[110] By the thirteenth century, we see a number of attempts to deal with these criticisms by using alternative models that employ devices consisting of uniformly rotating spheres, those of Ṭūsī, ʿUrḍī, and Shīrāzī being the most prominent; the proposal of alternative models continues for several centuries in Islam. It is important to emphasize that this historical development is sustained and traceable; Ṭūsī and his successors knew of earlier criticisms and alternative models, and they explicitly sought to build upon their predecessors. This long-term historical process is precisely what is missing in the accounts of those who advocate a "parallel development" in the Latin West. As we have seen, the Ṭūsī-couple appears there in fits and starts; we do not find a sustained discussion of the "equant problem" before Copernicus,[111] and we certainly do not see a sustained, historically coherent development of alternative models. Here, the evolution of Ṭūsī's various couples is instructive; from the initial discussion of the problem and announcement of a solution until he put forth his "final" versions, Ṭūsī took twenty-five years, during which he presented various models that he would later revise. But in the Latin case, there is no one about whom a story exists that can account for the rationale and development – indeed, the "logic" – for one or more versions of the Ṭūsī-couple. As we have seen, they just somehow appear. And no one after Ṭūsī claims to have independently discovered any of the versions of the couple, either in the Islamic world or in the Latin West.[112]

In their different scenarios, both Di Bono and Goddu have attempted to provide alternative "stories," but these are deeply flawed. Di Bono seeks to find the source for Copernicus's use of the Ṭūsī-couple in the Paduan Aristotelian-Averroist critiques of Ptolemy. But the problem here is that such critiques generally led to quite different homocentric modelling based on a variety of techniques that are quite distinct from those of Ṭūsī and his successors. In particular, Di Bono makes no attempt to explain how Copernicus could have used the epicycle-only modelling of Ibn al-Shāṭir if he had been so influenced by astronomers and natural philosophers adamantly opposed to epicycles and eccentrics.[113] In the case of an astronomer who did come out of that tradition and who did use one version of the Ṭūsī-couple, namely Amico, we have an astronomy quite different from that of Copernicus. As for Goddu's attempt to locate Copernicus's discovery and use of the Ṭūsī-couple in the Aristotelian environment of Cracow, here we have what amounts to a misunderstanding. As we have seen, Brudzewo, whom Goddu wishes to make the immediate predecessor for Copernicus's use of the couple, is in fact using Ibn al-Haytham's Eudoxan-couple. It is true that Brudzewo does mention it in the context of the motion of the epicyclic apogee due to the Moon's prosneusis point, which, interestingly enough, is one of the examples Ṭūsī uses to explain the need for the curvilinear version of his couple.[114] But again, neither Brudzewo nor anyone else adduced by Goddu proposes a Ṭūsī-couple device for dealing with the problem.[115] In sum, both Di Bono and Goddu depend on tenuous connections that would have us believe that their actors can move from model to model without clear agency or plausible historical context. And it is this stark contrast – between, on the one hand, Islamic astronomy's well-developed historical context for dealing with the irregular motions of Ptolemaic astronomy and, on the other hand, the Latin West's ad-hoc, episodic, and decontextualized "parallel" attempts – that in my opinion provides us with the most compelling argument for transmission of non-Ptolemaic models such as the Ṭūsī-couple from Islam to Europe before the sixteenth century.[116]

Given what we know, it seems that one possible scenario is that Copernicus was indeed influenced by Brudzewo's comments to pursue the problem of the Moon's epicyclic apogee. And perhaps he realized at some point that what was needed was a curvilinear oscillation on the epicycle's circumference, as Ṭūsī had before him. Then, while in Italy, he somehow encountered, through one of the routes outlined above, one or more versions of the Ṭūsī-couple that he would subsequently use. But it is also clear that he was not overly interested in the complexities of the models, which would account for his use of the apocopated two-sphere (as opposed to the full three-sphere) version in the *Commentariolus*. And by the time of composing *De revolutionibus*, he was willing to make a further simplification by using Ṭūsī's two-circle version even though it did not fulfil the need either for a full-scale physical model for rectilinear motion or for a version that could produce true curvilinear oscillation.

In summary, it seems that, as put so perceptively by Dobrzycki and Kremer, "We may be looking for a means of transmission both more fragmentary and widespread than a single treatise."[117] And certainly by the time Copernicus wrote *De revolutionibus*, one version or another of the Ṭūsī-couple would have been available in the Latin West for several centuries; in other words, it had become commonplace. So perhaps Copernicus, the man from Toruń, felt no need to worry about its origins, whether in Tūn or elsewhere, and could, without qualms, cross out the redundant remark in his holograph that "some people call this the 'motion along the width of a circle.'"[118]

8

Jews as Scientific Intermediaries in the European Renaissance

Robert Morrison

AT THIS POINT IN TIME, and by this point in the present volume, the circumstantial evidence that Nicholas Copernicus relied on European and Islamic astronomers whom he did not name is not news.[1] Noel Swerdlow has found that there were five, mostly uncited, sources for Copernicus's work, including Islamic sources.[2] More recently, F. Jamil Ragep has found that a proof about planetary motions, which became the foundation of Copernicus's transformation of a geocentric cosmos into a heliocentric one, had appeared earlier in the Islamic world.[3] Still, despite the numerous parallels that are difficult to explain by invoking independent discovery,[4] despite the existence of Byzantine Greek manuscripts containing theories from scholars in Islamic societies that appear in Copernicus's work,[5] and despite significant new findings of scientific exchange between Islamic societies and Europe around 1500,[6] some recent publications have continued to argue for independent discovery or for a solely European context for Copernicus's work.[7] But historians have not found a full European context for Copernicus's achievements.[8] Rather, recent research has found that commercial, intellectual, and personal contacts between Europe and Islamic societies proliferated in Copernicus's time and in the following centuries.[9] Arguments for independent discovery would mean that Copernicus was not affected by his historical context. This chapter discusses both another way that astronomers in Renaissance Europe and Islamic societies saw a need to criticize and modify Ptolemaic astronomy and another way that Copernicus could have learned of the achievements of astronomers from Islamic societies.

The locus for the chapter's discussion is the Ṭūsī-couple, a "hypothesis" (i.e., a building-block of an astronomical model)[10] produced by Naṣīr al-Dīn al-Ṭūsī (d. 1274), the director of the renowned Marāgha Observatory near Tabriz.[11] The Ṭūsī-couple combined uniformly revolving orbs to produce a linear oscillation and make Ptolemy's models physically coherent. A hypothesis identical to the Ṭūsī-couple was present in Copernicus's model in *De revolutionibus orbium coelestium* for trepidation – that is accelerations and decelerations in the motion of the equinoxes.

This chapter presents evidence of how an astronomy text composed originally in Judeo-Arabic, a dialect of Jews in the Arabic-speaking world written in Hebrew characters, interested Renaissance astronomers. This text contained, among other things, a hypothesis mathematically identical to the curvilinear Ṭūsī-couple (i.e., a Ṭūsī-couple for producing curvilinear oscillation on the surface of an orb) and physically equivalent to the truncated two-sphere Ṭūsī-couple. To argue for scholarly exchange, this chapter first points out remarkable parallels between that text, entitled *The Light of the World*, and works of European astronomers; the chapter then argues how scholarly exchanges could explain those connections. To provide a broader context for those exchanges, the chapter describes other ways that Jews functioned as intermediaries for Renaissance astronomers' knowledge of astronomy. Finally, this chapter points out that this volume's reassessment of Copernicus has had the effect of directing our attention to the full extent of the involvement of Jews in Renaissance astronomy and expanding our knowledge of inter-religious scientific cooperation. Although most of this chapter is about theoretical astronomy, which focused on physical models and hypotheses that could account for the available observations, the latter part discusses the production of tables of planetary positions that could be used for calendar calculations, timekeeping, and astrological forecasting.

As the topic of the chapter is scholarly cooperation and exchange between Jews and Christians, I should clarify that conversion to Christianity was a reality of Jews' lives, and since the conditions around such conversions were often fraught, this chapter's criteria for classification as a Jew are, unless there is evidence that a historical actor never identified as a Jew, either a knowledge of a Jewish language or membership in the Jewish religious community at one point in the historical actor's life.[12] Given negative perceptions of Islam in Renaissance Europe, Jews were more viable colleagues for European scientists.[13]

In fact, the transregional position of Jews in the fifteenth century rendered them an important means of communication, such as between the Republic of Venice and the Ottoman Empire following 1453.[14] The persecution and then final expulsion of the Jews from Spain in 1492 intensified existing connections among these various Jewish communities. The historical actors that this article describes travelled among and contributed to the intellectual lives of a number of regions. For instance, members of an Ibn Naḥmias family went from Castille to Albania, and then to Salonika, before moving to Venice by the 1600s.[15] Members of that family also established the earliest printing press in the Ottoman Empire by the end of the fifteenth century, probably in 1493.[16] Another prominent family in science, the Qusṭanṭīnī family, responsible for the copying of Johannes Regiomontanus's *Epitome of the Almagest* found in Paris MS Hébreu 1100, fled to Candia on Crete after the expulsion.[17] By the 1600s, members of the family started coming to Italy. These two brief anecdotes begin to show how intellectual mobility characterized the Jewish communities of the Iberian Peninsula, Italy, Greece, and the Ottoman Empire.

I now turn to *The Light of the World* by Joseph ibn Naḥmias (fl. ca. 1400).[18] In this text, Ibn Naḥmias attempted to represent all celestial motions as occurring on the surface of an orb (a hollow sphere) by theorizing a set of homocentric orbs, with the Earth at the precise centre of that orb or set of orbs (see figure 8.1). *The Light of the World* was an attempt to improve Nūr al-Dīn al-Biṭrūjī's (fl. 1200) *Kitāb fī al-hay'a* (On the Principles of Astronomy), a text that was translated into Latin by Michael the Scot (fl. 1217) and cited by Copernicus in *De revolutionibus.*[19] Most astronomers followed Claudius Ptolemy's (fl. 50–125 CE) *Almagest* and did not insist that the Earth had to be at the centre of all orbs; removing the Earth from the centre of the orbs facilitated achieving greater accuracy when accounting for the observed variations and irregularities in the planets' motions.

Unlike the models found in Ptolemy's *Almagest,* models composed of homocentric orbs (such as Biṭrūjī's and Ibn Naḥmias's) often could not "save the appearances," which meant that they could not account accurately for the planets' observed motions in longitude and latitude or predict accurately where the planets would be in the future.[20] Ibn Naḥmias's models did improve on the predictive accuracy of Biṭrūjī's models, although the positions predicted by Ibn Naḥmias's lunar model still diverged from observations to an extent that would be noticeable.[21] Perhaps this relatively poor predictive and retrodictive accuracy is why historians of science have not paid as much attention to models constructed out of homocentric orbs.[22] But Michael Shank's research has shown that important Renaissance astronomers were interested in these models of homocentric orbs due to their philosophic consistency.[23] Regarding Regiomontanus's (d. 1476) interest in models composed of homocentric orbs, Shank has pointed out that "models that yielded sound physical predictions would also improve mathematical predictions."[24] Peter Barker has argued that Copernicus

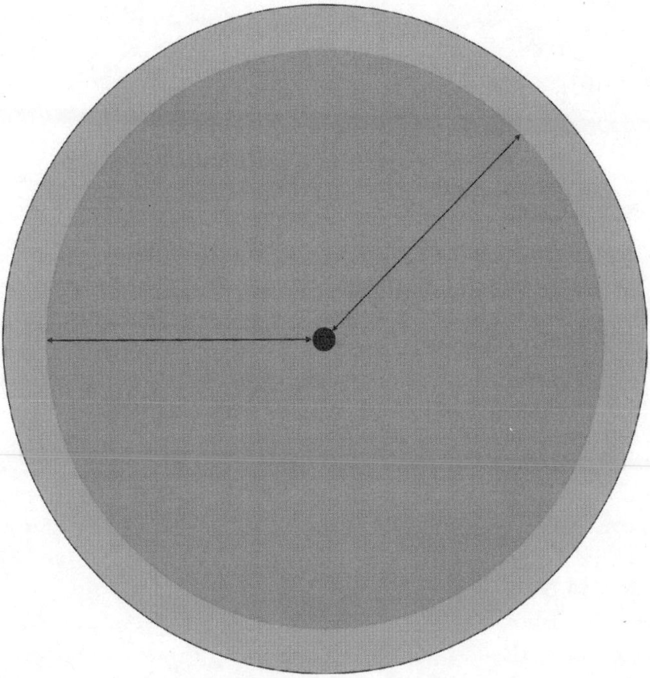

8.1 Cross-section of an orb with the black dot representing the Earth. The distance from the Earth to all points on the concave edge of the orb is the same.

paid attention to Averroist criticisms of Ptolemy (i.e., the criticisms of those astronomers and philosophers who posited a cosmos of wholly homocentric orbs),[25] and Bernard Goldstein has made a persuasive argument that Copernicus's hypothesis of heliocentricity solved the cosmological problem of planetary distances, not problems of physical consistency or mathematical astronomy.[26] Homocentric models, although they prioritized physical elegance over predictive and retrodictive accuracy, were an important part of fifteenth-century astronomy.

Ibn Naḥmias's *The Light of the World* had its roots in the intellectual life of twelfth-century Andalusia, when a fervent desire to adhere rigorously to a certain interpretation of Aristotle's natural philosophy occasioned a cluster of writings that A.I. Sabra has named "the Andalusian revolt against Ptolemaic astronomy."[27] Sabra identifies Biṭrūjī's *On the Principles of Astronomy* as "the final outcome of the Andalusian endeavor."[28] Ibn Naḥmias's work is certainly connected to, but not a continuation of, the revolt because Ibn Naḥmias disagreed with Biṭrūjī over whether motions that might appear to be in opposite directions could occur in the heavens.[29] Biṭrūjī and Ibn Naḥmias were participating in a long-running debate that originated with the question of whether Aristotle meant, in *De Caelo* 270a, that the categories of "up" and "down" could apply to celestial motions. Ibn Naḥmias's reinterpretation of the issue of opposite motion in the heavens facilitated his attempt to improve the predictive accuracy of the homocentric models found in Biṭrūjī's *On the Principles* because Biṭrūjī's attempt to exclude apparently opposite motions from his models was, to Ibn Naḥmias, an unnecessary imposition that complicated attempts to achieve a measure of predictive and retrodictive accuracy.[30] There was much at stake because the planets' mean motions in longitude appeared to be in the *opposite* direction of the cosmos's diurnal motion; if apparently opposite motions were permissible, a single orb could account for the planets' mean motions in longitude.

In addition to the original Judeo-Arabic version of *The Light of the World*, there exists a Hebrew recension that is due either to Ibn Naḥmias himself or to someone who agreed with him.[31] The author of the recension accepted a cosmos of homocentric orbs and, more important, tried to propose physical movers for some of the motions that the Judeo-Arabic version often explained through the motions of great circles on the surface of a single orb. The Hebrew recension would have been more linguistically accessible to scholars in Europe. Furthermore, key passages in the Hebrew recension addressed a central technical challenge faced by any astronomer who proposed models composed of uniformly revolving orbs with a common centre, namely how could combinations of uniformly revolving orbs with a common centre produce an oscillation on an arc on the surface of an orb that would help to account for observed variations in the planets' motions (such as how variations in the Sun's motion cause the seasons to be of unequal length)?[32] Both *The Light of the World* and European Renaissance astronomers found identical ways of producing such an oscillation on the surface of an orb. In the case of the Sun, the resultant oscillation could either keep the Sun in the ecliptic (i.e., the path of the Sun against the background of the zodiac) or account for the variations of the Sun's motion from the mean.

The contents of *The Light of the World*, given developments in European astronomy in the fifteenth century, would have interested European scholars working on homocentric astronomy. The Renaissance astronomer Regiomontanus presented his models of homocentric orbs in a letter dated to 1460.[33] His models featured a hypothesis that converted circular motion into a linear oscillation (see figure 8.2); scholarship on Regiomontanus has called this oscillation a reciprocal motion. Swerdlow has pointed out that the mechanism (or hypothesis) for reciprocal motion was similar to a slider-crank mechanism.[34] The oscillation is on the surface of an orb, so the oscillation is supposed to be along an arc. In the simplest case of the Sun, an outer orb causes the revolution of the Sun in the ecliptic at a uniform velocity, while the pole of the inner orb revolves along a small circle, causing, in turn, an oscillation of a point along the ecliptic on the outer orb. In figure 8.2, DHG is the small circle of the path of the pole of the inner orb, and LBM is the ecliptic arc on the uniformly revolving outer orb along which a point oscillates through the path of the pole of the inner orb on DHG. If the period of the revolution of the outer orb and the period of the revolution of the pole of the inner orb on DHG were the same, a point on LBM would speed up and slow down throughout a single revolution of the outer orb. The proposal of a revolving circle that moves a point back and forth does function geometrically, but physical problems remain. In particular, Swerdlow wonders how the oscillating point would know to remain in the ecliptic (LBM) without some sort of track.[35]

An equivalent hypothesis to produce a linear oscillation, faults and all (again, the matter of what keeps the oscillating point moving on an arc), had already appeared in the Hebrew recension of *The Light of the World*.[36] Swerdlow's argument that Regiomontanus's models of perfectly concentric orbs were his own work is, however, quite plausible.[37] Swerdlow cites internal evidence to explain how the hypothesis for the reciprocal motion arose from a modification of Biṭrūjī's own models, and there is no evidence for the presence of theories from *The Light of the World* in the Veneto as early as 1460. It is more significant that the Hebrew recension proposed improvements of the hypothesis (or mechanism) for reciprocal motion. A text that addressed a weakness of Regiomontanus's homocentric astronomy would certainly have interested other Renaissance astronomers aware of Regiomontanus's homocentric astronomy.

The improvement of the reciprocation mechanism preferred by the Hebrew recension of *The Light of the World* relies on a hypothesis of two simultaneous motions in seemingly opposite directions that keep a point oscillating on (approximately) an arc on the surface of an orb (see figure 8.3). This proposal allows the revolution of Z to P to E to B to occasion an oscillation from K to G to D all through the complete revolutions of orbs. As well, this preferred proposal does not rely on fixing the oscillating point in a track like the preliminary proposals in the Hebrew recension or like Regiomontanus's hypothesis for reciprocal motion. The Hebrew recension's hypothesis requires only two orbs, one with pole Z, and points D, G, and K, and another with pole A.[38] One presumes that each orb has a negligible thickness.

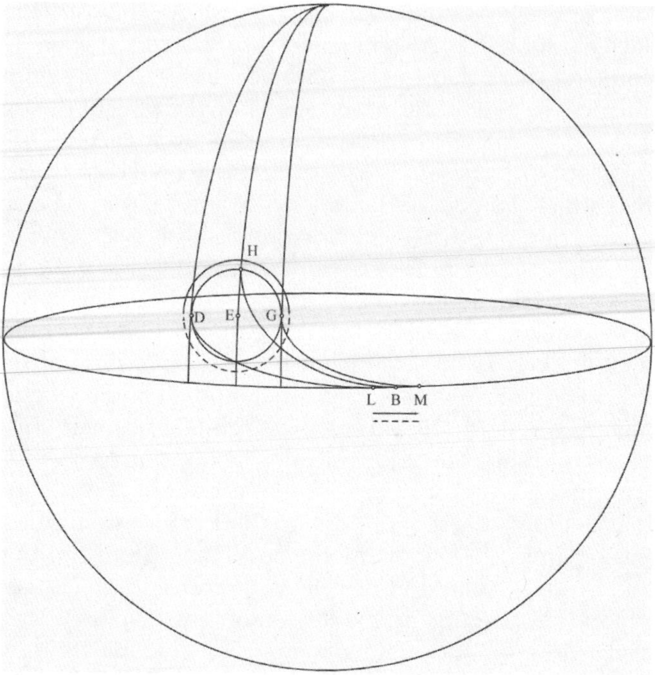

8.2 Regiomontanus's reciprocation mechanism.

The first orb is fixed in the second at a point (A') corresponding to pole A on the second orb. The second orb revolves about pole A, causing pole Z to revolve a certain number of degrees (45° in the figure) about pole A to arrive at point P. As a result, the position of point K would become point N. Simultaneously, the first orb whose pole is Z revolves about pole Z to bring point K from point N over to point Y. The Hebrew recension does not specify the increments of the orbs' motions; one should imagine the revolutions occurring by minimal increments, not by increments of 45°. Thus the point does not jump from K to Y. Rather, it oscillates, through all the intervening points, on an arc from K to Y and then to G. As the Hebrew recension acknowledges, this solution will not function precisely.[39] Y is close to but not precisely on arc KGJD.

Besides this development of the mechanism for reciprocal motion, both the Judeo-Arabic original and the Hebrew recension of *The Light of the World* contain another hypothesis that is mathematically equivalent to the curvilinear version of the Ṭūsī-couple in Ṭūsī's *Al-Tadhkira fī ʿilm al-hayʾa*.[40] Because Ibn Naḥmias's hypothesis is physically congruent

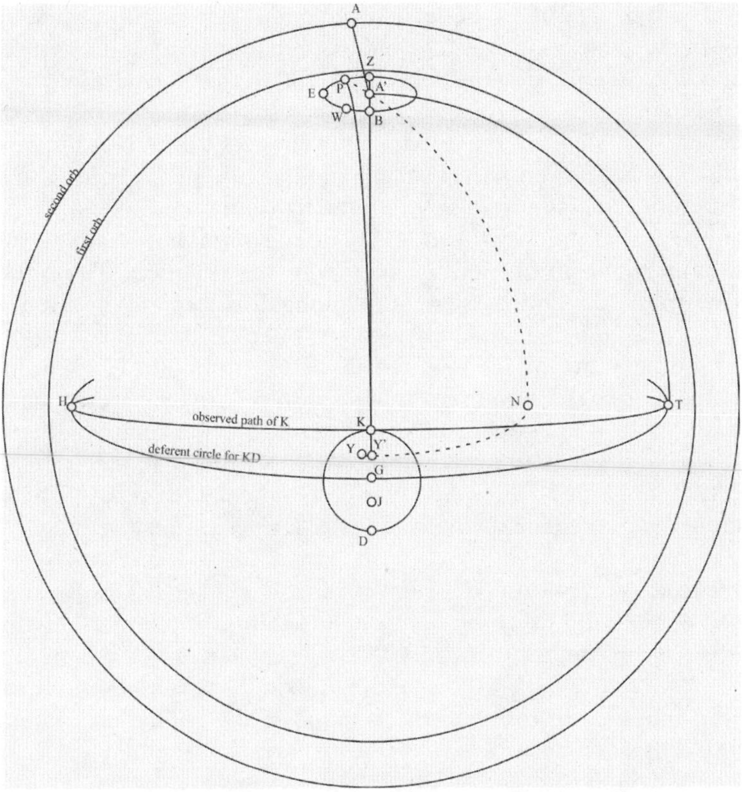

8.3 An improved reciprocation mechanism (a Eudoxan-couple).

only to what Ragep calls the truncated two-sphere version of the Ṭūsī-couple, I call Ibn Naḥmias's hypothesis the double-circle hypothesis. Ṭūsī developed the Ṭūsī-couple to resolve the physical inconsistencies of Ptolemy's astronomy, but Ibn Naḥmias used the double-circle hypothesis to improve predictive accuracy (see figure 8.4). Ibn Naḥmias had found that the motions of the circle of the path of the centre and the inclined circle carrying the circle of the path of the centre were insufficient, on their own, to account for the Sun's motions because they do not place the Sun in the ecliptic where it is observed. Searching for a solution, Ibn Naḥmias saw that a near-oscillation along an arc of an orb could be produced by the opposite revolutions of two equal, small circles on the surface of an orb, with the centre of one fixed on the circumference of the other. In figure 8.4, the centres of the small circles are C and A. One small circle on the surface of an orb revolved in one direction at a given

small circles (= math)

speed, and the centre of the second circle was carried on the circumference of that first circle and revolved twice the measure of the first revolution. Ibn Naḥmias described the centres of the small circles and points on their circumferences as "poles."[41]

poles

In figure 8.4, pole Z, the centre of the circle of the path of the centre of the Sun, has moved with the mean motion (45° in the figure), and A revolves with the motion (also 45° in the figure) of the circle of the path of the centre of the Sun.[42] Ibn Naḥmias has shown, however, that these two motions are insufficient to keep A in the ecliptic.[43] The double-circle hypothesis eliminates the outstanding displacement of A from the ecliptic. The upper small circle revolves θ (45° in the figure) about A, moving the centre of the lower small circle from point H to point B. Simultaneously, the lower small circle revolves 2θ (90° in the figure) about its pole, now at point B. Point X marks the Sun's final position after the revolution of 2θ from point A. Arc AN is the diameter of the circle HBC, and in the figure arc AN is perpendicular to arc HC. The Sun is supposed to oscillate on arc AN. There remains a small yet negligible displacement from AN.[44]

The Judeo-Arabic and Hebrew versions of *The Light of the World* proposed eliminating from the solar model both the circle of the path of the centre and the inclined circle carrying the circle of the path of the centre (see figure 8.5). In this model, the double circles would revolve on the ecliptic with the solar mean motion. The oscillation produced by the double circles would account for the solar anomaly (i.e., variation in the Sun's motion).[45] It turns out that such a solar model, with a double-circle hypothesis, appeared in the text on homocentric astronomy of Giovanni Battista Amico (d. 1538), who wrote in the 1530s at Padua.[46] Swerdlow's study of Amico's *De motibus corporum coelestium* shows that Amico used only double-circle hypotheses to account for the planets' anomalies.[47] Another astronomer of the period, Girolamo Fracastoro, produced an astronomy based on homocentric orbs, published at Venice in 1538. Although Fracastoro did refer to the double-circle hypothesis, he did not seem to incorporate that hypothesis into his astronomy.[48]

The key difference between the Ṭūsī-couple found in Ṭūsī's *Tadhkira* and his Persian *Ḥall-i mushkilāt-i Muʿīniyya*, on the one hand, and the double-circle hypothesis found in *The Light of the World* and in the theories of the Paduan astronomers, on the other hand, is that the two circles of this version of the Ṭūsī-couple (i.e., the double-circle hypothesis) were the same size and were not tangent.[49] This difference opens up a real possibility that Amico and Fracastoro could have learned of the double-circle hypothesis from *The Light of the World*. Even if one

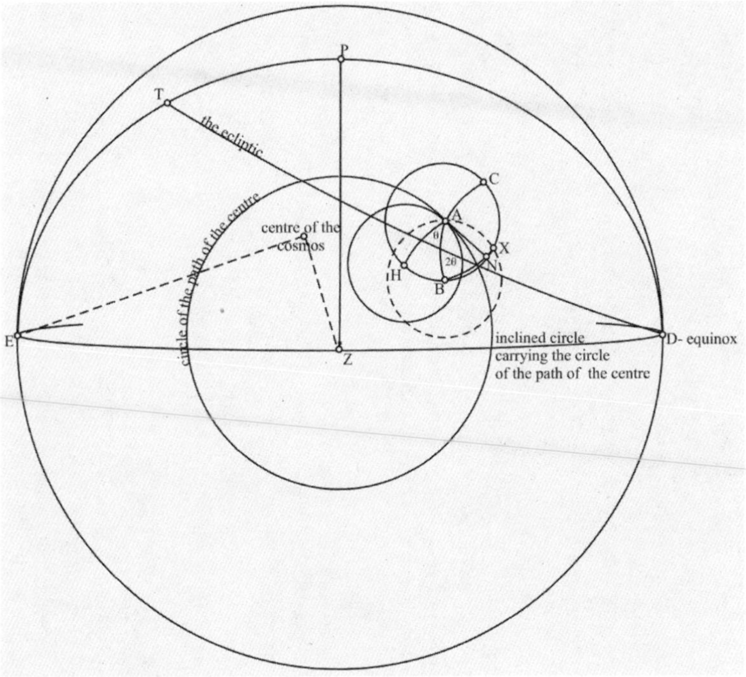

8.4 The double-circle hypothesis.

were to reject Ibn Naḥmias's role and accept Mario Di Bono's proposition of an independent European invention of a double-circle hypothesis, there are too many other parallels between Copernicus and astronomers of Islamic societies left to explain.[50] Moreover, according to Di Bono's reasoning, one could conclude that Ibn Naḥmias and the medieval Jewish scholar Avner de Burgos (ca. 1270–1340), who produced a version of the Ṭūsī-couple, worked independently of each other and in isolation from others.[51] Although it would be difficult to trace every occurrence of a double-circle hypothesis or Ṭūsī-couple back to Ṭūsī, and there were many ways that Copernicus could have learned of the Ṭūsī-couple, it would be equally difficult to conclude that every instance of the Ṭūsī-couple was a case of independent discovery. Fracastoro and Amico are better understood as evidence of how a serious interest in homocentric astronomy in Italy continued after Regiomontanus's career. Amico knew of Biṭrūjī's *Planetarum theorica*,[52] so an available improvement on Biṭrūjī's *On the Principles*, such as Ibn Naḥmias's *The Light of the World*, would certainly have interested Amico.

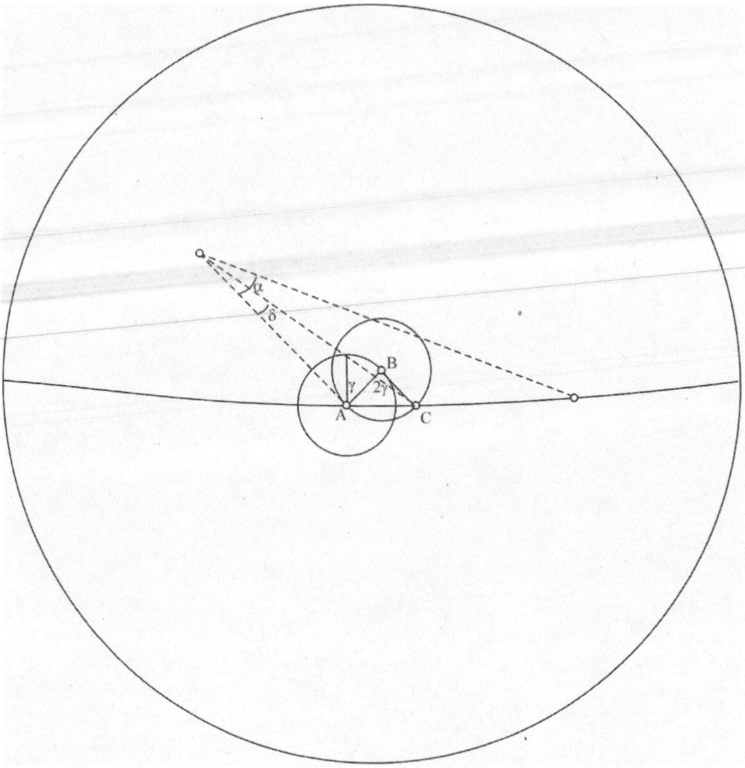

8.5 Ibn Naḥmias's solar model shared by Amico.

Let us now turn to the historical evidence to support the argument that astronomers in late-fifteenth-century Padua interested in Regiomontanus's astronomy could easily have found out about *The Light of the World* and its improvement on Regiomontanus's hypothesis for reciprocal motion. Tzvi Langermann has written two articles about the activities and ideas of a Jewish scholar named Moses ben Judah Galeano (Mūsā Jālīnūs, d. after 1542), who was present at the court of the Ottoman sultan Bāyazīd II (r. 1481–1512); Galeano is a key figure for understanding the passage of *The Light of the World* to Renaissance Italy.[53] Galeano composed a Hebrew text entitled *Ta'alumot ḥokmah* (Puzzles of Wisdom) around 1500, although he finalized the text in the 1530s. *Puzzles of Wisdom* mentioned, among other things, the astronomy of 'Alā' al-Dīn ibn al-Shāṭir (d. ca. 1375), whose models, it has been noted, parallel Copernicus's in many places.[54] *Puzzles of Wisdom* also explained that *The Light of the World* was a text about an astronomy of homocentric orbs, although Galeano identified the author as Joseph ibn Yaʿīsh.[55] Still, another text Galeano wrote about homocentric astronomy, the sole manuscript of which is present in the Topkapı Library in Istanbul, indicates his familiarity with the details of Ibn Naḥmias's theories.[56] *Puzzles of Wisdom* described Galeano's visit to Venice around 1500, during which he met with the prominent printer Gershom Soncino.[57] Figures such as Soncino would have been very interested in news of scientific developments.

Although Galeano was exceptional, his association with scientific activity at the Ottoman court was not an exception. Ilyās ibn Ibrāhīm al-Yahūdī (d. after 1512), known after his conversion to Islam as ʿAbd al-Salām al-Muhtadī or ʿAbd al-Salām al-Daftarī, came from al-Andalus to the court of Bāyazīd II and wrote a text in Hebrew about how to use an astronomical instrument, which he invented, known as al-Dābid.[58] Then he translated the text into Arabic at the sultan's request in 1502. *Puzzles of Wisdom* was also replete with criticisms of Galeano's contemporaries, some of whom were Jewish, at the sultan's court.[59] Earlier, an older relative of Galeano had worked with scholars at the court of Mehmed II (d. 1481) to produce a Hebrew version of Maḥmūd ibn Muḥammad ibn ʿUmar al-Jaghmīnī's *Al-Mulakhkhaṣ fī ʿilm al-hayʾa al-basīṭa.*[60] There is also evidence for the passage of *The Light of the World* from al-Andalus to Istanbul to Padua. First, linguistic evidence suggests that Galeano's own text on homocentric astronomy found in the Topkapı Library was translated from Hebrew or transcribed from Judeo-Arabic.[61] Thus it is plausible that the extant Arabic text by Galeano is a translation or transcription, carried out in Istanbul, of a now lost Hebrew or Judeo-Arabic version of *The Light of the World* probably made before Galeano left Istanbul for Venice. In any case, the contents of *The Light of the World*, if not a complete manuscript, clearly transited to Istanbul. Subsequently, Giulio Bartolocci's (1613–87) *Bibliotheca magna rabbinica de scriptoribus*, a bio-bibliographical dictionary of Jewish literature, contained a report of *The Light of the World* being seen at Padua.[62] The striking parallels between Ibn Naḥmias's theories and those of astronomers at Padua, Galeano's voyage to Venice, and the eventual report of *The Light of the World* being at Padua make it highly likely that scholars at Padua such as Amico and Fracastoro were aware of the contents of *The Light of the World.* The career of Moses ben Judah Galeano helps to explain, as well, the numerous parallels with Ibn al-Shāṭir's theories in Copernicus's work.

A significant question regarding scholarly exchange is whether any Jews knew of what contemporary European Renaissance astronomers were doing, for if Jews knew about recent developments in European Renaissance astronomy, there would be a broader context for contact between Galeano and scholars at Padua. A well-documented area of contact between Jews and Christians in Europe was the translation of Averroes's (Ibn Rushd's) corpus into Latin. All told, Jews translated three-fourths of Averroes's writings into Latin from Hebrew translations of the original Arabic.[63] And there is some evidence that the last Jewish Averroist, Elijah Delmedigo (d. 1493), knew of recent efforts to develop new theories in astronomy.

Delmedigo's brief life was full of scholarly exchange. He hailed from Candia on Crete, which had an educated Jewish populace (and was the city where Galeano died), but between 1480 and 1490, Delmedigo was actively sought out by Christian scholars.[64] He went to Venice and taught Christians in Padua, among them Giovanni Pico della Mirandola (d. 1494),[65] a scholar interested, in addition to philosophy, in the Qabbala and hermeticism.[66] Pico himself studied Hebrew and was familiar with the work on astronomy of Levi ben Gerson (Gersonides, d. 1344).[67] Much contact between Jews and Christians in philosophy occurred at the University of Padua, which graduated its first Jewish physician in 1409.[68] In 1501 the Polish legate to Rome reported that he knew "of six Jews of Polish origin who were attending the university under assumed names."[69] Copernicus spent time at Padua (1501–03)[70] studying medicine, and some holders of chairs in natural philosophy in northern Italy came from Poland. During Delmedigo's lifetime, Crete was under Venetian control. The connection between Crete and Venice endured throughout the 1500s. In 1508 Elijah Capsali (d. 1555), the chief rabbi of Candia, came to Padua to study.[71]

Delmedigo's commentaries on Averroes's Latin *Metaphysics* and on his *On the Substance of the Celestial Orb* (which referenced Latin texts) do not seem to contain explicit references to Ibn Naḥmias or even to Biṭrūjī's work. But Delmedigo's Hebrew commentary (which referenced Latin texts) on Averroes's *On the Substance of the Celestial Orb* made a clear connection between the dismissal of eccentrics and epicycles and Renaissance Averroism's interest in the physical world. He also made, in that same commentary on *On the Substance of the Celestial Orb*, a tantalizing reference to attempts to reform Ptolemy:

> All of these [i.e., Ibn Rushd, Ibn Ṭufayl, and Maimonides] said that the roots of natural science contradict what Ptolemy proposes regarding this [i.e., astronomy]. And it [i.e., natural science] is true and, without doubt, there are attached to Ptolemy's astronomy enormous gaps (*harḥaqot ʿaṣumot*) in his proofs since he did not have completely satisfactory evidence for the epicycle and eccentric he proposed, as Ibn Rushd explained in many places. Even the words of the modern astronomers and their like,[72] who thought to save Ptolemy, necessitate that there be a heavenly body without any function so that there will not be proposed any

void with a few of the planets. This body nearly resolves the difference, that is to say that a part of it is very thick and that a part is very thin and that it happens to move in a way agreeing with the rest of the bodies that are with it until no void occurs nor interpenetration of [celestial] bodies as is known to whomever looks at their words. All of this is a worthless fancy.[73]

This quotation begins with Delmedigo's citation of contemporaries of Averroes who questioned the existence of epicycles and eccentrics. Then Delmedigo may have alluded to Ibn Rushd's *Talkhīṣ al-Majisṭī*, or *Qiṣṣur al-Magisṭī* (Epitome of the *Almagest*), a text in which Ibn Rushd both accepted the existence of epicycles and eccentrics and referred to Abū ʿAlī al-Ḥasan ibn al-Haytham's *Shukūk ʿalā Baṭlamyūs* (Doubts against Ptolemy).[74] Subsequently, Delmedigo mentions astronomers close to his own generation and their like (*ha-aḥaronim ve-dimyoneihem*) who also thought to save Ptolemy and notes that they, too, accepted the existence of complementary bodies that prevented the existence of a void between, say, eccentric and parecliptic orbs. At the end of the cited passage, Delmedigo rejects the existence of these bodies, which did not serve as movers and which varied in thickness, suggesting that he favoured an astronomy of perfectly homocentric orbs. The tantalizing question, then, is the identity (or, failing that, the century) of the astronomers Delmedigo believed to be engaged in saving Ptolemy.

The would-be savers of Ptolemy to whom Delmedigo referred would not have been anyone who proposed models composed of perfectly homocentric orbs because complementary bodies with varying thicknesses could not be part of such models. It is perhaps possible that Delmedigo was thinking of Ibn al-Haytham (d. ca. 1040) or Jābir ibn Aflaḥ (twelfth century), critics of Ptolemy cited in Ibn Rushd's *Talkhīṣ al-Majisṭī*. The manuscript of Delmedigo's commentary that I have consulted indicates a composition date of 1485 and a copying date of 1492.[75] Thus the text was composed and copied before Delmedigo returned from Italy to Crete.[76] Given the young age at which Delmedigo left Crete for Italy, it is also possible that the attempts to save Ptolemy to which Delmedigo referred were attempts by European astronomers such as Regiomontanus, not the work of recent Islamic astronomers.[77] Delmedigo's comment is evidence that a prominent Jewish scholar may well have known of developments in fifteenth-century European astronomy, providing more indications that Galeano would have known that there were European astronomers interested in the news he was bringing from the Ottoman Empire, and/or it is evidence that another Jewish scholar in Galeano's milieu knew about important achievements in Islamic astronomy. But even if the referent were earlier critics of Ptolemy, this text would have alerted the reader to the interest of scholars in Europe, which is where Delmedigo wrote, in models based on perfectly homocentric orbs as solutions to the known problems of Ptolemaic astronomy.

That Jews, both from Andalusia and the Ottoman Empire, would be involved in these exchanges regarding theoretical astronomy makes a great deal of sense given what we know about the Jews' cooperation with European astronomers in the composition of astronomical tables. Originating in the Iberian Peninsula, the *Alfonsine Tables*, themselves developed by the Jewish astronomers Judah ben Moses ha-Cohen and Isaac ben Sid at the order of Alfonso X (d. 1284) of Castile, became the basis of many noted astronomical tables in the fifteenth and sixteenth centuries throughout Europe. In Italy, Mordechai Finzi (fl. 1440–75) of Mantua adapted the *Alfonsine Tables* to the geographical coordinates of Mantua.[78] Finzi's extensive contacts with Christian scholars are well known.[79] He was in fact assisted by an astrologer and mathematician of the Gonzagas.[80]

The *Alfonsine Tables* were one of the sources that Copernicus relied upon in his *De Revolutionibus*.[81] Giovanni Bianchini's (d. ca. 1469) tables aimed to simplify the use of the *Alfonsine Tables* and were an unacknowledged source for Copernicus's work on planetary latitudes.[82] Bianchini's tables were also important for scholars writing in Hebrew.[83] Copernicus's use of Bianchini's tables constitutes more evidence of his connection to his scientific contemporaries and of the importance of further exploration of Copernicus's context.[84] Bianchini, in addition, had received a letter from Regiomontanus explaining what Regiomontanus saw to be the most pressing astronomical problems of the day.[85] For example, the *Alfonsine Tables* could not predict certain planetary conjunctions and aspects of lunar eclipses. There was no satisfactory theory for variations in the obliquity of the ecliptic and for the motions of the fixed stars.[86] On the one hand, these were problems that Copernicus did not solve with his heliocentric models, meaning that the motivations for his theories did not stem only or even mainly from a drive for increased predictive accuracy. On the other hand, there is no question that astronomers (including Jewish astronomers) working on tables were engaged with significant problems and that the existence of such problems indicated that the available theories were not satisfactory since those theories could not account for certain observations.[87]

Jews' cooperation with Christians in the production of astronomical tables was also a feature of the cultural life of the Byzantine Empire. Tables entitled *The Six Wings* by Immanuel Bonfils, written in the mid-fourteenth century at Tarascon in France, led to a Greek version in 1435 by Michael Chrysococces,[88] and Mark Eugenicos translated *The Cyclic*

Tables of the Perpignan Jewish astronomer David Bonjorn (fl. 1361) into Greek.[89] By the end of the fourteenth century, a Jew in Salonika, Sharbiṭ ha-Zahab̲ (Solomon ben Elijah, fl. 1374–86), had translated into Hebrew the *Persian Tables* of George Chrysococces. Chrysococces was a Byzantine scholar who wrote an explanation of the tables from Ṭūsī's *Zīj-i Īlkhānī* (The Īlkhānid Astronomical Handbook), brought to Byzantium by Gregory Chioniades.[90] In Constantinople, Rabbi Mordechai Khomṭiano (or Comtino, d. 1482) wrote a *Commentary on the Persian Tables*.[91] Khomṭiano's comments addressed a Christian critic and evinced a deep familiarity with the material.[92] Khomṭiano's proficiency with astronomical instruments brought him to the attention of a *kadiasker* (military judge) to whom Khomṭiano donated an instrument. Khomṭiano's student, Rabbi Elijah Mizrahi (d. 1526), also the teacher of Moses ben Judah Galeano, wrote, in Hebrew, on mathematics and astronomy.[93] Subsequently, *The Six Wings*, the astronomy of Sharbiṭ ha-Zahab̲, *The Paved Way* of Isaac al-Ḥadīb ha-Sep̄ardi, and Khomṭiano's *Defense of the Persian Tables* were all sources for George Gemistos Pletho's *Manual of Astronomy*.[94] An adversary of Pletho, George Scholarios, alleged that Pletho had studied with a Jew![95]

Khomṭiano's comments in his *Commentary on the Persian Tables* indicate that he understood the parameters and positions provided by the tables to be intimately connected to astronomy's physical models.[96] Another text by Khomṭiano related theoretical astronomy and astronomy's physical models to Khomṭiano's religious thought.[97] This connection between Khomṭiano and the *Persian Tables* is fascinating because Chrysococces also reported that Chioniades acquired in Tabriz a number of astronomy texts that he then brought back to the Byzantine Empire.[98] In three manuscripts containing the *Zīj al-Sanjarī* (The Sanjarī Astronomical Handbook) exists a separate Greek text, attributed to Chioniades,[99] with non-Ptolemaic models that incorporated the Ṭūsī-couple.[100] Thus, through the introduction to the *Persian Tables*, readers of Hebrew in the Ottoman Empire were linked to Chioniades's travels to Tabriz, a conduit for the introduction into the Byzantine Empire of hypotheses of Islamic astronomy that appeared in *De Revolutionibus*.[101] Anne Tihon has concluded, "The fifteenth century was a period of intense intellectual exchanges, which left their traces in manuscripts, so that one finds notes in Latin or Hebrew in Greek astronomical manuscripts of the period."[102]

The production of new tables was neither mechanical nor simply a question of computation nor divorced from other areas of Jewish intellectual life. The translation of tables and these numerous contacts between Jewish astronomers and Christian (and Muslim) astronomers in

Renaissance Europe, the Byzantine Empire, and the Ottoman Empire mean that contact between astronomers on matters of theoretical astronomy is more plausible than a presumption of no contact. Our emerging knowledge of the Jews' position as facilitators of the exchange of theoretical astronomy makes sense in the context of their work with Christian astronomers on tables.

This chapter has used the activities of Jewish scholars to help answer two of the most puzzling questions in Copernican studies. The first question is how Copernicus's major overhaul of Ptolemy emerged in a context where there was no history of a sustained critique of Ptolemy, whereas such a critical tradition had existed in the Islamic world for several centuries. This chapter has shown that the production of astronomical models of homocentric orbs, a tradition that tied together Jewish, Muslim, and European astronomers, was a locus for criticisms of Ptolemy. The second question is how Copernicus would have learned of details of Islamic astronomy, such as the Ṭūsī-couple and the models of Ibn al-Shāṭir, that played crucial roles in his own theories. This chapter has proposed a possible historical connection between Renaissance Italy and the Ottoman Empire, via Crete, facilitated by Moses ben Judah Galeano and Elijah Delmedigo. The parallels between *The Light of the World* and European Renaissance astronomy point to homocentric astronomy as a locus for intellectual exchange. The work of Renaissance astronomers, including Copernicus, should be understood as a continuation of astronomy in Jewish and Islamic civilization (and in late-medieval Europe), not as a radical disjuncture with the past.

Notes

INTRODUCTION

1 Peter Barker has recently referred to the lack of a "generally agreed and historically respectable answer to the question of why Nicholas Copernicus adopted heliocentrism" as a "scandal." Barker, "Why Was Copernicus a Copernican?" 203.

2 Arthur Koestler coined this phrase and portrayal in his popular account *The Sleepwalkers*.

3 Recent books on Copernicus have, for the most part, continued this pattern. See Omodeo, *Copernicus in the Cultural Debates*; and Neuber, Rahn, and Zittel, *Making of Copernicus*.

4 For extended reviews, see Westman, "Two Cultures or One?"; and Swerdlow, "Essay."

5 Here, we project backward from Kuhn's later, and more famous, *The Structure of Scientific Revolutions*. Kuhn's *The Copernican Revolution*, no doubt, provided an important "paradigm" for the latter work.

6 Neugebauer, *Exact Sciences in Antiquity*, 205–6.

7 Swerdlow, "Derivation and First Draft." A succinct summary of Swerdlow's reconstruction may be found in Swerdlow and Neugebauer, *Mathematical Astronomy*, part 1, 55–60.

8 Swerdlow, "Pseudodoxia Copernicana"; Rosen, "Reply to N. Swerdlow."

9 A sampling includes Aiton, "Celestial Spheres and Circles"; Jardine, "Significance of the Copernican Orbs"; and Westman, "Astronomer's Role." For a more recent view, see Goddu, "Response to Peter Barker."

10 For Swerdlow's view, see Swerdlow and Neugebauer, *Mathematical Astronomy*, part 1, 41–8.

11 Swerdlow, in part, addressed this criticism in his book with Neugebauer, distinguishing between the "how" and "why" of the Copernican transformation. Ibid., 59. Nevertheless, he has remained firmly convinced that the

reasons for Copernicus's decision must be sought within his analysis of the planetary models. As Swerdlow put it more recently, "My own opinion, for what it is worth, is that neither of these [causes proposed by Kuhn] created conditions pertinent to Copernicus's motivation, which lay entirely within the domain of theoretical and physical astronomy and was entirely technical, namely, a concern with the nonuniform rotation of spheres required by the physical forms of Ptolemy's planetary models." Swerdlow, "Essay," 83.

12 Di Bono, *Le sfere omocentriche*; Di Bono, "Copernicus, Amico, Fracastoro."

13 Goddu, *Copernicus.*

14 Westman, *Copernican Question*, esp. 76–105.

15 This effect has been particularly true of Westman's book. See Swerdlow's review, "Copernicus and Astrology"; and Westman's response, "Copernican Question Revisited." See also Shank's review, "Made to Order"; Westman's response, "Reply to Michael Shank"; and Shank's "Rejoinder." Goddu's *Copernicus* was also subject to a long essay review by Barker and Vesel, "Goddu's Copernicus," to which Goddu's "Response to Peter Barker" was the reply.

16 This is glaringly the case with Westman, *Copernican Question*, which in 681 pages has only one minor footnote dealing with the rich and extensive literature on the possibility of an Islamic influence on Copernicus (531n136). Di Bono, "Copernicus, Amico, Fracastoro," 143–9, and Goddu, *Copernicus*, 476–86, refer to this possibility, only to seek to undermine it. In their translation of Copernicus, *De revolutionibus (Des révolutions)*, in a section entitled "La vaine recherche d'autres précurseurs ou inspirateurs au xxe siècle," Michel-Pierre Lerner and Alain-Philippe Segonds devote several pages to debunking any Islamic influence on Copernicus, often with long-discredited secondary sources (vol. 1, 551–60). It is telling that in their notes on the quite complex Mercury model in *De revolutionibus (Des révolutions)*, they fail to note its striking mathematical equivalence to that of Ibn al-Shāṭir's model (vol. 3, 394–409), something long known and remarked upon. Kennedy and Roberts, "Planetary Theory," 232–3. One should say that Jews and Byzantines fare little better in any of these works.

17 Swerdlow, "Derivation and First Draft," 435.

18 Barker, "Copernicus and the Critics," 352, argues that Copernicus was likely exposed to Averroist critiques of Ptolemy, but he concludes, "At Bologna he [Copernicus] may well have learned about proposals by Regiomontanus to establish a homocentric astronomy that would be equally successful in calculating planetary positions. However, Copernicus's own later work strongly suggests that he found any such proposals unsuccessful or unpersuasive."

19 Swerdlow and Neugebauer, *Mathematical Astronomy*, part 1, 47.

20 Swerdlow, "Essay," 86, clearly makes this distinction: "However, one caution is in order in evaluating the preface [of *De revolutionibus*] as an explanation

of Copernicus's motivation, namely, that it was written in 1542, some thirty years after he had developed his new theory and explained it in the *Commentariolus* (ca. 1510–14). There he makes no reference to the incoherence of the separate models, the absence of a common measure or unification of the planetary system, and concentrates entirely on the violation of uniform circular motion ... Hence, although this is an argument from silence in the *Commentariolus*, it appears that his later indictment of earlier planetary theory as incoherent did not occur to him until *after* he had developed the heliocentric theory and saw that it did indeed give an unambiguous account of the order and distances of the planets."

21 Goldstein, "Copernicus and the Origin." Westman, "Copernican Question Revisited," 106–8, clearly depends heavily on Goldstein to justify his own interpretations. Goddu, "Reflections on the Origins," also depending on Goldstein, provides a speculative reconstruction of how Copernicus arrives at heliocentrism, but there is little in the historical record or the *Commentariolus* to support the view that Copernicus arrived at his heliocentric hypothesis before first engaging with the problem of Ptolemy's violations of uniform circular motion, which is, after all, what Copernicus highlights in his introduction to that work. Goddu also underestimates the complexity of Copernicus's models, reducing them to "the so-called Tusi couple" (45).

22 On the *hayʾa* tradition, see F.J. Ragep, *Naṣīr al-Dīn al-Ṭūsī's Memoir*, vol. 1, 33–41; and S.P. Ragep, chapter 6, this volume. On the connection of *hayʾa* with Copernicus, see Swerdlow and Neugebauer, *Mathematical Astronomy*, part 1, 41–8.

23 This finding is based on new research on Copernicus's Mercury models undertaken by F. Jamil Ragep and Sajjad Nikfahm-Khubravan, which indicates that Copernicus adapted Ibn al-Shāṭir's models in one way in the *Commentariolus*, resulting in a defective model for Mercury because of Copernicus's desire to have the Sun at the centre of Mercury's orbit, but adapted them in a different way in *De revolutionibus*, where he accepts the need for eccentrics and multiple centres. For details, see F.J. Ragep, "Ibn al-Shāṭir and Copernicus"; and Nikfahm-Khubravan and F.J. Ragep, "Ibn al-Shāṭir and Copernicus."

24 For more on these contexts, see Chen-Morris and Feldhay, chapter 5, this volume. See also F.J. Ragep, "Ṭūsī and Copernicus."

25 Swerdlow, "Essay," 85.

26 Koestler, *Sleepwalkers*.

CHAPTER ONE

1 Celenza, *Lost Italian Renaissance*, xiii, 147–9.

2 On the latter point, see McLaughlin, *Literary Imitation*.

3 On fifteenth-century philosophy as missing from canonical accounts, see Celenza, "What Counted as Philosophy."
4 See Celenza, *Lost Italian Renaissance*, 147–8, for literature.
5 See Celenza, "End Game," "Petrarch, Latin," and "What Counted as Philosophy." All three articles assume the historical importance and centrality of the five-generation period coterminous with what I have called the "long fifteenth century."
6 Most recently, see Hunt, *Inventing Human Rights*. For earlier antecedents, see Skinner, *Foundations*; Skinner, *Liberty before Liberalism*; and Pocock, *Machiavellian Moment*.
7 For background, see Mulsow and Stamm, *Konstellationsforschung*.
8 See Connell, "Republican Idea."
9 On the experience of traders, see Trivellato, *Familiarity of Strangers*. On a period later than Copernicus but implicated in the same set of developments, see Jacob, *Strangers Nowhere in the World*.
10 See, for example, Shapin, *Social History of Truth*; and Daston and Park, *Wonders and the Order*, 215–53.
11 Lines, "Natural Philosophy," 134.
12 Ibid. Lines refers to Possevino, *Bibliotheca selecta*, book 13, ch. 25, para. 2, 109–10.
13 Kepler, *Gesammelte Werke*, vol. 7, 253, cited in Westman, "Two Cultures or One?" 111: "They [universities] are established in order to regulate the studies of the pupils and are concerned not to have the rules of teaching change very often: in such places, because it is a question of the progress of the students, it frequently happens that the things which have to be chosen are not those which are the most true but those which are most easy."
14 In addition to Celenza, "End Game," "Petrarch, Latin," and "What Counted as Philosophy," see also Celenza, "Humanism and the Classical Tradition."
15 See Lines, "Humanism and the Italian Universities."
16 See Daston, "Preternatural Philosophy."
17 See ibid.; and Daston and Park, *Wonders and the Order*.
18 Daston, "Preternatural Philosophy," 20; Pomponazzi, *De naturalium effectuum*. On the diffusion of this text, see Zanier, *Ricerche sulla diffusione*. See also Pine, *Pietro Pomponazzi*; Nardi, *Saggi sul pensiero inedito*; and Nardi, *Studi su Pietro Pomponazzi*.
19 This lack of resolution is noted by Daston, "Preternatural Philosophy," 24, who points out that demons, representing in effect agents of chance, were not included in most treatises of natural philosophy during this period, even if they remained staples of witchcraft writings. See also Clark, *Thinking with Demons*; and Stephens, *Demon Lovers*.
20 Guicciardini, *Ricordi*, no. 211, 132: "Io credo potere affermare che gli spiriti siano; dico quella cosa che noi chiamiamo spiriti, cioè di quelli aerei che

dimesticamente parlano con le persone, perché n'ho visto esperienzia tale che mi pare esserne certissimo; ma quello che siano e quali, credo lo sappia sì poco chi si persuade saperlo, quanto chi non vi ha punto di pensiero. Questo, ed el predire el futuro, come si vede fare talvolta a qualcuno o per arte o per furore, sono potenzie occulte della natura, overo di quella virtù superiore che muove tutto; palesi a lui, segreti a noi, e talmente, che e' cervelli degli uomini non vi aggiungono." This passage is partially cited in Maggi, *In the Company of Demons*, viii, 169n5.

21 He goes on in that maxim: "Astrologers don't know what they're talking about and they are never right except by chance. This is the case to such an extent that if you take a prediction from any astrologer and another from a man chosen at random, the latter will have the same likelihood of being true as the former." Guicciardini, *Ricordi*, no. 207: "Della astrologia, cioè di quella che giudica le cose future, é pazzia parlare; o la scienza non é vera, o tutte le cose necessarie a quella non si possono sapere, o la capacità degli uomini non vi arriva; ma la conclusione é, che pensare di sapere el futuro per quella via é uno sogno. Non sanno gli astrologi quello dicono, non si appongono se non a caso; in modo che se tu pigli uno pronostico di qualunque astrologo, e di uno di un altro uomo fatto a ventura, non si verificherà manco di questo che di quello."

22 Ficino, *De vita libri tres*, 306–7, cited in Celenza, "Late Antiquity," 95.

23 Westman, "Two Cultures or One?" 89. For an extended discussion, see Westman, *Copernican Question.*

24 In them, by the way, Novara names a number of Arabic astronomers, although they are by and large earlier than Ṭūsī. Bònoli, Colavita, and Mataix, "L'ambiente culturale bolognese," 874.

25 Blair, *Too Much to Know*, "Reading Strategies," "Note-Taking as an Art."

26 Johns, *Nature of the Book*; McKitterick, *Print, Manuscript.*

27 This increase in reading material is a key point in the argument of Eisenstein, *Printing Press.* See also Grafton, "Importance of Being Printed."

28 See Eisenstein, Grafton, and Johns, "Forum." On the issue of unbound quires, see Rizzo, *Il lessico filologico*, 42–7.

29 Habermas, *Strukturwandel der Öffentichkeit.*

30 Garin, ed., *Prosatori latini del Quattrocento*, 806: "Legimus saepe ego et noster Politianus quascumque habemus tuas aut ad alios, aut ad nos epistolas; ita semper prioribus certant sequentia et novae fertiliter inter legendum efflorescunt veneres, ut perpetua quadam acclamatione interspirandi locum non habeamus."

31 For a critical edition of this text, see Poliziano, *Lamia.* On this edition, see Celenza, "Petrarch, Latin." For an English translation and introductory studies, see Celenza, *Angelo Poliziano's Lamia*, where this anecdote is reported at pages 14–16.

32 See Lombardus, *Sententiae*; Colish, *Peter Lombard*; Stegmüller, *Repertorium commentariorum*; and Doucet, *Supplément au Répertoire*.

33 Bonaventure, *Opera Omnia*, vol. 1, 14–15 (from his *Commentaria in Sententias*): "Aliquis enim scribit aliena, nihil addendo vel mutando, et iste mere dicitur *scriptor*. Aliquis scribit aliena, addendo sed non de suo, et iste *compilator* dicitur. Aliquis scribit et aliena et sua, sed aliena tamquam principalia, et sua tamquam annexa ad evidentiam, et iste dicitur *commentator*, non auctor. Aliquis scribit et sua et aliena, sed sua tamquam principalia, aliena tamquam annexa ad confirmationem, et talis debet dici *auctor*."

34 Chenu, "Auctor, actor, autor."

35 There is an excellent discussion of medieval conceptions of authorship in Coxon, "Introduction."

36 See Hankins, *Plato in the Italian Renaissance*, vol. 1, 18–26.

37 See Stock, "Minds, Bodies, Readers," and the papers accompanying it in the journal *New Literary History* 37, no. 3 (2006).

38 Copernicus, *De revolutionibus*, preface: "Quare hanc mihi operam sumpsi ut omnium philosophorum quos habere possem libros relegerem, indagaturus an ne ullus unquam opinatus esset alios esse motus sphaerarum mundi quam illi ponerent qui in scholis mathemata profiterentur."

39 On memory and associated reading and graphic practices, see Carruthers, *Craft of Thought*; Bolzoni, *La stanza della memoria*; Bolzoni, *Gallery of Memory*; and Yates, *Art of Memory*.

40 Petrucci, *Writers and Readers*, 204.

41 See Quintilian, *Institutio oratoria*, book 1, ch. 2, sec. 15, in vol. 1, 88–9, and book 2, ch. 5, sec. 4, in vol. 1, 302–3.

42 John of Salisbury, *Metalogicon*, ch. 24: "Qui ergo ad philosophiam aspirat, apprehendat lectionem, doctrinam, et meditationem cum exercitio boni operis, nequando irascatur Dominus, et quod videbatur habere, auferatur ab eo. Sed quia legendi verbum aequivocum est, tam ad docentis et discentis exercitium quam ad occupationem per se scrutantis scripturas – alterum, id est quod inter doctorem et discipulum communicatur (ut verbo utamur Quintiliani) dicatur praelectio, alterum quod ad scrutinium meditantis accedit – lectio simpliciter appelletur" (punctuation altered).

43 Westman, "Two Cultures or One?" 106n87. The intermediary source was Thomas Streete's *Astronomia Carolina* (1661), which Newton read in 1664.

44 See Celenza, *Lost Italian Renaissance*, xiii, 142, 147–9; and Celenza, "Petrarch, Latin," 528.

45 Petrarch, *Epistolae seniles*, book 9, epistle 1, edited in Casamassima, *L'autografo Riccardiano*, 116. See also Rizzo, *Ricerche sul latino umanistico*, 37.

46 See Celenza, "Petrarch, Latin," 513–15; Hadot, *Philosophy*; and Hadot, *What Is Ancient Philosophy?*

47 See Celenza, "End Game." The most thorough treatments of this problem can be found in Tavoni, *Latino, grammatica, volgare*; Coseriu and Meisterfeld,

Geschichte der romanischen Sprachwissenschaft, vol. 1, 149–237; and Ramminger, "Humanists and the Vernacular." The entire issue of the journal *Renaessanceforum* 6 (2010), in which Ramminger's article appears, is quite valuable in understanding key problems and questions as neo-Latin moved north. See http://www.renaessanceforum.dk.

48 Much of what follows is drawn from Celenza, "End Game."

49 IJsewijn, *Companion to Neo-Latin Studies*, vol. 1, 246.

CHAPTER TWO

1 See Shank, chapter 4, this volume.

2 Swerdlow and Neugebauer, *Mathematical Astronomy*; F.J. Ragep, "Copernicus."

3 F.J. Ragep, "Copernicus," 67–8.

4 See Morrison, chapter 8, this volume.

5 F.J. Ragep, "ʿAlī Qushjī and Regiomontanus," 360; F.J. Ragep, "Copernicus," 76; Swerdlow and Neugebauer, *Mathematical Astronomy*, part 1, 47–54.

6 These are not uncomplicated examples, as both Pico and Postel were condemned by the Catholic Church for their religious ideas. In addition, despite Postel's admiration for aspects of Ottoman society, he still advocated a strong and swift military confrontation against the Ottomans. See Housley, *Later Crusades*, 383; and Bisaha, *Creating East and West*, 171–2, 182. On the rise of Arabic studies in Europe, see Dannenfeldt, "Renaissance Humanists"; and Burnett, "Second Revelation."

7 One notable exception is Francesco Petrarch (1304–74), who roundly dismissed all Arab learning, possibly due to his own anti-scholastic bent. See Bisaha, "Petrarch's Vision."

8 See Mavroudi, "Exchanges with Arab Writers"; and Meserve, *Empires of Islam*, ch. 5.

9 A growing body of scholarship examines the ways that premodern authorship differed significantly from modern notions. See Ranković, ed., *Modes of Authorship*; Bjørnstad, ed., *Borrowed Feathers*; and Bolens and Erne, eds, *Medieval and Early Modern Authorship*. On the practice of compiling extracts as a form of authorship, see Sønnesyn, "Obedient Creativity," 113–14. On the sixteenth-century scientific compilations, see Carlino, "*Kunstbüchlein* and *Imagines Contrafactae*"; and Grafton, *Footnote*, which reminds us how recent the practice of careful attribution is to scholarship.

10 Recent essays by Anna Contadini and Palmira Brummett speak to the complexity of perceptions and exchange with the Ottomans at this time. Contadini examines material objects that originated in the Muslim East while acknowledging that we cannot know how European consumers of these goods actually viewed the culture of origin. Contadini, "Sharing a Taste?" Brummett speaks to a variety of exchanges but also notes that "fear and conversation … were often tightly intertwined." Brummett, "Lepanto Paradigm Revisited," 71.

11 Nabil Matar notes a similar practice by Portuguese sailors and pirates who regularly raided North Africa and carried off inhabitants in the sixteenth and seventeenth centuries. Matar, *Europe through Arab Eyes*, 13–18. For an overview of the Ottoman advance and its impact on Europe, see Housley, *Later Crusades*.

12 See Brummett, *Ottoman Seapower*; and Dursteler, *Venetians in Constantinople*.

13 See Schwoebel, *Shadow of the Crescent*; and Cardini, *Europa e Islam*.

14 This is certainly the case in the fourteenth century and especially in the fifteenth century following 1453. Bisaha, *Creating East and West*; Hankins, "Renaissance Crusaders." There is, however, some softening in attitude and a greater desire to gather reliable information about the Ottomans, mixed with vigorous calls for crusade, in sixteenth-century accounts. Meserve, *Empires of Islam*.

15 Much earlier, Ibn Jubayr (1145–1217) spoke of the grave dangers of travelling in Christian lands north of Sicily. In the fifteenth century, Bertrand de la Broquière related that he was refused lodging by the innkeepers in Vienna when he arrived dressed as a Turk until he managed to convince them he was Christian. Schwoebel, *Shadow of the Crescent*, 212. Matar also discusses the difficulties and suspicions that prevented all but a few Muslims from travelling in western Europe during the Renaissance. Matar, *Europe through Arab Eyes*. The French alliance with the Ottomans under Francis I and Suleyman the Magnificent is an important exception, but it is worth noting that the French felt it necessary to clear the port of Toulon of all its inhabitants so that the Ottoman fleet could winter there in 1543–44. Isom-Verhaaren, *Allies with the Infidel*, 134–5.

16 A good introduction to Aeneas's life and works is Mitchell, *Laurels and the Tiara*. Several collected essays on Aeneas have been published. Among them, see Von Martels and Vanderjagt, eds, *Pius II*; and Tarugi, ed., *Pio II*. See also Maffei, ed., *Enea Silvio Piccolomini*, an older but excellent collection of essays.

17 Nederman, "Aeneas Sylvius Piccolomini."

18 Izbicki and Nederman, eds and trans, *Three Tracts on Empire*, 100–5.

19 See, for instance, his characterization of the "common people" who overthrew the elite leadership of the Ambrosian Republic of Milan in Aeneas, *Europe*, 226–7.

20 Aeneas's gradual change of attitude and mounting frustration can be seen in the thoughtfully selected letters found in Thomas M. Izbicki, Gerald Christianson, and Philip Krey's edited and translated text of Aeneas, *Reject Aeneas*.

21 Aeneas had already called for a crusade in an oration delivered at Basel. His letters written during his service to Frederick also closely follow events like the Crusade at Varna and various military engagements and negotiations between Christian powers to ally against the Ottomans.

22 I discuss some of Aeneas's letters in 1453 and 1454 in Bisaha, "Discourses of Power and Desire."

23 Aeneas, *Der Briefwechsel*, vol. 68 (1918), 200. See also Aeneas, *Reject Aeneas*, for a translation of this letter. For a fuller discussion of reports on the loss of books, see Bisaha, *Creating East and West*, 65ff.

24 Aeneas, *Der Briefwechsel*, vol. 68 (1918), 209.

25 Ibid.

26 See Weinig, *Aeneam suscipite, Pium recipite.* The Frankfurt oration was printed in Aeneas's *Opera Omnia.* See Aeneas, *Commentaries* (2003), vol. 1, 396n105.

27 Hay, *Europe*, 96. Denys Hay does note Aeneas's great importance in articulating an influential concept of Europe and ascribes an important innovation to him: his role in bringing the adjective "European" into widespread use and significantly expanding the rhetorical uses of the term. Ibid., 86.

28 On this subject, see Hay, *Europe*; Jordan, "'Europe' in the Middle Ages"; and Akbari, "From Due East."

29 Aeneas, *Europe*, 51. See my introduction to this work for a fuller discussion of Aeneas's influences, goals, and role within the larger fields of Renaissance geography and history.

30 See Casella, "Pio II"; and Baldi, "Enea Silvio Piccolomini."

31 Felix V, formerly Duke Amedeo of Savoy, could not attract enough powerful supporters to his side and ended up conceding in 1449.

32 Aeneas, *Europe*, 307.

33 Aeneas does acknowledge, however, that Vladislas IV of Hungary broke his ten-year truce with Murad II in 1444, precipitating the massive defeat at Varna. Aeneas, *Europe*, 83.

34 See Casella, "Pio II." Nicola Casella also provides a list of the many early printings of *Europe* and *Asia*, giving a sense of the works' popularity and dissemination. *De Asia* may be found in Aeneas, *Opera quae extant omnia.*

35 Vollmann, "Aeneas Silvius Piccolomini," 54.

36 Meserve, *Empires of Islam*, 225. This may complicate the notion that Aeneas began to generalize all Muslims as enemies and barbarians, as I argue in Bisaha, *Creating East and West*, 78. However, in the *Commentaries*, Aeneas ends the story of Uzun Ḥasan with serious doubts about the legitimacy of the delegates who claimed to represent him, stating "[f]rom that time the Pope was suspicious of any communications from the East, especially when they were brought by men who were poor and unknown." Aeneas, *Commentaries* (1937–57), vol. 3, 374.

37 Aeneas, *Epistola*, 38.

38 Ibid., 101. For other examples of Aeneas's use of the term "Europe" as pope, including his often copied and printed oration at the Congress of Mantua (1459), which he summoned to plan a crusade, see Hay, *Europe*, 85.

39 "Letter to Leonardo Benvoglienti" (1453), in Pertusi, *La caduta di Costantinopoli*, vol. 2, 262. Greek writers Doukas and Michael Kritoboulos

detailed Mehmed's efforts to spare portions of the city and its populace in 1453 and to rebuild it after the conquest. However, such accounts were not circulated in the Latin West in the fifteenth century.

40 Bisaha, "Pope Pius II's Letter."

41 Aeneas announced his intention to lead the crusade and "wipe out the Turkish race" in early 1462 at the same time that he was revising the letter. Ibid., 196.

42 Hay, *Europe*, 84.

43 Aeneas, *Epistola*, 13.

44 Ibid., 74.

45 Ibid., 91

46 Ibid., 91–2. Earlier, Aeneas also accuses the Prophet Muhammad of prohibiting dispute of Islam given its nonsensical, irrational base. Ibid., 88. This criticism seems a bit one-sided given the limits of debate on Christian orthodoxy in Europe, as recently seen in the burning of Jan Hus in 1414.

47 Aeneas, *Opera quae extant omnia*, 313; Meserve, *Empires of Islam*, 225.

48 In 1473 Filelfo mourned the loss of Uzun Hasan to Ottoman forces, stating that people of Persian stock, "which was in no way wild or uncivilized," were preferable to the Turks. Meserve, *Empires of Islam*, 228.

49 The terms "Christendom" and "Europe" were interchangeable until the eighteenth century, when the latter term began to dominate. See Hay, *Europe*.

50 Said, *Orientalism*.

51 See, for example, Wells, *Barbarians Speak*; von Staden, "Liminal Perils"; and Gruen, *Culture and National Identity*.

52 Sabra, "Appropriation and Subsequent Naturalization," 225.

53 See Jardine and Brotton, *Global Interests*, which I have questioned elsewhere.

54 On Bessarion's trip to Vienna and relationship with Peurbach and Regiomontanus, see Shank, chapter 4, this volume. On the theory of Bessarion as a transmitter of central Asian astronomical models, see F.J. Ragep, "ʿAlī Qushjī and Regiomontanus," 360.

55 Shank, "Classical Scientific Tradition." For more on Bessarion's perceptions of the Ottoman Empire and his role as a crusade advocate, see Bisaha, *Creating East and West*. See also Monfasani, *Byzantine Scholars*.

56 Vast, *Le cardinal Bessarion*, 454–5.

57 Bisaha, "Barbarians or Intellectual Peers?" It has been posited that Bessarion's teacher in Mistra, George Gemistos Pletho, was directly acquainted with Ottoman learning. See Kafadar, *Between Two Worlds*, 90.

58 One is reminded of Aeneas's questionable boast that Latin scholars had surpassed the Byzantines based on one debate that took place at the Council of Ferrara-Florence. See Aeneas, *Europe*, 235–6.

59 Both these individuals, incidentally, wrote against the Ottoman "barbarians." On Tignosi, see Sensi, "Niccolò Tignosi da Foligno." On Ficino, see Bisaha, *Creating East and West*, 74–5, 171.

60 Brann, "Humanism in Germany."
61 Andrews and Kalpaklı, *Age of the Beloveds*, 10.
62 Feingold, "Decline and Fall."

CHAPTER THREE

1 The *Commentariolus* was not published in Copernicus's lifetime. The three extant copies are all descended from a copy in the possession of Tycho Brahe. For my purposes, the English translation by Edward Rosen in Copernicus, *Three Copernican Treatises*, is not reliable because he does not translate "orb" in a consistent way. A better English translation may be found in Swerdlow, "Derivation and First Draft." I sometimes emend Swerdlow's translations by using the Latin texts in Copernicus, *Erster Entwurf seines Weltsystems*, or Copernicus, *Das neue Weltbild*.
2 Swerdlow, "Derivation and First Draft," 441; Copernicus, *Das neue Weltbild*, 8.
3 Peurbach, *Theoricae novae planetarum*; Aiton, "Peurbach's *Theoricae novae planetarum*."
4 Swerdlow, "Derivation and First Draft," 424.
5 "Finit commentariolum super theoricas novas Georgii Purbatii," in Brudzewo, *Commentaria utilissima*; and in Brudzewo, *Commentariolum super Theoricas*.
6 Brudzewo, *Commentaria utilissima*, fol. a5r; Brudzewo, *Commentariolum super Theoricas*, 16.
7 Brudzewo, *Commentaria utilissima*, fol. a5r; Brudzewo, *Commentariolum super Theoricas*, 16.
8 Brudzewo, *Commentaria utilissima*, fol. a5r; Brudzewo, *Commentariolum super Theoricas*, 16–17.
9 A work that Brudzewo frequently cites, the *Almagestum parvum*, may be a compilation from works in Arabic. Lorch, "Some Remarks"; Zepeda, "Medieval Latin Transmission."
10 Brudzewo, *Commentariolum super Theoricas*, 26–7: "sicut testatur Albeon, in prima parte sui [Instrumenti] capitulo decimo, dicens, 'Non quod in coelestibus sunt huiusmodi ecentrici aut epicycli sicut imaginatio mathematica sibi fingit, quod nullus disciplinatus potuit verisimiliter putare, sed quia sine huiusmodi imaginationibus mathematicis, de stellarum motibus regularis ars tradi non potest, quae sic earum loca ad quodvis momentum certificet, quod a nostris aspectibus non discordent.' Haec ille [Richard of Wallingford]. Debemus ergo de hoc modo fore contenti, cum – ipso mediante – artem perfectam astrorum in motibus complectamur."
11 Swerdlow, "Derivation and First Draft," remarks in more than one place that Copernicus in the *Commentariolus* does not seem to understand the mathematical models that he was using, such as on page 469: "One may seriously wonder whether he understood the fundamental properties of his model for the first anomaly." See also Swerdlow and Neugebauer,

Mathematical Astronomy, part 1, 410: "The models in the *Commentariolus* were not intended for practical application – at least not with the crude and incomplete parameters supplied in the text – and at the time of its composition Copernicus was evidently not secure in constructing a model for Mercury."

12 Translation modified from Swerdlow, "Derivation and First Draft," 436.

13 Copernicus, *Letter against Werner*, 99–100, as translated by Rosen. I have made a change in the translation where the Latin is inserted.

14 Aiton, "Peurbach's *Theoricae novae planetarum*," 9–10.

15 John of Sicily, *Scriptum super canones Azarchelis*, para. J285, 134: "Oportet autem ad hoc ut apparentia salventur et contra naturalem philosophiam inconvenientia non sequantur, in unoquoque planetarum ad minus 3 orbes sphaericos assignare, quorum unus in concavitate alterius statuantur: ita quod inferior sit concentricus terrae quantum ad superficiem concavam, excentricus autem quantum ad convexam; secundus autem sit excentricus terrae quantum ad utrumque superficiem, concentricus tamen superficiei convexae orbis inferioris, ita quod omnino sit super idem centrum, supra quod est orbis inferioris convexa superficies constituta; tertius autem superior orbis sic excentricus terrae ex parte suae concavitatis, sit tamen ex eadem parte concentricus secundo; ex parte vero suae convexitatis sit omnino concentricus terrae ... Quilibet etiam istorum orbium proprium habet motorem ... Duo tamen extremi, licet diversis motibus et inaequalibus revolvantur, habent nihilominus proportionales motus, ita quod determinata pars inferioris semper est sub determinata parte superioris, hoc est latior pars inferioris sub strictiore superioris et strictior sub latiore. Et hii duo dicuntur orbes revolventes seu deferentes augem, eo quod aux mediani orbis excentrici movetur ad motus istorum duorum."

16 The features of the *theorica orbium* will become clearer when Ibn al-Haytham's *On the Configuration of the World* is described.

17 Grant, *Planets, Stars, and Orbs*. More broadly, Grant did not regularly make use of works with the title *Theorica planetarum* in his research. He credited the idea of three-part orbs to Ptolemy's *Planetary Hypotheses* rather than to Ibn al-Haytham's *On the Configuration of the World*.

18 F.J. Ragep, *Naṣīr al-Dīn al-Ṭūsī's Memoir*, vol. 1, 100.

19 Grant, *Planets, Stars, and Orbs*, 38.

20 This characterization of Roger Bacon's final view is questionable, as Bacon argues first one side and then the other. Bacon, *De celestibus*, 437–9.

21 Grant, *Planets, Stars, and Orbs*, 278.

22 A Latin passage from John of Sicily's work is included above in note 15.

23 Grant, *Planets, Stars, and Orbs*, 279. In his note 30, Grant writes, "De quadam ymaginatione modernorum. See Bacon, *Opus tertium*, 1909, 125 ... Bacon, *De celestibus*, fasc. 4, 1913, 438 ... Although the 'modern' theory seems

ultimately derived from Ptolemy's *Hypotheses of the Planets*, Bacon's immediately preceding discussion ... may have derived from an earlier attempt to materialize eccentrics on the basis of Ptolemy's description in the *Almagest*." Here, Grant suspects what is true, with the earlier attempt being that of Ibn al-Haytham.

24 It is almost funny to see how Grant seems determined to omit Ibn al-Haytham from his genealogy of the three-orb system.

25 Grant, *Planets, Stars, and Orbs*, 281.

26 See Barker, "Albert of Brudzewo's *Little Commentary*"; and Malpangotto, "Original Motivation."

27 Brudzewo, *Commentaria utilissima*, fol. a3r; Brudzewo, *Commentariolum super Theoricas*, 7–8. In an appendix to one Arabic manuscript of Ibn al-Haytham's *On the Configuration of the World*, there is a list of principles: "The premises on which the construction of the orbs of the stars and all the bodies which move around the world are based are four. The first of them is that the natural body does not by its nature move with more than one natural motion. The second is that the simple natural body does not move with a varying motion, that is, it always covers equal distances in equal times during its revolution. The third is that the body of the heavens does not admit of being acted upon. The fourth is that the vacuum does not exist." Langermann, *Ibn al-Haytham's "On the Configuration,"* 231.

28 Albert of Saxony, *Quaestiones subtilissime*, fol. E2vb. Also in Albert of Saxony, *Questiones et decisiones*, book 2, Q. 6.

29 Brudzewo provides this same reference to Aristotle's *Metaphysics*.

30 Albert of Saxony, *Quaestiones subtilissime*, fol. E3rb-va; Albert of Saxony, *Questiones et decisiones*, fol. 105vb.

31 D'Ailly, *Questions on the Sphere*, Q. 2, "Utrum sint precise novem sphere celestes et non plures nec pauciores," fol. Q2ra.

32 Swerdlow, "Derivation and First Draft," 433–6. Here, I have started with, but modified, Swerdlow's translation, for instance changing the translation of "modus" to "way," rather than "model," because "model" carries too much philosophical baggage.

33 Langermann, *Ibn al-Haytham's "On the Configuration."*

34 Taub, *Ptolemy's Universe*, 113–18.

35 F.J. Ragep, "Hay'a," 1061.

36 Ibid., 1062.

37 Morrison, "Quṭb al-Dīn al-Shīrāzī's Hypotheses," 21. See also Hugonnard-Roche, "Problèmes Méthodologiques."

38 Langermann, *Ibn al-Haytham's "On the Configuration,"* 40.

39 Ibid., 34–6.

40 See, for example, Duhem, *To Save the Phenomena*, ch. 6; and Grant, "Late Medieval Thought."

41 See Westman, "Melanchthon Circle"; and Westman, *Copernican Question*, ch. 5, which incorporates most of the earlier paper.
42 Jardine, *Birth of History*.
43 Hoenen, "*Via Antiqua*." See also Sylla, "John Buridan."
44 See Hugonnard-Roche, "Problèmes Méthodologiques," 62: "Il ne s'agit plus de déduire *a priori* des axiomes de la physique les hypothèses géometriques de l'astronomie, mais, au rebours, de trouver *a posteriori* un 'modèle physique,' c'est-à-dire une interprétation physique, des schémas mathématiques ptoléméenes, qui soit compatible avec les principes reçus de la doctrine aristotélenne … Les modèles bien connus permettant cette réconciliation sont ceux des sphères célestes emboîtées que Ptolémée avait décrites dans ses *Hypothèses des planètes,* et que les médiévaux ont redécouvertes par l'intermédiaire des traductions latines de la *Configuration du monde* de Ibn al-Haytham (d. ca 1041) … cette 'modélisation' de l'astronomie ptoléméenne est la méthode couramment utilisée, dès la fin du XIIIe siècle, par tous les savants."
45 See Sylla, "Status of Astronomy"; and Sylla, "John Buridan."
46 Many of the views I put forth in this section of the chapter were formulated before the appearance of André Goddu's *Copernicus and the Aristotelian Tradition*. Nevertheless, Goddu makes a strong case for the relevance of the Aristotelian context. See Barker and Vesel, "Goddu's Copernicus"; and Goddu, "Response to Peter Barker."
47 De Groot, "Modes of Explanation"; De Groot, *Aristotle's Empiricism*.
48 Murdoch, "Thomas Bradwardine."
49 See Glasner, *Averroes' Physics*.
50 Sylla, "Averroes and Fourteenth-Century Theories."
51 Sylla, "John Buridan," 222. According to Laird, "*Scientiae Mediae*," 140, 194, Aristotle, Robert Grosseteste, Thomas Aquinas, Walter Burley, John Buridan, Albert of Saxony, and others do not subalternate perspective (or optics) to natural philosophy, whereas Aegidius Romanus and Paul of Venice do.
52 Grosseteste, *Commentarius*, 99–100; Burley, *Quaestiones super librum posteriorum*, 88–9 (Latin text). See Sylla, "Status of Astronomy," 273n29. It should be noted that in defining science, Grosseteste has in mind not just any claimed science but true science.
53 In the quotation from Copernicus's *Letter against Werner*, cited at note 13, Copernicus reflects this same idea.
54 For supporting documentation, see Sylla, "John Buridan."
55 On Avicenna and Averroes, as reported by Galileo Galilei, see Galileo, *Tractatio de praecognitionibus*, 98–101.
56 Sylla, "Status of Astronomy."
57 Sylla, "John Buridan"; Maier, *Metaphysische Hintergründe*, 387.
58 Grellard, *Croire et Savoir*; Biard, *Science et nature*.

59 Buridan, *In Metaphysicen Aristotelis*, book 2, Q. 2, fol. BB3v; Maier, *Metaphysische Hintergründe*, 388.

60 Aristotle, *Prior Analytics*, 46a17–24. The Latin translation uses the word "declarare."

61 This is the thesis of Sylla, "John Buridan," 216–19.

62 Livesey, "*Metabasis*."

63 Sylla, "Political, Moral."

64 Aristotle, *Meteorologica*, 344a5–7: "We consider that we have given a sufficiently rational explanation of things inaccessible to observation by our senses if we have produced a theory that is possible." Quoted in Martin, "Conjecture, Probabilism," 271.

65 Sylla, "A Posteriori Foundations."

66 Bacon, *De celestibus*, 443.

67 Ibid., 444. I question whether this text is reliable.

68 Ibid., 441.

69 Ibid., 442.

70 Ibid., 445.

71 See Sylla, "John Buridan," 219–20.

72 Sabra, *Optics of Ibn al-Haytham*, book 1, ch. 1, sec. [5], 5, cited in Sylla, "John Buridan," 220.

73 Ibid., book 1, ch. 1, sec. [6], 5, cited in Sylla, "John Buridan," 220.

74 See Sylla, "John Buridan," 219–21.

75 See Hugonnard-Roche, "Problèmes Méthodologiques."

76 Sylla, "Averroes and Fourteenth-Century Theories."

77 See Sylla, "Oxford Calculators' Middle Degree."

78 Glasner, *Averroes' Physics*.

79 Copernicus, *Letter against Werner*, 98–9 (in accord with the *Posterior Analytics* and the *moderni*), describes how astronomers work *from* measured positions to their theories.

80 I discuss the work of Albert of Brudzewo and John of Głogów in more detail in Sylla, "Astronomy at Cracow University."

81 Ibid.

82 Głogów, *Quaestiones super libros Analyticorum posteriorum*, fol. 16v.

83 Ibid., fols 17v–18r.

84 According to Albertus Magnus, once the intellect acquires knowledge by the senses, it is shaped (becoming the *intellectus adeptus* or "acquired intellect") and has no further need for the senses. Hasse, "Soul's Faculties," 318.

85 Ibid., 319.

86 On this issue of the fallibility of scientific knowledge, see Pasnau, "Science and Certainty."

87 Głogów, *Quaestiones super libros Analyticorum posteriorum*, fol. 71v. This point about proper principles had been made in question 16 of Versor's commentary.

88 Ibid., fol. 72r.
89 Ibid., fols 72v–73r.
90 Ibid., fol. 73v.
91 Ibid., fol. 74r.
92 Ibid., fols 83r, 85r–v.
93 Ibid., fol. 86r.
94 Ibid., fol. 86v.
95 Głogów, *Introductorium compendiosum*, fol. B2v.
96 Ibid., fol. D1v.
97 Ibid., fol. K3r: "Quilibet autem planeta tres habet circulos preter solem, scilicet equantem, deferentem, et epiciclum. Equans quidem lune est circulus concentricus cum terra et est in superficie eclyptice."
98 Ibid., fol. K3v: "Notandum etiam pro intelligentia de equante qui et in luna ponitur et etiam ibi determinat de aliis tribus. Equans est circulus imaginatus cuius imaginatio ab astronomis sic est inventa, quia enim planete non equaliter moventur super centro mundi, nec semper moventur equaliter super centro deferentium suorum, astronomi imaginati sunt orbem aliis planetis a sole per quem illa difformitas reduceretur ad uniformitatem. Omne enim difforme reducendum est ad uniforme, et omnis inequalitas ad equalitatem. Unde propter hoc nomen equantis habet. Quomodo autem hoc contingat resolutissime Magister Johannes Danckonis in suis theoricis et Erhardus Cremonensis et ceteri theoriste describunt."
99 Ibid., fol. K4r–v.
100 Ibid., fol. K4r–v: "Notandum pro intelligentia istius quod in quinque planetis equans et deferens sunt ecentrici et in eadem superficie. Quod Erhardus Cremonensis quem Magister Georgius Purbachius vir dignus, et *Dialogus* Magistri Joannis de Monteregio sepius reprehendi et confutat."
101 Ibid., fol. K4r.
102 This section of the chapter was originally written before Peter Barker's "Albert of Brudzewo's *Little Commentary on George Peurbach's 'Theoricae Novae Planetarum.'*" Although I agree in many respects with his interpretations of Brudzewo's work, I retain here some partly repetitious passages that are relevant to the thesis of the chapter, namely that Copernicus's attitudes toward the status of astronomy as a science as they appear in the *Commentariolus* are very like what he would have learned from Brudzewo. This similarity comes out more clearly in my translations than in Barker's.
103 Brudzewo, *Commentariolum super Theoricas*, 4–5.
104 Ibid., 5–7.
105 Averroes, *Aristotelis De Caelo*, vol. 5, book 2, comment 67, fol. 144rb. Note that at least in the Latin translation, Averroes speaks of "the modern Arabs" (*moderni Arabes*), perhaps fitting with Bacon's use of the phrase

"imaginatio modernorum," which I suppose refers to the Arabs. See Averroes, *Large Commentary*.

106 Averroes, *Aristotelis De Caelo*, vol. 5, book 2, fol. 118r–v.

107 Ibid., vol. 5, book 2, fol. 118v.

108 Brudzewo, *Commentaria utilissima*, fols a5v–a6r; Brudzewo, *Commentariolum super Theoricas*, 18–20.

109 Brudzewo, *Commentaria utilissima*, fols b1v–b2r; Brudzewo, *Commentariolum super Theoricas*, 34: "CIRCULUS ITAQUE ECENTRICUS. Tertia pars principalis, in qua volens declarare terminos quibus utuntur tabularii, praemittit definitionem circuli ecentrici tenentis vicem in similitudine sui orbis. Et ita refert Magister dispositionem orbis realis, corpus solare deferentis, ad circulum imaginarium, quomodo et in plano figurari. Consueverunt enim Theoristae, orbium realium dispositionem intelligentes, circulos imaginarios, illis similes, in eorum locis subordinare, tandem singula, quae talem dispositionem sequantur, in plano oculis subiicere. Sensus enim saepius adiuvat intellectum speculari."

110 Brudzewo, *Commentaria utilissima*, fol. a5r–v; Brudzewo, *Commentariolum super Theoricas*, 17–18.

111 Brudzewo, *Commentariolum super Theoricas*, 28.

112 Brudzewo, *Commentaria utilissima*, fol. a8v; Brudzewo, *Commentariolum super Theoricas*, 30–1. See Sylla, "A Posteriori Foundations."

113 Brudzewo, *Commentariolum super Theoricas*, 31.

114 Ptolemy, *Ptolemy's Almagest*, 153.

115 F.J. Ragep, *Naṣīr al-Dīn al-Ṭūsī's Memoir*, vol. 1, 144: "Ptolemy chose the former model – there being no necessity to do so – because it is simpler." See also Averroes, *Aristotelis De Caelo*, vol. 5, where he says there is no formal demonstration from observations to the eccentrics and epicycles supposed to explain them.

116 Aiton, "Peurbach's *Theoricae novae planetarum*," 23: "It follows from the motion of these orbs that the center of the orb of the deferent of the epicycle likewise describes in the same time a certain circumference of a small circle."

117 Brudzewo, *Commentaria utilissima*, fol. d7v; Brudzewo, *Commentariolum super Theoricas*, 85. It might be noted that the equant point really does help to describe the motion of the planets because it corresponds to the empty focus of the elliptical path of the planet following Kepler's laws. If a planet sweeps out equal areas in equal times around the Sun in one focus of its elliptical orbit according to Kepler's second law, then it will sweep out equal angles in equal times about the other, empty focus of the ellipse.

118 Copernicus, *Letter against Werner*, 99.

119 Brudzewo, *Commentariolum super Theoricas*, 13–15. Brudzewo's answers as well as questions come from Albertus Magnus, *Metaphysica*, book 11,

ch. 26, lines 19–63, 516. In Albertus Magnus, the editors read *astronomorum*, rather than *astrologorum*, at the end.

120 Brudzewo, *Commentariolum super Theoricas*, 96.

121 Ibid., 134.

122 Ibid., 62–3.

123 See ibid., 96–7.

124 Oxford University, Bodleian Library, MS Canon Misc. 45, fols 1a–57a (forty-two figures); Langermann, *Ibn al-Haytham's "On the Configuration,"* 41.

125 Although some of the extant copies of the Latin translations of *On the Configuration* are not illustrated, there are numerous manuscript copies of Hebrew translations of *On the Configuration* now found in European libraries that do have illustrations. Langermann, *Ibn al-Haytham's "On the Configuration,"* 34–6.

126 See Pedersen, "'Theorica Planetarum.'"

127 Edward Rosen, in his translation of the *Commentariolus*, confused the issue because he very frequently translated "orb" in the Latin as "circle." Copernicus, *Three Copernican Treatises*.

128 See Benjamin and Toomer, eds, *Campanus of Novara*. On the common acceptance of physical orbs in the fourteenth century, see Hugonnard-Roche, "Problèmes Méthodologiques."

129 John of Sicily, *Scriptum super canones Azarchelis*, para. J285, 134, para. J318, 156, para. J344, 168–9.

130 John of Jandun, *In Duodecim Libros Metaphysicae*, book 12, Q. 20, fol. 141r–v.

131 John Marsilius Inguen, *Quaestiones subtillissime*, fols 55rb–55va.

132 This subject is discussed in S.P. Ragep, chapter 6, this volume.

133 See Brudzewo, *Commentariolum super Theoricas*, index.

134 See Barker, "Albert of Brudzewo's *Little Commentary*"; and Malpangotto, "Original Motivation."

135 See Shank, "Regiomontanus and Homocentric Astronomy"; Swerdlow, "Regiomontanus's Concentric-Sphere Models"; and Shank, "Regiomontanus as a Physical Astronomer." See also Malpangotto, *Regiomontano*.

CHAPTER FOUR

For their criticisms and suggestions, I thank the editors, as well as Rich Kremer, Ron Numbers, Noel Swerdlow, and two anonymous referees.

1 In addition to the literature in the notes below, see Sylla, chapter 3, this volume.

2 The spelling of "Peurbach" is an erroneous English-language convention that ignores the name's origins in the town of Peuerbach. I prefer

"Peuerbach," so spelled in German, but have followed my colleagues in this volume in using the anglicized form.

3 Ptolemy's *Planetary Hypotheses,* which takes such matters seriously, was not known in the Latin world and did not circulate widely in Arabic either. One notable exception is Abū ʿAlī al-Ḥasan ibn al-Haytham, for which see Langermann, *Ibn al-Haytham's "On the Configuration,"* 12–25.

4 Rheticus, *Narratio prima,* fol. A2r: "in every kind of learning and in expertise of astronomy, he is not inferior to Regiomontanus" (my translation). See Swerdlow, "Annals of Scientific Publishing," esp. 273–4.

5 Zinner, *Leben und Wirken,* 19.

6 Shank, "Mechanical Thinking," esp. 22–6; Shank, "Geometrical Diagrams."

7 Zinner, *Leben und Wirken,* 13–16; Zinner, *Der deutsche Kalender,* 7.

8 Geoffrey Chaucer (d. 1400) claimed to have written his *Treatise on the Astrolabe* at the request of his ten-year-old son. Benson, *Riverside Chaucer,* 661–2. On the Leipzig context, see Wussing, "Regiomontanus als Student in Leipzig."

9 See Georg Tanstetter's introduction to Peurbach and Regiomontanus, *Tabulae eclypsium,* fol. aa3v.

10 Grössing and Stuhlhofer, "Versuch einer Deutung"; Shank, "Academic Consulting." For an overview of the Viennese background, see Byrne, "Stars, the Moon." We know more about Frederick III's later years than his earlier ones. Hayton, *Crown and the Cosmos,* esp. 22–4.

11 The scant literature on Nihil is cited in Shank, "Academic Consulting," 262–4.

12 Uiblein, "Die Wiener Universität," 395.

13 Aeneas, *De Bohemorum origine,* ch. 70, 159–61.

14 Vienna, Österreichische Nationalbibliothek (ÖNB), cod. 5203.

15 Andalò di Negro of Genoa (d. ca. 1340) entitled his work "Nova theorica planetarum" in a 1421 manuscript in Baldassare Boncompagni's possession in the late nineteenth century and in another lost one. See Boncompagni's appendix to Cornelio De Simoni's "Intorno alla vita ed ai lavori de Andalò di Negro," esp. 351, 358. The "newe theorik of planetis" that Derek de Solla Price mentions in Cambridge, Trinity College, MS O.5.26 (ca. 1375), is evidently an English translation of Andalò di Negro's work. See Price, *Equatorie of the Planetis,* 197; and Eagleton, *Monks, Manuscripts, and Sundials,* 173. Johannes de Fundis had also called his own work "Theorica novae planetarum" in the early fifteenth century. See Pedersen, "'Theorica planetarum,'" esp. 71–8; Pedersen, "Decline and Fall," esp. 162–8; and Lerner, *Le Monde des sphères.* Peurbach may have learned about Negro and Fundis during his Italian travels.

16 Vienna, ÖNB, cod. 5203, 2r: "Theorica nova realem sperarum habitudinem atque motum cum terminis tabularum declarans." The same title appears in

the copy that Peurbach dedicated to Johannes Vitéz (Cracow, Biblioteca Universitaria, 599). The first folio is reproduced in Nagy, "Ricerche cosmologiche," third folio of illustrations between pages 80 and 81. Curiously, this title resembles that of a treatise that Frederick of Drosendorf (d. 1404), canon of St Stephen's in Vienna and *astrologus Austriae*, left to the chapter library a generation before Peurbach: "Tres sexternos de theorica nova, que tabularum terminis dimissis realem sperarum et motuum habitudinem declarant." In Gottlieb, *Mittelalterliche Bibliothekskataloge Österreichs*, vol. 1, 281–2.

17 This concern is particularly clear in Peurbach's chapter on the deflection of the inferior planets' eccentrics, in which concentric spheres enclose epicycles or eccentrics. Aiton, "Peurbach's *Theoricae novae planetarum*," 34–5; Mancha, "Ibn al-Haytham's Homocentric Epicycles," esp. 74–7, 80. Additional grounds for novelty include the description of the curve traced by the centre of Mercury's epicycle as "ovalis" and the much improved, albeit qualitative, description of latitude theory. Pedersen, "Decline and Fall," 162–8; Lerner, *Le Monde des sphères*, vol. 1, 122–3. See also Sylla, chapter 3, this volume.

18 Regiomontanus, *Disputationes contra deliramenta cremonensia* (Nuremberg, c. 1475), fol. 3v–4r, in *Joannis Regiomontani Opera collectanea*, 518–19; Shank, "Regiomontanus on Ptolemy," esp. 192–7.

19 The emphasis on three-dimensional spheres sets Peurbach's *Theoricae novae planetarum* apart from the *Theorica planetarum communis* but not from Campanus of Novara's *Theorica planetarum*. In the tradition of Alfraganus's (al-Farghānī's) *Elements*, Campanus discussed the internal and outer dimensions of the various planetary spheres. Benjamin and Toomer, *Campanus of Novara*, 25ff.; Lerner, *Le Monde des sphères*, vol. 1, 121–6.

20 Lerner, *Le Monde des sphères*, vol. 1, 310n58, 311n64; Shank, "Academic Consulting," 253.

21 Aiton, "Peurbach's *Theoricae novae planetarum*," 10. More recently, see Barker, "Reality of Peurbach's Orbs," esp. 11–12; and Malpangotto, "L'univers auquel."

22 A late-fourteenth-century copy of Sacrobosco's *On the Sphere of the World* (perhaps by Nicholas of Erfurt, who copied another text in the same hand) shows such "filled-in" diagrams for the Sun and a superior planet. Naples, Biblioteca Nazionale, ms VIII.D.31, 18v. Similar illustrations appear in Viennese manuscripts in the early fifteenth century. Grössing, *Humanistische Naturwissenschaft*, 98, has suggested plausibly that the diagrams in the *Theorica planetarum communis* preserved in Vienna, ÖNB, cod. 5266 (ca. 1434), influenced Peurbach's *Theoricae novae planetarum*. For illustrations, see Shank, "Zwischen Berechnung und Experiment," esp. 200–1.

23 Pantin, "First Phases."

24 Vienna, ÖNB, cod. 5203.

25 Kren, "Homocentric Astronomy"; Kren, "Medieval Objection"; Mancha, "Ibn al-Haytham's Homocentric Epicycles," 73–8.

26 Goldstein, *Astronomy of Levi ben Gerson,* 105–6; Mancha, "Latin Translation," esp. 28. Since Levi ben Gerson's work was known in mid-fourteenth-century Parisian circles, he seems a likely source of these critiques.

27 Zinner, *Leben und Wirken,* 107–8; Swerdlow, "Regiomontanus on the Critical Problems," 172–4. In one striking passage, Regiomontanus proposes to keep the solar distance invariable by giving the Sun a nonuniformly moving concentric sphere centred on the Earth. Regiomontanus, *Defensio,* 156v.

28 Aristotle, *Metaphysics,* book 12. See also Dicks, *Early Greek Astronomy,* 200–2. As usually understood, the Eudoxean planetary model for Saturn, for example, had four spheres: one for the twenty-four-hour daily east-to-west motion, one for Saturn's thirty-year west-to-east motion through the ecliptic, and two to generate the hippopede that simulated its retrograde loop. To obtain from this a "clean" once-in-twenty-four-hours motion for Jupiter, it was necessary to "undo" each of the last three motions, which were peculiar to Saturn. Thus, below the Saturn model, three counteracting spheres were needed that each moved at the same rate as, but in the opposite direction of, the counterpart it cancelled. The mechanical result left the daily motion intact to account for Jupiter's daily motion, followed by three spheres to account for its twelve-year west-to-east motion, its hippopede, and so on.

29 Biṭrūjī, *De motibus celorum,* 71, 150; Goldstein, ed. and trans., *Al-Biṭrūjī*; Sabra, "Eleventh-Century Refutation."

30 Avi-Yonah, "Ptolemy vs. al-Biṭrūjī," esp. 137–44; Grant, "Celestial Motions," esp. 141–7; Shank, "Rings in a Fluid Heaven," esp. 185–9.

31 Vienna, ÖNB, cod. 5203.

32 Shank, "'Notes on al-Biṭrūjī'"; Shank, "Rings in a Fluid Heaven," 181–5.

33 Swerdlow, "Regiomontanus's Concentric-Sphere Models"; Shank, "Regiomontanus and Homocentric Astronomy."

34 Ptolemy's final lunar model implied a four-fold variation in observed area, whereas he treated the Sun as observationally equidistant from the Earth (31′) even though his model predicted a parallax. Venus had a very large epicycle, from which one would expect enormous variations in brightness.

35 Aschbach, *Geschichte der Wiener Universität,* vol. 1, 538, which quotes the manuscript of the Acta Facultatis Artium (Archive of the University of Vienna), vol. 3, fols 117, 136, 144; Zinner, *Leben und Wirken,* 78.

36 Monfasani, *George of Trebizond,* 104–9; d'Alessandro and Napolitani, *Archimede Latino,* 70–2.

37 George of Trebizond deemed Plato a traitor to Athens, a besmircher of rhetoric, an advocate of pedophilia, and a pagan who lent aid and comfort to Greek Christianity. See Hankins, *Plato in the Italian Renaissance,* vol. 1, 167–73; and Bisaha, *Creating East and West,* 152–7.

38 Monfasani, *George of Trebizond*, 131–2.
39 Lorch, "Some Remarks," esp. 422–3.
40 Zinner, *Leben und Wirken*, 54, 130, 297.
41 Swerdlow, "Derivation and First Draft," esp. 426. In the letter that appears in some copies of the 1496 edition of the *Epitome*, Giambattista Abioso noted that he preferred the latter to the *Almagest*. Thorndike, *History of Magic*, vol. 5, 341.
42 Regiomontanus, *Epytoma in Almagestum Ptolemaei*, fol. a7r: "Praefationem autem Ptolemei ad litteram exprimere libuit, tum propter crebras in ea sententias scitu dignissimas, tum propter auctoritatem Ptolemei, quo etiam imitatio nostra fidelior redderetur."
43 Venice, Biblioteca Nazionale Marciana (BNM), lat. 327.
44 *Imitatio* can have multiple meanings, from "imitation" through "likeness" to the rearrangement of materials into a certain order, as per Macrobius. *Imitatio* (*mimesis*) was a classical rhetorical stance that was undergoing transformations in the fourteenth and fifteenth centuries. The Roman rhetorician Quintilian had already hinted that imitation should strive to go beyond the model by not merely reproducing but instead emulating it. Petrarch would make a similar point. Ruthven, *Critical Assumptions*, 103–4. On Macrobius, see Pelttari, *Space that Remains*, 25–32. Peurbach was certainly hewing close to the *Almagest* in book 1 of the *Epitome*.
45 In his *Disputationes*, Regiomontanus addressed the responsibility of the commentator: "*Cracovian*: … I see so many commentaries on his work, all of which undertake to explain this *Theorica* while noting nothing mistaken, indeed nothing unreasonably stated. *Viennese*: If the author himself has written something obscure, or perhaps taught something unwisely, it is the commentator's job to clarify the former, and to point out the latter with proper restraint – otherwise the commentator deserves to be considered lazy and thoughtless." Regiomontanus, *Disputationes contra deliramenta cremonensia* (Nuremberg, c. 1475), fol. 2v, in *Joannis Regiomontani Opera collectanea*, 516 (my translation).
46 In 1463 Regiomontanus was writing his *Problemata Almagesti*, now lost, which also followed the thirteen-book structure of the *Almagest*. The few surviving references to its early mathematical sections suggest amplification, leaving in doubt the extent to which any criticism of the *Almagest* may have appeared in later books. Zinner, *Leben und Wirken*, 118–21, 324–5.
47 Regiomontanus, *Epitome*, book 5, prop. 22, in *Joannis Regiomontani Opera collectanea*, 145. The argument prompted one annotator of the first edition (Venice, BNM, Inc V 119) to doubt the reality of the Ptolemaic devices and to suggest that Aristotle's (homocentric) approach (following Eudoxus of Cnidus and Calippus of Syracuse) in *Metaphysics*, book 12, deserved another look.

48 Shank, "Regiomontanus as a Physical Astronomer," esp. 336–42.

49 The astronomical portion of Theon of Smyrna's work is now in Venice, BNM, Z Gr. 303 (534). This manuscript, which also contains a heavily annotated *Almagest* and Proclus's *Hypotyposis*, appears in the 1468 and the 1474 inventories of Bessarion's library (numbers 257 and 573, respectively). Labowski, *Bessarion's Library*, 167, 222, 467; Huxley, "Theon of Smyrna."

50 I have recently discovered that Bessarion at least once conflated the two Theons (of Smyrna and of Alexandria). His *In calumniatorem Platonis* attributes merely to "Theon" what is clearly an allusion to Theon of Smyrna's book *Theon summus ille mathematicus, cum opus de quattuor mathematicis disciplinis edidisset, compendium in Platonis libros inscripsit.* Mohler, *Kardinal Bessarion*, vol. 2, 74–5. If Regiomontanus did so as well, the rationale for "defending Theon" becomes more complicated.

51 Dupuis, *Théon de Smyrne*, 269–87. A serviceable English translation appears in Theon of Smyrna, *Mathematics*, 108–15.

52 Dupuis, *Théon de Smyrne*, esp. 279, 303–5; Theon of Smyrna, *Mathematics*, 121–2.

53 Proclus, *Procli Diadochi Hypotyposis astronomicarum positionum*, ch. 3, §76–§83, 76–82; Pingree, "Teaching of the *Almagest*," esp. 79.

54 Regiomontanus, *Defensio*, 254r, 262r; at 262r–v, Regiomontanus has written out a page and half from the Greek. See http://collections.dartmouth.edu/zoom/zoom.php?p=/regio/tiff_pyr/262v.tif&t=Regiomontanus%20-%20262v. For the seventh entry in the printing prospectus, entitled "Procli sufformationes astronomicae," see Zinner, "Die wissenschaftlichen Bestrebungen Regiomontans," esp. 94.

55 Nuremberg, Stadtbibliothek, MS Cent V 57. See Chabás and Goldstein, *Astronomical Tables*, esp. ch. 1.

56 Swerdlow, "Regiomontanus on the Critical Problems," esp. 173–4, 184–7. Levi ben Gerson also believed that Mars varied far less than Ptolemy claims. Goldstein, *Astronomy of Levi ben Gerson*, 188.

57 Zinner, *Leben und Wirken*, 96–100.

58 Ibid., 118–21, 324–5; Rigo, "Bessarione, Giovanni Regiomontano," esp. 90–7. See the list of post-Regiomontanus references to the whereabouts of the *Problemata* in Zinner, *Leben und Wirken*, 324–5; and Malpangotto, *Regiomontano*, 205–6.

59 Regiomontanus, *Oratio Iohannis de Monteregio*, 43–53, esp. 46, 47, 51. Paraphrases and partial translations in Swerdlow, "Science and Humanism"; and Byrne, "Humanist History of Mathematics?"

60 Regiomontanus, *Oratio Iohannis de Monteregio*, 47; Pingree, "Gregory Chioniades"; Pingree, *Astronomical Works*, vol. 1, 23–9; Neugebauer, *History*, part 1, 11–13; Swerdlow, "Science and Humanism," 147; Labowski, *Bessarion's Library*, nos 230, 233, 250, 271, at 166–8. These manuscripts are

still in Venice, BNM, MG 323, 327, 333, 328 respectively. Manuscript group 327 was added later (inventory of 1474). Richard Kremer and Jerzy Dobrzycki have speculated that Marāgha-like techniques (i.e., double epicycles) stand behind the new tables on which Peurbach was working at his death. Dobrzycki and Kremer, "Peurbach and Marāgha Astronomy?"

61 Birkenmajer, "Marcin Bylica," 533; Vargha and Both, "Astronomy in Renaissance Hungary," esp. 279.

62 Mercati, "Le due lettere," esp. 68 and the letters on 85–99. On George's trip to Constantinople and back, see Monfasani, *George of Trebizond*, 132, 188n.

63 Mercati, "Le due lettere," 85, 91–2. On Geber's criticisms of Ptolemy, see Lorch, "Astronomy of Jābir ibn Aflaḥ," esp. 94–9. The letters express George's eschatological vision and his identity as a prophet. See John Monfasani's translation of "On the Divinity of Manuel" in Monfasani, *Collectanea Trapezuntiana*, 564–9.

64 The enmity between George and Regiomontanus intersects with a complex social landscape of sympathies, friendships, employments, and prospective and actual patronage. Monfasani, *George of Trebizond*, 196–8.

65 Ibid., 194–5. The king's splendid copy of George's translation is in Vienna, ÖNB, cod. 24 (17 March 1467).

66 Monfasani, *George of Trebizond*, 194–200; Monfasani, *Collectanea Trapezuntiana*, 564–74.

67 My transcription of Regiomontanus's draft dedication to the king is inserted at 37r-39r of his *Defensio* and appears as chapter 0. See http://regio.dartmouth.edu/diplomatic/00.html.

68 Rome, *Commentaires*, vol. 1, vi.

69 For example, Regiomontanus held that, contrary to what the *Almagest* plainly states at the beginning of book 12, Ptolemy actually held that either eccentric or epicyclic models could account for the second anomaly of both the inferior and superior planets. Accordingly, Regiomontanus supplied the proof, which he took Ptolemy to have omitted (more below). Shank, "Regiomontanus as a Physical Astronomer," 336–40.

70 Shank, "Regiomontanus on Ptolemy," esp. 200–3.

71 Shank, "Regiomontanus and Homocentric Astronomy"; Swerdlow, "Regiomontanus's Concentric-Sphere Models"; Regiomontanus, *Defensio*, 156v, 226v–227r.

72 Shank, "Regiomontanus on Ptolemy," 190–200, quotation at 200. Regiomontanus did not use his homocentric models for the Sun and Moon in his *Tabula primi mobilis*. Swerdlow, "Regiomontanus's Concentric-Sphere Models," 5. Nonetheless, in book 12 of the *Defensio*, 225r, he attacks George by criticizing partial orbs: "a kind of monstrosity, that there are in the heavens bodies of such varied thickness, i.e., here so skinny, there however

swelling greatly by some extraordinary assemblage – such a defective picture that nature could not possibly be pleased by it." And at 226v–227r, he tells George, "the defender of eccentrics," that retrograde motion can be generated with concentrics.

73 Kuhn, *Structure of Scientific Revolutions.*

74 George of Trebizond, *Commentary on the Almagest*, book 9, as quoted in Regiomontanus, *Defensio*, book 9, 152r: "For, since the parallax of Venus and Mercury is effectively imperceptible and makes an angle much less than perceptible, we cannot easily discover the distances by these means, but transposing proportionally from the distances that he [Ptolemy] demonstrated, we will demonstrate that Mercury can be located, in immediately ascending [order], after the Moon, and Venus after Mercury, and then the Sun. It is therefore necessary, lest there be a vacuum, that the greatest distance of the Moon be the least distance of either Venus or Mercury. Not that of Venus, however, therefore that of Mercury. For Mercury is recorded [scribitur] to have gone below Venus, and is faster than Venus. And it is necessary that, in circular motion, the inferior be faster" (my translation).

75 Theon of Smyrna and Proclus (to say nothing of Plato) had proposed planetary orders at variance with Ptolemy's. Without accepting his comments on heliocentrism, see Siorvanes, *Proclus*, 304–11. Regiomontanus's more recent predecessors also had different ideas, including Geber, who placed Mercury and Venus above the Sun on grounds of no parallax; Biṭrūjī, who placed the Sun between Mercury and Venus; and Levi ben Gerson. On Geber, see Lorch, "Astronomy of Jābir ibn Aflaḥ," esp. 94–9. Two manuscripts of Geber are in, or annotated by, Regiomontanus's hand. Ibid., 91; Zinner, *Leben und Wirken*, 310–11.

76 Prosdocimo de' Beldomandi had once made a similar point. Markowski, "Die kosmologische Anschauungen," esp. 270.

77 "Tres superiores per epicyclum Soli colligantur; tres vero inferiores, non per epicyclum sed per motum longitudinalem; hec esse potest una ratio rethorica situs planetarum et ordinis. Cur non quinque retrogrados in una parte natura locavit? Cur non animadvertit sexus rationem?" Regiomontanus, *Defensio*, 153v. This last remark may not be snide, for Regiomontanus makes the case in the *Defensio* for a tighter spatial fit if Mercury is above Venus.

78 Goldstein, "Copernicus and the Origin." Regiomontanus's concerns show why Westman, *Copernican Question*, chs 1–3, is unconvincing when tracing Copernicus's concern about planetary order to his putative attempt to answer Giovanni Pico della Mirandola's passing mention of the uncertainty in the latter's critique of astrology, entitled *Opuscula, Disputationes adversus astrologos* (1496). For my review, Westman's response, and my rejoinder, see Shank, "Made to Order"; Westman, "Reply to Michael Shank"; and Shank,

"Rejoinder." Swerdlow translated the *Epitome*, book 9.1, from the best manuscript and published it in the online appendix to his review of Westman's book. Swerdlow, "Copernicus and Astrology."

79 The so-called "Letter to Christian Roder" appears in Curtze, "Der Briefwechsel Regiomontans," 324–36. Menso Folkerts has shown that, contrary to Curtze and everyone else, Regiomontanus's correspondent in Erfurt was not Roder (an inference) but Gottfried Wolack. Folkerts, "Conrad Landvogt," esp. 234n32.

80 Zinner, *Leben und Wirken*, 161.

81 Vargha and Both, "Astronomy in Renaissance Hungary," esp. 282n4.

82 The prospectus is now edited and discussed in Malpangotto, *Regiomontano*, 149–54, 184–209. See also the analysis in Zinner, "Die wissenschaftlichen Bestrebungen Regiomontans."

83 Regiomontanus's first edition had not given the work a title. This one is Erhard Ratdolt's, diffused in the second and third editions of the work (Venice, 1482 and 1485).

84 Pedersen, "Decline and Fall," 162–8; Zinner, *Leben und Wirken*, 34; Aiton, "Peurbach's *Theoricae novae planetarum*," esp. 6–7.

85 Shank, "Geometrical Diagrams," 31–3.

86 Pantin, "First Phases," 5–7.

87 Shank, "Geometrical Diagrams." A notable example occurs in Regiomontanus's manuscript of Langenstein's *De reprobatione ecentricorum et epicyclorum* (1364). Vienna, ÖNB, cod. 5203.

88 The brass arms still survive in some copies. Their appearance and that of the volvelles can be gauged from illustration 12–13, plates xliii–iv, in Wattenberg, "Johannes Regiomontanus," preceding page 345.

89 These corrections will be documented in detail in my forthcoming edition of the *Disputationes* and in my forthcoming articles on Regiomontanus as a printer.

90 Swerdlow, "Annals of Scientific Publishing."

91 Poulle, "L'horloge planétaire de Regiomontanus."

92 This so-called "Letter to Christian Roder" (see note 79 above) is summarized in Zinner, *Leben und Wirken*, 163–9, esp. 164–5.

93 Kremer, "War Bernard Walther"; Steele and Stephenson, "Eclipse Observations"; Kremer, "Bernard Walther's Astronomical Observations," esp. 174–7, 185–6; Kremer, "Use of Bernard Walther's"; Swerdlow, "Regiomontanus on the Critical Problems," 171–4.

94 Babicz, "Die exakten Wissenschaften"; Goddu, *Copernicus*, 162ff.

95 Czartoryski, "Library of Copernicus," 366; Goddu, "Copernicus's Annotations," esp. 207–8.

96 Goddu, *Copernicus*, 159–67.

97 Johannes Werner seems to have seen some of Regiomontanus's autographs before 1514. Zinner, *Leben und Wirken*, 313. Alive in Nuremberg during each of Copernicus's trips was Bernard Walther (d. June 1504), who was Regiomontanus's collaborator as observer and the guardian of his *Nachlass*, including the *Defensio* and the *Problemata Almagesti. De revolutionibus* mentions three of Walther's unpublished observations of Mercury, usually explained by Georg Joachim Rheticus's visit to Copernicus. Kremer, "War Bernard Walther," 161; Kremer, "Text to Trophy"; Kremer, "Use of Bernard Walther's," 125.

98 Zinner, *Leben und Wirken*, 240–1.

99 Although listed in Kristeller, *Iter Italicum*, vol. 1, part 1, 403, the manuscript (Naples, Biblioteca Nazionale, ms VIII C 40) has been overlooked. Copied by an unreliable scribe, it was owned, corrected, and annotated (in book 1, primarily) by a specialist who also owned and had annotated a copy of the *Almagest* ("vide glosam super hac figura in meo libro Almagesti") (15v). A likely owner is Giovanni Battista Abioso (or Abiosus, 1463 to ca. 1524?), who saw the 1496 Venice edition of Regiomontanus's *Epitome* through the press. Abioso, from Bagnoli (Naples), was in Ferrara between 1490 and 1492 at least. Having dedicated his *Dialogus in astrologiae defensionem* ... (Venice, 1494) to King Alfonso II of Aragon, he returned to Naples in the sixteenth century. These biographical data are consistent with the manuscript's colophon, annotations, and present location. Pastore-Stocchi, "G.B. Abioso," esp. 21–4. Of the other manuscripts of the *Epitome*, two were in Bessarion's library (Venice, BNM, Z.328 [1760] and Z.329 [1843]), one in King Matthias Corvinus's library (Vienna, ÖNB, cod. 44), and one in Florence dated 1477, probably in Martin Bylica of Ilkusch's library (Florence, Biblioteca Nazionale, Magl. XI, 144).

100 In *De revolutionibus*, Copernicus used his Bologna observation of the Moon's occultation of Aldebaran on 9 March 1497 as a test of the Moon's parallax. Ludwik Antoni Birkenmajer has suggested that it was a test of the Ptolemaic lunar theory's predicted, but unverified, four-fold increase in the Moon's disk at quadratures (see Regiomontanus, *Epitome*, book 5), but this was a gross discrepancy that did not require careful observation to document. Copernicus, however, may well have been interested in documenting lunar size at various points in the circuit. In his discussion of lunar parallax in *De revolutionibus*, book 4, 27, Copernicus used this observation as empirical evidence for his theory, which predicted a modest parallax, against Ptolemy's, which predicted a larger one. Swerdlow has shown that Copernicus's claims and computations are filled with problems, noting that Copernicus's conclusion does not follow from his

corrected arguments. Birkenmajer, *Mikolaj Kopernik*, 20–6; Swerdlow and Neugebauer, *Mathematical Astronomy*, part 1, 266–71.

101 Birkenmajer, *Mikolaj Kopernik*, ch. 1; Swerdlow, "Derivation and First Draft," 425–6.

102 Rosen, "Copernicus and al-Biṭrūjī." Rosen's other comments about Regiomontanus and Biṭrūjī no longer stand. Shank, "'Notes on al-Biṭrūjī.'"

103 *Epitome*, book 9, prop. 1, fol. k2r, translation appended to Swerdlow, "Copernicus and Astrology."

104 Swerdlow and Neugebauer, *Mathematical Astronomy*, part 1, 54–7. On the slight additional motion of the mean Sun in Copernicus's theory, see Neugebauer, "On the Planetary Theory," esp. 96.

105 As F. Jamil Ragep has recently shown, ʿAlī Qushjī proved the equivalence perhaps as much as thirty years earlier, but he did so as an extensive critique of Ptolemy – a reading that is less puzzling than that of Regiomontanus. Ragep notes that, in Qushjī and the published *Epitome* of 1496, the diagrams of the proofs are remarkably similar (including analogous lettering and orientation of the epicycle). Ragep, "'ʿAlī Qushjī and Regiomontanus." Note that the diagrams Regiomontanus drew in the *Defensio* and the manuscripts of the *Epitome* in Bessarion's library are less similar in orientation to the Qushjī drawing than that in the 1496 printed *Epitome*, which Regiomontanus obviously never saw. One must therefore be cautious here. A similarity between diagrams can have more causes than direct copying, including layout constraints, attempts to save space, and conventions, both idiosyncratic and scribal. Ragep's is a tantalizing hypothesis about possible borrowing rather than evidence of borrowing. The diagrams and proofs in Qushjī and Regiomontanus illustrate contradictory beliefs about Ptolemy's views. Qushjī proves that Ptolemy is *wrong*, whereas Regiomontanus in the *Defensio* (wrongly) claims to represent Ptolemy *correctly* when proving George's summary of Ptolemy wrong. On similar diagrams and transmission, see Blåsjö, "Critique of the Arguments," esp. 185–6.

106 Shank, "Regiomontanus as a Physical Astronomer," 336–42.

107 Swerdlow, "Derivation and First Draft," 471–6, summarized in Swerdlow and Neugebauer, *Mathematical Astronomy*, part 1, 55–7; Swerdlow, "Astronomy in the Renaissance." E.J. Dijksterhuis seems to have been the first to draw attention to Regiomontanus's derivation in relation to Copernicus. Dijksterhuis, *Mechanization of the World Picture*, 291–2.

108 Swerdlow, "Derivation and First Draft," 471–6, summarized in Swerdlow, "Astronomy in the Renaissance."

109 One can visualize the eccentric model as equivalent to inverting the sizes and positions of epicycles and deferents. In the epicyclic model, a large deferent roughly centred on the Earth and moving at the rate

characteristic of each planet (i.e., thirty years for Saturn, etc.) carries a small epicycle that moves at the same rate as the mean Sun; in the eccentric model, a small deferent centred on the Earth and moving at the same rate as the mean Sun carries a large epicycle that in turn carries each planet at its characteristic rate (i.e., thirty years for Saturn, etc.).

110 Swerdlow, "Derivation and First Draft," 426–9, 471–8; Dobrzycki, "Notes on Copernicus's Early Heliocentrism."

111 Goldstein, "Copernicus and the Origin"; Goddu, "Reflections on the Origins." Swerdlow's geometrical and textual analysis stands on its own merits, whether or not one believes that Copernicus had to eliminate a "Tychonic" planetary arrangement before moving to heliocentrism.

112 Swerdlow, "Derivation and First Draft," 426.

113 Dobrzycki, "Notes on Copernicus's Early Heliocentrism," 223; Swerdlow, "Derivation and First Draft," 426.

114 Swerdlow and Neugebauer, *Mathematical Astronomy*, part 1, 84–5.

CHAPTER FIVE

1 Copernicus, *Commentariolus*: "Etenim quibus physiologi stabilitatem eius astruere potissime conantur, apparentijs plerumque innituntur: quae omnia hic inprimis corruunt, cum etiam praeter apparentiam versemur eandem" (our translation). The deviation from standard translations was suggested to us by an anonymous reviewer, and we are grateful for the insight. Noel Swerdlow provides this translation: "And in fact, [the evidence] by which natural philosophers attempt so very hard to confirm the immobility of the earth depends for the most part upon appearances. All [their evidence] falls apart here in the first place since we overthrow the immobility of the earth also by means of an appearance." Swerdlow, "Derivation and First Draft," 439. On the following page, Swerdlow writes, "He [Copernicus] … shifts the issue to his own domain by stating that their arguments for the immobility of the earth depend mostly upon appearances while he will show that the immobility of the earth is only an appearance." See also Edward Rosen's translation: "For, the principal arguments by which the natural philosophers attempt to establish the immobility of the earth rest for the most part on appearances. All these arguments are the first to collapse here, since I undermine the earth's immobility as likewise due to an appearance." Copernicus, *Complete Works*, vol. 3, 82.

2 Westman, *Copernican Question*, 101.

3 Daston and Lunbeck, eds, *Histories of Scientific Observation*.

4 Park, "Observation in the Margins," 20.

5 Ibid., 32.

6 Pomata, "Observation Rising," 69.

7 Copernicus, *Complete Works*, vol. 2, 21–3, emphasis added: "Quae omnia ratio ordinis, quo illa sibi inuicem succedunt, et mundi totius harmonia nos docet, si modo rem ipsam ambobus (utaiunt) oculis inscipiamus." See also Goddu, "Logic of Copernicus's Arguments," 57. Copernicus may have found the latter expression in Georgio Valla's *De expectendis rebus* (1501), known for including a reference to the moving Earth and for speaking on "Sphaera ambobus spectate" in the optical part of his work. Tucci, "Giorgio Valla"; Goddu, *Copernicus*, 229. On Copernicus's acquaintance with Valla's work, see commentary by Rosen in Copernicus, *Complete Works*, vol. 2, 368–82.

8 The *locus classicus* for the various uses of "appearances" is Duhem, *To Save the Phenomena*, even though the thesis is outdated. For a critique, see Lloyd, "Saving the Appearances."

9 The Greeks used sight as a metaphor for intelligizing. Plato makes clear the distinction between seeing, or *aestheta* (*sensa*), and intelligizing, or *noeta* (*intellecta*), in the Divided Line parable. In the *Phaedrus*, the soul is able to fly beyond the outer rim of the heavens and can gaze for a while at the ideas. Such concrete-abstract entities may have their origins in the Parmenidean poetic tradition. However, when Socrates speaks about "gazing at the ideas," he probably thinks of the dialectical process of active thinking, examining different and often contradictory possibilities, raising objections to common opinions, provoking interlocutors with paradoxes, and bringing in new hypotheses. Later on, however, Cicero spoke of Plato's own "eyes of the soul" (*oculi animi*), with which Plato could see that workshop by means of which he makes the world to be constructed by God. Neo-Platonists interpreted nonintellectual gazing at the ideas as contemplation, a view that was appropriated by Christian theologians and became particularly popular among fifteenth-century philosopher-theologians. Dr Ivor Ludlam, an expert on ancient Greek philology and philosophy, communicated these insights to us. For the discourse concerning sight and instruments, see Gal and Chen-Morris, *Baroque Science*, 53–78.

10 On the evidence of direct contact and acquaintance between Leon Battista Alberti and Nicholas of Cusa, see Santinello, "Nicolo Cusano e Leon Battista Alberti," appendix to *Leon Battista Alberti*, 265–96; and on the intellectual relationship between Alberti's theory of perspective and Cusa's metaphysical speculation, see Koenigsberger, *Renaissance Man*. More recently, with an emphasis on vision and notions of infinitude, see Harries, "On the Power and Poverty"; and the extensive comparative analysis of these two thinkers in Carman, *Leon Battista Alberti*. Our analysis diverts from these illuminating studies by stressing the mathematical context of Alberti's and Cusa's redefinitions of the demarcating line between the visible and the invisible.

11 See Cusa, *On Learned Ignorance*, part 1, ch. 3, 8.
12 On Regiomontanus, see Shank, chapter 4, this volume.
13 Rose, *Italian Renaissance*, ch. 5.
14 Swerdlow, "Derivation and First Draft," 471–2.
15 Ibid., emphasis added.
16 Aristotle, *On the Parts of Animals*, vol. 1, 1003–4.
17 Ptolemy, *Ptolemy's Almagest*, 35–6. See also Taub, *Ptolemy's Universe.*
18 Alberti, *On Painting and On Sculpture*, 36. We have preferred the older translation of Cecil Grayson over Rocco Sinisgalli's new one for stylistic reasons. Although Sinisgalli provides a more extensive and up-to-date critical edition and convincing arguments that Alberti's Italian version predated the Latin one, Grayson's translation serves our purposes better. Sinisgalli, *Il nuovo De Pictura*; Alberti, *On Painting: A New Translation.* The literature on Alberti's theory of painting is extensive to say the least; Grafton, *Leon Battista Alberti*, is the best available intellectual biography of Alberti. For different interpretations, see Jarzombek, *On Leon Battista Alberti*; and Kircher, *Living Well.* For Alberti's theory of perspective, see Field, *Invention of Infinity*, esp. 2–42; and Kemp, *Behind the Picture*, esp. 79–101. On the question of Renaissance painting and visibility, see Summers, *Vision, Reflection.*
19 Alberti, *On Painting and On Sculpture*, 37.
20 Ibid.
21 Ibid., 49, emphasis in original.
22 Ibid., 68.
23 Ibid., 98.
24 Unlike medieval Neo-Platonists who contemplated visual objects to negate material visibility and to go beyond it in order to capture abstract and divine reality with their mind's eye (e.g., see the discussion of Abbot Suger of St Denis), Alberti aspires to materialize the notion of beauty as a sensuous experience per se, or as he formulates it, with "la piu` grassa Minerva." See Sinisgalli, *Il nuovo De Pictura*, 94.
25 Alberti, *On Painting and On Sculpture*, 98.
26 Ibid., 102: "Sed et naturae dotes industria, studio atque exercitatione colendae, augendaeque sunt, et praeterea nihil quod ad laudem pertineat, negligentia praetermissum a nobis videri decet."
27 Cusa, *Nicolai Cusae Cardinalis: Opera*, ch. 7, 184, r: "Volumus autem ipsum ut principium indivisibile videre"; Hopkins, trans., *Nicholas of Cusa*, 794. For further discussion, see Gal and Chen-Morris, *Baroque Science*, 53–78.
28 Cusa, *Nicolai Cusae Cardinalis: Opera*, ch. 2, 184, v; Hopkins, trans., *Nicholas of Cusa*, 792–3.
29 Panofsky, *Abbot Suger*, 63–5.
30 "The power of our reason" can be interpreted as the measuring power.
31 Cusa, *Nicolai Cusae Cardinalis: Opera*, ch. 1; Cusa, *Complete Philosophical and Theological Treatises*, vol. 2, 792.

32 Cusa, *Nicolai Cusae Cardinalis: Opera*, ch. 5, 184, r; Cusa, *Complete Philosophical and Theological Treatises*, vol. 2, 794.
33 Cusa, *Nicolai Cusae Cardinalis: Opera*, ch. 5, 184, r; Cusa, *Complete Philosophical and Theological Treatises*, vol. 2, 794.
34 Cusa, *Nicolai Cusae Cardinalis: Opera*, ch. 36, 192, v; Cusa, *Complete Philosophical and Theological Treatises*, vol. 2, 824.
35 Cusa, *De docta ignorantia*, 2–3: "Omnes autem investigantes in comparatione certi proportionabiliter incertum iudicant, comparativa igitur est omnis inquisitio medio proportionis utens ... Omnis igitur inquisitio in comparativa proportione facili vel deficili existit ... Proportio vero cum convenientiam in aliquo uno simul et alteritatem dicat absque numero intelligi nequit. Numerus ergo omnia proportionnabilia includit." We have followed the translation of Jasper Hopkins in Cusa, *On Learned Ignorance*, 50.
36 Cusa, *Nicolai Cusae Cardinalis: Opera*, ch. 32, 190, r; Hopkins, trans., *Nicholas of Cusa*, 818.
37 Cusa, *Nicolai Cusae Cardinalis: Opera*, ch. 37, 192, r; Hopkins, trans., *Nicholas of Cusa*, 825.
38 Alberti, *On Painting and On Sculpture*, 52–4.
39 Cusa, *Nicolai Cusae Cardinalis: Opera*, ch. 27, 189, v; Hopkins, trans., *Nicholas of Cusa*, 812.
40 Cusa, *Nicolai Cusae Cardinalis: Opera*, ch. 16, 186, v; Hopkins, trans., *Nicholas of Cusa*, 801.
41 Cusa, *Nicolai Cusae Cardinalis: Opera*, ch. 16, 189, v; Hopkins, trans., *Nicholas of Cusa*, 811.
42 Cusa, *Nicolai Cusae Cardinalis: Opera*, ch. 19, 187, v; Hopkins, trans., *Nicholas of Cusa*, 804.
43 Cusa, *On Learned Ignorance*, part 1, ch. 13, 21, emphasis added.
44 Plato, "Republic," book 6, 509D–513E.
45 *Wisdom of Solomon* 11:20.
46 Cusa, *On Learned Ignorance*, part 2, ch. 13, 99.
47 Clagett, *Archimedes*, vol. 3, 310n18 (for calculation of pi).
48 Cusa, *Nikolaus von Kues*, 7–46.
49 Nicolle, *Nicolas de Cues*, 78–9.
50 Ibid., 80: "Semidiameter circuli isoperimteri trigono inscripto, se habet ad lineam a centro circuli cui trigonus inscribitur, ad quartam lateris ductam, in proportione sesquiquarta." See also ibid., 81–3, esp. notes 5 and 6.
51 See Cusa, *Nikolaus von Kues*, 191, 207, 216, 232, 243.
52 Ibid., 10–21; Clagett, *Archimedes*, vol. 3, ch. 1; and Nicolle, *Nicolas de Cues*, 9–35.
53 Nicolle, *Nicolas de Cues*, 79.
54 Notable are Hofmann, Clagett, and Nicolle.
55 Ibid., 76.

56 Ibid., 80.
57 Ibid., 76.
58 Ibid.
59 Ibid., 128.
60 Ibid., 36.
61 Ibid., 134, emphasis added.
62 Ibid., 154.
63 Ibid., 284.
64 Ibid., 396.
65 Ibid., 434.
66 We follow Hopkins's translation in his *Nicholas of Cusa*, vol. 2, 842–76.
67 Ibid., 842
68 Ibid., 857.
69 Ibid.
70 Ibid., 858.
71 Ibid., 848.
72 Ibid., 859.
73 Ibid., 847.
74 Regiomontanus, *Joannis Regiomontani Opera collectanea*, 415–510.
75 "De quadratura circuli secundum Nicolaum Cusensem, Dialogus Ioan. De Monteregio Aristophilus. Critias," in ibid., 438–44.
76 We follow Clagett's translation in his *Archimedes*, vol. 3, 367.
77 Ibid., vol. 3, 369.
78 Ibid., vol. 3, 371.
79 Ibid., vol. 3, 373.
80 It is interesting to note that Hofmann rejects this critique of Regiomontanus. He justifies this conclusion by computing the value of pi using Cusa's methods and comes up with a rather good approximation, within Archimedes's bounds.
81 Clagett's translation, in Clagett, *Archimedes*, vol. 3, 378.
82 Curtze, "Der Briefwechsel Regiomontans," 192–292.
83 Swerdlow, "Regiomontanus on the Critical Problems," 190.
84 Ibid., 171.
85 Ibid., 170–1.
86 Ibid., 183. The publication was Johannes Schoener's *Scripta clarissimi mathematici M. Joannis Regiomontani*, fols 27, 36–42, quoted in ibid., 192n4.
87 For the problematics of observation in the fifteenth century, see Kremer, "Bernard Walther's Astronomical Observations"; and Kremer, "Use of Bernard Walther's." On the notion of "testing," see F.J. Ragep, "Islamic Reactions."
88 Swerdlow, "Regiomontanus on the Critical Problems," 183.
89 Swerdlow, "Science and Humanism."

90 Curtze, "Der Briefwechsel Regiomontans," 324–36. See also Swerdlow, "Regiomontanus on the Critical Problems."

91 Ibid.

92 See letter in ibid.; and discussion in Swerdlow, "Science and Humanism."

93 See Labowski, *Bessarion's Library*; and Monfasani, *Byzantine Scholars.*

94 Uzielli and Celoria, *La vita e i tempi.*

95 See the end of Regiomontanus, *De Triangulis omnimodis*; and also Regiomontanus, *Oratio Iohannis de Monteregio.*

96 See Aristotle, *Posterior Analytics*, vol. 2, 19.

97 See, for one specific example, Proclus commenting on astronomy as a science of visibility in his *A Commentary on the First Book of Euclid's Elements*, 34: "There remains astronomy, which enquires into the cosmic motions, the sizes and shapes of the heavenly bodies, their illuminations and distances from the earth, and all such matters. This art draws heavily on sense perception."

98 Quoted in Heath, *Mathematics in Aristotle*, 14–15. See also Diels, ed., *Commentaria in Aristotelem Graeca*, vol. 10, 290–3.

99 Ptolemy, *Ptolemy's Almagest*, 141.

100 One should note that Ptolemy initially vouchsafes his observations and claims that his predecessors, especially Hipparchus of Nicaea, were unable to save planetary motions because they had no reliable observations at their disposal to achieve such a theory.

101 Ibid., 480.

102 Peurbach, *Theoricae novae planetarum* (1569), 45: "Unde et punctus ille centrum aequantis dicitur, et circulus super eo ad quantitatem deferentis secum in eadem superficie imaginatus, eccentricus aequans appellatur."

103 Alberti, *On Painting and On Sculpture*, 60–1: "absentes … present es esse faciat."

104 Our translation; see note 1.

105 Ibid., 436, fifth, sixth, and seventh postulates.

106 Ibid., 435.

107 Ibid., 441.

108 Ibid., 480.

109 Copernicus, *De revolutionibus*, book 5, ch. 1, 134: "Patet igitur, quod Saturni, Iouis, & Martis uera loca tunc tantum modo nobis conspicua fiunt, quando fuerit αωρονυκτυ, quod accidit fere in medio repedationum. Coincidunt enim tunc medio loco Solis in lineam rectam, illa commutatione exuti." Translation in Copernicus, *On the Revolutions of the Heavenly Spheres*, 233.

110 Swerdlow, "Derivation and First Draft," 461.

111 Copernicus, *On the Revolutions of the Heavenly Spheres*, 22.

112 Copernicus, *De revolutionibus*, book 1, ch. 4, 3: "Consentaneum est aequales illorum motus apparere nobis inaequales, uel propter diuersos

illorum polos circulorum, siue etiam quod terra non sit in medio circulorum, in quibus illa uoluuntur, & nobis à terra spectantibus horum transitus syderum accidat ob inaequales distantias propinquiora seipsis remotioribus maiora uideri, (ut in opticis demonstratum)." Translation in Copernicus, *On the Revolutions of the Heavenly Spheres*, 39.

113 Copernicus, *De revolutionibus*, book 1, ch. 4, 3: "[N]e dum excelsissima scrutari uolumus, quae nobis proxima sunt, ignoramus, ac eodem errore quae telluris sunt attribuamus coelestibus." Translation in Copernicus, *On the Revolutions of the Heavenly Spheres*, 39–40. This statement is a take on the ancient scornful tale of the astronomers who gaze at the sky without noticing the pit under their feet; thus Diogenes Laertius narrates of Thales that, while contemplating the stars, he fell into a ditch. An old woman reprimanded him that "[t]hou art like, indeed, to discover what is above at such a distance in the Sky, that can'st not see a Ditch just before thy nose." Diogenes Laertius, *Lives, Opinions*, 26.

114 Copernicus, *De revolutionibus*, book 1, ch. 5, 4; translation in Copernicus, *On the Revolutions of the Heavenly Spheres*, 40.

115 Cusa, *Vision of God*, 3.

116 Ibid., 15.

117 See the discussion of the artistic context of Cusa's speculations in Koerner, *Moment of Self-Portraiture*, esp. 127–38.

118 Cusa, *Vision of God*, 70.

119 See also the discussion of "De vision dei" in Levao, *Renaissance Minds*, esp. 76–84.

120 Copernicus, *De revolutionibus*, book 1, ch. 6, 4; translation in Copernicus, *On the Revolutions of the Heavenly Spheres*, 41.

121 Copernicus, *De revolutionibus*, book 1, ch. 6, 4; translation in Copernicus, *On the Revolutions of the Heavenly Spheres*, 42.

CHAPTER SIX

1 See Bagheri, "Newly Found Letter," 242–3. This letter supplements another letter written by Kāshī to his father about Ulugh Beg's circle of scholars in Samarqand, a city he depicts as having no parallel in Fārs (a province in southern Iran) in the teaching and learning of mathematics. See also Kennedy, "Letter of Jamshīd al-Kāshī"; and Sayılı, *Uluğ Bey*. Aydın Sayılı provides the Persian text and a Turkish translation with a commentary in both English and Turkish.

2 Keep in mind that the period after the Muslim theologian Abū Ḥāmid Muḥammad al-Ghazālī (d. 505/1111) is often depicted as being one of scientific decline or stagnation within Islamic lands. According to Edward Sachau, "But for Al Ashʿarî and Al Ghazâlî the Arabs might have been a

nation of Galileos, Keplers, and Newtons." Quoted in Sayılı, *Observatory in Islam*, 408. Christopher Beck assumed that classical Arabic science "did not really outlive Averroës" (d. 1198). Beck, *Warrior of the Cloisters*, 140. And the Nobel Laureate Steven Weinberg asserted that after Ghazālī, "there was no more science worth mentioning in Islamic countries." Weinberg, "Deadly Certitude." For a rebuttal, see F.J. Ragep, "When Did Islamic Science Die?"

3 Research indicates the Tīmūrids built over sixty-nine madrasas. In addition, Ulugh Beg established many civil institutions and *külliyyes* (complexes of buildings that surrounded mosques and could include libraries, hospices, and madrasas). See Fazlıoğlu, "Samarqand Mathematical-Astronomical School," 9–10; and Eshenkulova, "Timurlular Devri Medrese."

4 Ḥasan al-Jabartī's (d. 1188/1774–75) circle of scholars provides us an example of earlier theoretical astronomical works still being studied in eighteenth-century Cairo at Al-Azhar Mosque. According to his famous son, the historian ʿAbd al-Raḥmān al-Jabartī (d. 1241/1825–26), Ḥasan was a member of the *ʿulamāʾ* (scholars of the religious sciences) and attracted students from all parts of the world; and his instruction included Jaghmīnī's early thirteenth-century *Mulakhkhaṣ* and Qāḍīzāde al-Rūmī's fifteenth-century commentary on it. See İhsanoğlu, ed., *History of the Ottoman State*, vol. 2, 586–7; and Murphy, "Improving the Mind," 97–100. See also Rosenfeld and İhsanoğlu, *Mathematicians, Astronomers*, 410; Şeşen and İzgi, eds, *Osmanlı astronomi*, vol. 2, 479, no. 19; and İzgi, *Osmanlı Medreselerinde İlim*, vol. 1, 386, ç8. Almost a century after Ḥasan al-Jabartī, the Muslim Ottoman scholar Ṣadr al-Dīn al-Qūnawī (fl. 1857) attempted to reconcile the traditional science and the new (*jadīd*) science, which contained a version of the heliocentric system within the context of a traditional astronomical treatise for madrasa scholars. Morrison, "Reception of Early Modern."

5 Christopher Celenza raises culture as a factor in creating a willingness to question long-respected authority and to entertain divergent ideas. Celenza, chapter 1, this volume.

6 For our purposes, "Islamic" is here defined as encompassing those authors living under the umbrella of Islamic civilization – an area stretching from Spain and North Africa to central and south Asia – with a shared intellectual heritage; these authors included Jews, Christians, and others, in addition to Muslims. As well as Arabic, "Islamic" works were written in Persian, Turkish, Hebrew, and so on.

7 See Sayılı, *Uluğ Bey*, 36. Kāshī also reports that some 500 scientists witnessed his success in resolving a mathematical problem related to levelling the ground and determining the meridian at the Samarqand Observatory. Kennedy, "Letter of Jamshīd al-Kāshī," 198–9 (reprint 729–30). In fact, Kāshī's mathematical achievements have been highly lauded. See Schmidl,

"Kāshī"; and Sayılı, *Observatory in Islam*, 260–89. See also Kennedy, "Exact Sciences"; and Kennedy, *Planetary Equatorium.*

8 F. Jamil Ragep originally compiled these data in an unpublished paper entitled "Astronomy in the Fifteenth-Century Islamic World: More of the Same or the Beginnings of Modernism?" His findings have been supplemented here with astronomical works from Rosenfeld and Ihsanoğlu, *Mathematicians, Astronomers*, 258–319; Şeşen and İzgi, eds, *Osmanlı astronomi*; King, *Survey of the Scientific Manuscripts*; and King, "Appendix: A List of Mamluk Astronomers," in "Astronomy of the Mamluks," 553–5.

9 Since my primary focus is theoretical astronomy, I should mention that Edward S. Kennedy is not alone in considering *zījes* as "the most significant and historically rewarding subclass of the whole" Islamic astronomical corpus. Kennedy views *zījes* as more than numerical tables with accompanying explanations – that is, not just as end products enabling the astronomer/astrologer to solve a variety of planetary problems but also as gateways to understanding the underlying mathematics behind the numbers. On Arabic and Persian *zījes* from the eighth through fifteenth centuries, see Kennedy, "Survey," 123. For a supplement to Kennedy's survey with additional tables that are not contained within *zījes*, see King and Samsó, "Astronomical Handbooks."

10 David King points out that although the Mamluks wrote surprisingly little on astrology, "[a]ll *zījes* contain some astrological material." King, "Astronomy of the Mamluks," 550–1. See also Kennedy, "Survey," 144–5.

11 Ulugh Beg's *Zīj* consisted of four sections: "On Calendars," "On Times," "On the Positions of the Stars," and "On Astrology." Fazlıoğlu, "Samarqand Mathematical-Astronomical School," 16–17.

12 George Saliba details the pros and cons of astrology in his overview of the social status of the astrologer between the ninth and eighteenth centuries. Saliba, *"Role of the Astrologer."*

13 See Sabra, "Configuring the Universe," 289. For a discussion of astrology as a scientific discipline and some of the accepted methods of argumentation, see Burnett, "Certitude of Astrology."

14 On court astrologers, see Charette, "Locales," 124–8. For a comparison of the close relationship between the university and the court, especially the Habsburg dukes, see Shank, "Academic Consulting."

15 Although there is no direct evidence, some have inferred that Ulugh Beg's astrological motivations were similar to the patronage behind the building of the Marāgha Observatory. A.I. Sabra, for example, assumes that the Mongol "interest in the work of the observatory was undoubtedly astrological. The immediate goal was to produce a new set of astronomical tables, based on new observations." Sabra, "Situating Arabic Science,"

666–7. The narrative that Ulugh Beg believed in horoscopes and genethlialogy (true or not) has gained currency by the repeated anecdote that an astrological prediction had supposedly foreordained his actual assassination by his son ʿAbd al-Laṭīf. See Sayılı, *Observatory in Islam*, 277–8; and Fazlıoğlu, "Samarqand Mathematical-Astronomical School," 14–15.

16 For a scathing critique of astrology, see Ibn Khaldūn, *Muqaddimah*, vol. 3, 258–67. Ibn Sīnā (Avicenna) also wrote a well-known treatise entitled *Essay on the Refutation of Astrology*. See F.J. Ragep and S.P. Ragep, "Astronomical and Cosmological Works," 5, 8.

17 See, for example, Bīrūnī, *Al-Qānūn*, vol. 3, 1469, lxviii. See also Pines, "Semantic Distinction," 348–9; and Krause, "Al-Biruni," 10. For references to the *Qānūn* being studied at Samarqand, see Bagheri, "Newly Found Letter," 245, 253; Kennedy, "Letter of Jamshīd al-Kāshī," 197, 203, 209 (reprint 728, 734, 740); Sayılı, *Uluğ Bey*, 61–2, 71 (Persian), 99, 107 (Eng. trans.); and Fazlıoğlu, "Samarqand Mathematical-Astronomical School," 20, 52, 59.

18 According to Bīrūnī, he wrote the primer to provide definitions of astronomical and astrological terms in order to help facilitate their further application elsewhere. He informs us, "I have begun with Geometry and proceeded to Arithmetic and the Science of Numbers, then to the structure of the Universe, and finally to Judicial Astrology, for no one is worthy of the style and title of Astrologer who is not thoroughly conversant with these four sciences." Bīrūnī, *Kitāb al-Tafhīm*, 1; Kennedy, "Al-Bīrūnī," 155–6.

19 See King, *Astronomy in the Service*, 245–62.

20 François Charette emphasizes the need to re-examine the "utilitaristic interpretation" of texts on instruments. He points out, "There exists a plentiful Mamluk literature of instruments of a definitely *didactic* character (in the strong sense of the word)" that is concerned with "*training the minds of students*," in addition to teaching them how to use the instruments. Charette, "Locales," 130, 131, emphasis in original.

21 According to Aydın Sayılı, "The titles given to the mathematicians include the terms: *riyâḍî* (mathematician), *ḥâsib* (calculator), *muhandis* and *handasî* (probably two kinds of geometrician, or engineer and geometrician), and *ʿadadî* (arithmetician). Likewise, there were different titles corresponding to different fields of astronomy such as *falakî* (astronomer), *munajjim* (astrologer), *râṣid* (observer), and *usṭurlâbî* (instrument designer)." Sayılı, *Observatory in Islam*, 251.

22 King, *Astronomy in the Service*, ch. 1, 245; King, "On the Role of the Muezzin," 323.

23 King, "On the Role of the Muezzin"; King, "Astronomy of the Mamluks," 534–5. See also Sabra, "Situating Arabic Science," 668–9.

24 Listed chronologically from 1957 to 2007, see Roberts, "Solar and Lunar Theory"; Kennedy and Roberts, "Planetary Theory"; Abbud, "Planetary Theory"; Roberts, "Planetary Theory"; Kennedy, "Late Medieval Planetary Theory"; Swerdlow, "Derivation and First Draft"; Swerdlow and Neugebauer, *Mathematical Astronomy*, part 1, 41–8; and F.J. Ragep, "Copernicus." See also Saliba, "Theory and Observation," "Astronomical Tradition," "Arabic Planetary Theories," and *Islamic Science*.

25 This excerpt is from Noel Swerdlow and Otto Neugebauer's statement regarding evidence for the transmission of Arabic astronomy and the Marāgha School on Copernicus, the full assertion being, "The question therefore is not whether, but when, where, and in what form he [Copernicus] learned of Marāgha theory." Swerdlow and Neugebauer, *Mathematical Astronomy*, part 1, 47.

26 See F.J. Ragep, "Ibn al-Shāṭir and Copernicus"; and Nikfahm-Khubravan and F.J. Ragep, "Ibn al-Shāṭir and Copernicus."

27 See F.J. Ragep, "ʿAlī Qushjī and Regiomontanus," 360–1.

28 See Sayılı, *Observatory in Islam*, 250–1; and Sayılı, *Ulugh Bey*, 70–1 (Persian), 107 (Eng. trans.). See also, Kennedy, "Letter of Jamshīd al-Kāshī," 203–4 (reprint 734–5).

29 See Fazlıoğlu, "Samarqand Mathematical-Astronomical School," 25–34; F.J. Ragep, "Ḳāḍīzāde Rūmī"; and F.J. Ragep, "Qāḍīzāde al-Rūmī."

30 See Sayılı, *Ulugh Bey*, 71 (Persian), 107 (Eng. trans.). See also, Kennedy, "Letter of Jamshīd al-Kāshī," 203 (reprint 734).

31 According to a recent listing, Qāḍīzāde composed five works on theoretical astronomy, three on instruments, and three on timekeeping. Rosenfeld and Ihsanoğlu, *Mathematicians, Astronomers*, 272–4, no. 808.

32 Eshenkulova, "Timurlular Devri Medrese," 130–42.

33 See Sayılı, *Ulugh Bey*, 57, 61–2, 67, 71 (Persian), 95, 99–100, 104, 107 (Eng. trans.). See also Bagheri, "Newly Found Letter," 243, 245, 248, 253; and Kennedy, "Letter of Jamshīd al-Kāshī," 193–4, 197, 198, 203, 209 (reprint 724–5, 728, 729, 734, 740). İhsan Fazlıoğlu adds Ṭūsī's recensions *(taḥārīr)* and works by Muʾayyad al-Dīn al-ʿUrḍī and Ibn Sīnā to this list. Fazlıoğlu, "Samarqand Mathematical-Astronomical School," 19–20, 57–9.

34 Here, we have an example of a Sunnī student reading an astronomical text with a Shīʿī scholar at the shrine of the eighth Imām.

35 Fazlıoğlu, "Samarqand Mathematical-Astronomical School," 36–55.

36 Ibid., 41–6 (Arabic), 46–9 (Eng. trans.), 55. George Makdisi has argued that the "new organizational structure" of teaching that occurred within the Western college, namely the introduction of the basic *scholastic* method of teaching students, which consisted of the elements of *sic et non*, dialectic, and disputation, as well as the need to obtain the licence to teach, had

Islamic antecedents. Makdisi, "Baghdad, Bologna, and Scholasticism," 146–9, 151; Makdisi, "Madrasa and University."

37 See Ibn al-Akfānī's (d. 749/1348) *Kitāb irshād al-qāṣid ilā asnā al-maqāṣid*, edition by Jan Just Witkam, *De Egyptische Arts Ibn al-Akfānī*, 55–7, 408–7 (Arabic); and the Ottoman Aḥmad ibn Muṣṭafā Ṭāshkubrīzāde's (d. 968/1561) *Miftāḥ al-sa'āda wa-miṣbāḥ al-siyāda*, vol. 3, 348–9. They differ only in that Ṭāshkubrīzāde includes the *Mulakhkhaṣ* (under "famous abridgements") and also lists the four *Mulakhkhaṣ* commentaries of Faḍl Allāh al-'Ubaydī, Kamāl al-Dīn al-Turkmānī, al-Sayyid al-Sharīf al-Jurjānī, and Qāḍīzāde al-Rūmī. See also Fazlıoğlu, "Samarqand Mathematical-Astronomical School," 23.

38 Fazlıoğlu, "Samarqand Mathematical-Astronomical School," 8n13.

39 For example, the family of the Cairene Sibṭ al-Māridīnī (fl. ca. 1460), the extremely prolific fifteenth-century timekeeper of Al-Azhar Mosque, was originally from Damascus. It is noteworthy that Sibṭ al-Māridīnī's grandfather, Jamāl al-Dīn al-Māridīnī (d. 1406), was the timekeeper at the Umayyad Mosque in Damascus, and possibly a disciple of Ibn al-Shāṭir (d. ca. 1375) before travelling to Cairo. Rosenfeld and Ihsanoğlu, *Mathematicians, Astronomers*, 293–8, attribute fifty astronomical works and twenty-seven mathematical works to him. See also King, *Survey of the Scientific Manuscripts*, 80–2, C96.

40 İhsanoğlu, ed., *History of the Ottoman State*, vol. 2, 371.

41 See F.J. Ragep, "Astronomy in the Fanārī-Circle."

42 Ekmeleddin İhsanoğlu provides an overview of the formal teaching programs of the madrasas, which he calls "the most indigenous institutions of learning in Islam." İhsanoğlu, "Institutionalisation of Science," 265. For more sweeping surveys, see İhsanoğlu, ed., *History of the Ottoman State*, vol. 2, 368–90, esp. 383–7; and İzgi, *Osmanlı Medreselerinde İlim*. See also Inalcık, *Ottoman Empire*, 165–72, 173–8.

43 Inalcık, *Ottoman Empire*, 173–5. See also Topdemir, "'Abd al-Wājid."

44 Another possible spur for Qāḍīzāde would have been al-Sayyid al-Sharīf al-Jurjānī (d. 816/1413), who composed his own commentary on the *Mulakhkhaṣ* in 811/1409, three years earlier than Qāḍīzāde. It is striking that Qāḍīzāde's commentary led to (at least) twenty-five derivative works, whereas Jurjānī's inspired two. S.P. Ragep, *Jaghmīnī's Mulakhkhaṣ*, appendix 2, nos 10, 10a–b, 12, 12a–x.

45 It is said that Qāḍīzāde parted company with Jurjānī when he found his teacher deficient in the mathematical sciences. This oft-repeated account (true or not) is indicative of Samarqand's allure for those seeking a very high proficiency in a subject. Notwithstanding Qāḍīzāde's comment, Jurjānī's mathematical skills were "at such a high level that he could enter discussions with Jamshīd Kāshī, one of the greatest mathematicians and astronomers in the history of science." Fazlıoğlu, "Samarqand

Mathematical-Astronomical School," 35. Jurjānī also wrote several important astronomical commentaries on Ṭūsī's *Tadhkira*, Shīrāzī's *Tuḥfa*, and Jaghmīnī's *Mulakhkhaṣ*. See Ragep, "Qāḍīzāde al-Rūmī."

46 Qāḍīzāde al-Rūmī was certainly not alone in bringing the scientific knowledge of the Fanārī circle and Marāgha to Samarqand. Jurjānī was also an important bridge. He was a classmate of Fanārī in Bursa and later resided eighteen years in Samarqand (789–807/1387–1405). See Fazlıoğlu, "Samarqand Mathematical-Astronomical School," 34–6.

47 See Kennedy, "Late Medieval Planetary Theory," 365.

48 For example, Swerdlow and Neugebauer associate both fourteenth-century Ibn al-Shāṭir and fifteenth-century Copernicus with the thirteenth-century "Marāgha School": "In a very real sense, Copernicus can be looked upon as, if not the last, surely the most noted follower of the 'Marāgha School.'" Swerdlow and Neugebauer, *Mathematical Astronomy*, part 1, 294–5.

49 See F.J. Ragep, "Persian Context"; and F.J. Ragep, *Naṣīr al-Dīn al-Ṭūsī's Memoir*, vol. 1, 55–6.

50 The Marāgha Observatory became the model for similar, big-science initiatives throughout the world. It was said (true or not) that a childhood visit to the Marāgha Observatory site inspired Ulugh Beg to build the Samarqand Observatory. See Bagheri, "Newly Found Letter," 245–6; Kennedy, "Letter of Jamshīd al-Kāshī," 196 (reprint 727); and Sayılı, *Uluğ Bey*, 60 (Persian), 98 (Eng. trans.). See also Fazlıoğlu, "Samarqand Mathematical-Astronomical School," 7; and Sayılı, *Observatory in Islam*, 274.

51 These works included "Euclid's *Elements*, Ptolemy's *Almagest*, and the 'Middle Books' of mathematics and astronomy with treatises by Euclid, Theodosius, Hypsicles, Autolycus, Aristarchus, Archimedes, Menelaus, Thābit ibn Qurra, and the Banū Mūsā." F.J. Ragep, "Naṣīr al-Dīn al-Ṭūsī," 758.

52 See F.J. Ragep, "Shīrāzī's *Nihāyat al-idrāk*," 51 (Arabic), 55 (Eng. trans.).

53 For further reading, see F.J. Ragep, "*Hay'a*"; F.J. Ragep, "Astronomy"; and F.J. Ragep, *Naṣīr al-Dīn al-Ṭūsī's Memoir*, vol. 1, 29–41. See also the exchange on the subject of *hay'a* between A.I. Sabra and George Saliba: Sabra, "Configuring the Universe"; Saliba, "Arabic versus Greek Astronomy"; Sabra, "Reply to Saliba."

54 Naṣīr al-Dīn al-Ṭūsī provides us with what would become the classical definition of the discipline: "The subject of astronomy is the simple bodies, both superior and inferior, with respect to their quantities, qualities, positions, and intrinsic motions." F.J. Ragep, *Naṣīr al-Dīn al-Ṭūsī's Memoir*, vol. 1, 90–1, book 1, intro., para. 2. See also ibid., vol. 1, 38, "All Simple Bodies as the Subject Matter of Astronomy."

55 Although one finds topics dealing with the inhabited world included within Greek astronomical works, indeed a prominent example being Ptolemy's *Almagest*, book 2, it is significant that Islamic astronomers saw themselves as

doing something new and considerably expanded. F.J. Ragep, *Naṣīr al-Dīn al-Ṭūsī's Memoir*, vol. 1, 38.

56 George Saliba has maintained that the motivation to demarcate astronomy from the taint of astrology within a strictly Islamic context was what gave rise to the discipline of *hayʾa*. Saliba, "Islamic Astronomy in Context," 25–7, 42; Saliba, "Development of Astronomy"; Saliba, "Arabic versus Greek Astronomy," 328–9, 330.

57 For example, Jaghmīnī's *Mulakhkhaṣ* falls into this category. By contrast, in the tenth century *ʿilm al-nujūm* was still being used in Islamic reference books as the general term for "astronomy." Examples include al-Fārābī (Alpharabius) in his *Enumeration of the Sciences*, Abū ʿAbd Allāh al-Khwārizmī in his *Mafātīḥ al-ʿulūm*, and the Ikhwān al-Ṣafāʾ (Brethren of Purity) in epistle 3 of their encyclopaedic *Rasāʾil*, with the latter two designating *ʿilm al-hayʾa* as a branch of *ʿilm al-nujūm*. F.J. Ragep, *Naṣīr al-Dīn al-Ṭūsī's Memoir*, vol. 1, 34–7.

58 For example, Ṭūsī explicitly states in his introduction to the *Tadhkira* that there is a demarcation of subject matter between the disciplines and that the science of *hayʾa* relies on principles that are "proved in another science and are taken for granted in this science." F.J. Ragep, *Naṣīr al-Dīn al-Ṭūsī's Memoir*, vol. 1, 90–1, book 1, intro., para. 1.

59 For a general survey of Ancient and Islamic summary accounts of theoretical astronomy prior to the early thirteenth century, see S.P. Ragep, *Jaghmīnī's Mulakhkhaṣ*, Introduction, §I.3.

60 See F.J. Ragep, *Naṣīr al-Dīn al-Ṭūsī's Memoir*, vol. 1, 13, 21. Ibn al-Haytham informs us in his *Almagest* commentary, "Most commentators on the *Almagest* ... were more interested in proposing alternative techniques of computation than in clarifying obscure points for the beginner." Sabra, "Ibn al-Haytham," 199.

61 Ibn al-Haytham's *Al-Maqāla fī ḥarakat al-iltifāf* is not extant but has been reconstructed from later accounts. F.J. Ragep, "Ibn al-Haytham and Eudoxus," 787.

62 This is Tzvi Langermann's view in his *Ibn al-Haytham's "On the Configuration,"* 25–34.

63 Shank, chapter 4, this volume, points out that Regiomontanus lectured on Farghānī in Padua around 1464. See also Chen-Morris and Feldhay, chapter 5, this volume.

64 For more on Farghānī's *Jawāmiʿ* in general and on his parameters, see S.P. Ragep, *Jaghmīnī's Mulakhkhaṣ*. See also Sabra, "Al-Farghānī"; Abdukhalimov, "Aḥmad al-Farghānī"; and Farghānī, *Jawāmiʿ*, which is Fuat Sezgin's reprint of Jacobus Golius's 1669 Arabic printed edition with Latin translation.

65 This is discussed further in Shank, chapter 4, this volume; and Sylla, chapter 3, this volume.

66 See Langermann, *Ibn al-Haytham's "On the Configuration,"* for an edition of the text, along with an English translation and notes.

67 Tzvi Langermann suggests that Ibn al-Haytham may have had Farghānī's *Jawāmiʿ* in mind when he criticized his predecessors for producing works that "fall short" of putting forth "an explicit enunciation of the way in which the motions of the stars take place on the various spheres." Langermann, *Ibn al-Haytham's "On the Configuration,"* 26. See also F.J. Ragep, *Naṣīr al-Dīn al-Ṭūsī's Memoir*, vol. 1, 30–3.

68 Kharaqī (d. 553/1158) explicitly cites Ibn al-Haytham in three of his *hayʾa* works, namely his *Muntahā*, *Tabṣira*, and *ʿUmda*; and he credits him with being an important influence for motivating him to consider solid spheres as opposed to imaginary circles in astronomy, as well as to work toward reconciling physics with the mathematical models. All this would have an important influence on the Marāgha scholars. Naṣīr al-Dīn al-Ṭūsī devotes a chapter to the configuration of the epicycle orbs of the planets that is due to Ibn al-Haytham in an appendix to his Persian *Risālah-i Muʿīniyya*. F.J. Ragep, "Ibn al-Haytham and Eudoxus." See also F.J. Ragep, chapter 7, this volume.

69 See Sylla, chapter 3, this volume. The influence of Ibn al-Haytham on Peurbach and on other Renaissance scholars is noted by Aiton, "Peurbach's *Theoricae novae planetarum*," 7–8; and Hartner, "Mercury Horoscope," 122–7.

70 Extant copies of this treatise in Arabic are rare, and these are devoid of illustrations. For his edition, Tzvi Langermann also consulted Hebrew and Latin translations, with some figures contained in the Hebrew copies, not the Latin. Langermann, *Ibn al-Haytham's "On the Configuration,"* 34–41. However, when these were drawn and by whom remain unclear. However, the closing statements of the chapters on the orbs of the Sun, Moon, Mercury, Venus, the upper planets, the fixed stars, and the highest orb seem to indicate figures were intended. Ibid., 131 (para. 209), 150 (para. 272), 177 (para. 321), 196 (para. 337), 206 (para. 359), 215 (para. 374), 223–4 (para. 382) (English), 37, 46, 54, 57, 60, 63, 65 (Arabic). See also Sylla, chapter 3, this volume.

71 See F.J. Ragep, *Naṣīr al-Dīn al-Ṭūsī's Memoir*, vol. 1, 33.

72 F. Jamil Ragep points out that Ibn al-Haytham "seems to go out of his way to indicate that previous [astronomical] work has assumed the existence of solid spheres." Ibid., vol. 1, 30–3, quotation at 31. Tzvi Langermann also states that "it is quite clear … that Ibn al-Haytham does not regard himself to be the first person to address the problem of the physical description of the heavens." Langermann, *Ibn al-Haytham's "On the Configuration,"* 25.

73 For the Arabic edition of the *Muntahā* with a Persian translation, see Ghalandari, "Survey of the Works of '*Hayʾa*.'"

74 See ibid., abstracts 7 and 8; and İzgi, *Osmanlı Medreselerinde İlim*, vol. 1, 405–6, no. 7. Extant copies of Kharaqī's *Tabṣira* far outnumber his *Muntahā*. I am also currently unaware of any commentaries on the *Muntahā*, but there are a few on the *Tabṣira* (all seemingly composed in the thirteenth and fourteenth centuries) that indicate the treatise disseminated widely, was studied in Yemen and in eighteenth-century Egypt, and was translated into Hebrew. See also Fazlıoğlu, "Kamāl al-Dīn al-Turkmānī"; Langermann, "Kharaqī"; and Schmidl, "ʿUrḍī."

75 For an Arabic critical edition and English translation of, and commentary on, this important and influential treatise, see S.P. Ragep, *Jaghmīnī's Mulakhkhaṣ*. Appendix 2 lists sixty-one commentaries and supercommentaries on, and translations of, the *Mulakhkhaṣ*.

76 See Morrison, "Role of Oral Transmission."

77 My survey findings indicate that well over 50 per cent of the fifteenth-century authors who wrote on theoretical issues of astronomy had derivative works on the *Mulakhkhaṣ*: ʿAbd al-Wājid, Qāḍīzāde al-Rūmī, Humām al-Ṭabīb, al-Sayyid al-Sharīf al-Jurjānī, Sinān Pāshā, Fakhr al-Dīn al-ʿAjamī, Muḥyī al-Dīn al-Niksārī, al-Ḥaqq al-Kubnawī, Ḥusām Dellākoğlu, Muḥyī al-Dīn Akhawayn, and Fatḥ Allāh al-Shīrwānī all wrote commentaries or supercommentaries on Jaghmīnī's *Mulakhkhaṣ*. Shīrwānī would also compose a *Tadhkira* commentary (completed 879/1475) after having studied those of Nīsābūrī (composed 711/1311) and Jurjānī (composed 811/1409).

78 For example, a student of Muḥammad ʿAbduh (d. 1905) reported that *Sayyid* Jamāl al-Dīn al-Afghānī read the *Mulakhkhaṣ* in nineteenth-century Cairo with his students. Hildebrandt, "Waren Ǧamāl ad-Dīn al-Afġānī," 215n22.

79 F.J. Ragep, "New Light on Shams," 234–5.

80 According to David King, Cairo and Damascus were "leading centers of astronomy in the Islamic world, perhaps in the world in general … Mamluk astronomers worked in each of the major branches of astronomy: theoretical and computational planetary astronomy, spherical astronomy and timekeeping, instrumentation, and folk astronomy and astrology." King, "Astronomy of the Mamluks," 531, 551.

81 For the impact of the Tīmūrids on science, see Fazlıoğlu, "Samarqand Mathematical-Astronomical School," 20–2. For an extensive survey of Ottoman works, see Şeşen and İzgi, eds, *Osmanlı astronomi*; and Şeşen and İzgi, eds, *Osmanlı matematik*. As for India, keep in mind that Ẓahīr al-Dīn Bābur (1483–1530), the founder of the great Mughal Empire of India in 1526, was a Tīmūrid ruler and originally from Uzbekistan. Arabic and Persian scientific texts were translated into Sanskrit especially during the Sultanate period (twelfth-fifteenth centuries) and the Mughal period (sixteenth-eighteenth centuries). See also Ansari, "On the Transmission"; and Ansari, "Transmission of Islamic Exact Science."

82 See Fazlıoğlu, "Samarqand Mathematical-Astronomical School," 15–16. See also Sayılı, *Observatory in Islam*, 271; and van Dalen, "Ulugh Beg."

83 See Şeşen and İzgi, eds, *Osmanlı astronomi*, vol. 1, 37, no. 9.

84 See F.J. Ragep, "Freeing Astronomy," "ʿAlī Qushjī and Regiomontanus," and "Copernicus," 72–5. See also Saliba, "Al-Qushjī's Reform"; Saliba, "Reform of Ptolemaic Astronomy"; and Şeşen and İzgi, eds, *Osmanlı astronomi*, vol. 1, 27–38, no. 11 (Ali Kuşçu).

85 As mentioned earlier, Qāḍīzāde's *Sharḥ al-Mulakhkhaṣ* was particularly popular and studied well into the nineteenth century. To get a sense of the numbers of manuscripts involved, there are about 300 copies of his commentary in Istanbul libraries alone. Şeşen and İzgi, eds, *Osmanlı astronomi*, vol. 1, 8–20.

86 See Şeşen and İzgi, eds, *Osmanlı astronomi*, vol. 1, 30–3, no. 2, for *Al-Risāla dar ʿilm-i hayʾa*; and vol. 1, 33–5, no. 3, for *Al-Risāla al-Fatḥiyya*. See also İzgi, *Osmanlı Medreselerinde İlim*, vol. 1, 392–5, no. 2; and Fazlıoğlu, "Qushjī."

87 The view of the discipline of *hayʾa* as "the most noble of sciences" as a way to glorify God was explicitly articulated in Kharaqī's *Tabṣira* (twelfth century), Quṭb al-Dīn al-Shīrāzī's *Nihāya* (thirteenth century), and Qāḍīzāde's *Sharḥ al-Mulakhkhaṣ* (fifteenth century). See F.J. Ragep, "Shīrāzī's *Nihāyat al-idrāk*," 49 (Arabic), 54 (Eng. trans.); and F.J. Ragep, "Freeing Astronomy," 51, 64.

88 This point is developed in S.P. Ragep, *Jaghmīnī's Mulakhkhaṣ*, 115–54. My current research further examines evidence establishing that the mathematical sciences were being studied in Islamic institutions.

89 Fazlıoğlu, "Samarqand Mathematical-Astronomical School," 33–4; Fazlıoğlu, "Qushjī."

90 For a view that challenges this standard approach as it applies to Islamic scientific education, see S.P. Ragep, "Teaching of Theoretical Astronomy."

91 Brentjes, "Mathematical Sciences," 329.

92 According to Sonja Brentjes, in contrast to "rich evidence for Timurid courtly patronage of the mathematical sciences, there is little information known so far that illustrates the study of these disciplines and their works at madrasas and other endowed teaching institutes in the Timurid realm. One exception is the well-known case of the *madrasa* sponsored by Uluġ Beg in Samarqand where Qāżīzāda Rūmī and Ġiyāṯ al-Dīn Kāšī taught and ʿAlī Qušjī first studied and then taught." Ibid. 329–33, quotation at 332.

93 This sentiment, articulated by A.I. Sabra over thirty years ago, asserts that in medieval Islam a scientific education was "largely an individual affair in which individual students made special arrangements with individual teachers." And further, "insofar as the madrasa had anything like what we call a curriculum, the study of the 'ancient sciences' was not part of it." Sabra, "Science, Islamic," 85, 86.

94 Makdisi, *Rise of Colleges*, 9, 77–8; Makdisi, "Muslim Institutions," 15–16.

95 See Wisnovsky, "Nature and Scope"; Dallal, *Islam, Science*; Griffel, *Al-Ghazālī's Philosophical Theology*; F.J. Ragep, "Freeing Astronomy"; and Saliba, *Islamic Science*. See also Brentjes, "On the Location," "Reflections on the Role," and "Mathematical Sciences."

96 Grant cites Makdisi, *Rise of Colleges*, as support for his assertion that "[b]oth civilizations [Islam and the West] taught, studied, and wrote about logic, natural philosophy, and the sciences. But in contrast with Islam, the West taught, studied, and wrote about these disciplines in universities that fully supported them … In Islam, the foreign sciences, which comprised the analytic subjects derived ultimately from the Greeks, were rarely taught in religious schools such as the *madrasas*, which formed the core of Islamic higher education." Grant, "Fate of Ancient Greek," 523.

97 According to Huff, "the preceding legal, institutional and intellectual developments … made the Copernican innovation possible. That is, long before Copernicus and Galileo, there was an intellectual tradition established in Europe, above all in the *universities*, that, yes, *institutionalized* the study of natural phenomena, particularly by placing the corpus of Aristotle, along with a number of Arabic works and commentaries, at the centre of the university curriculum. This occurred in the twelfth and thirteenth centuries. In short, the Copernican revolution was a product of the educational system put in place by Europeans several hundred years earlier. As is well-known, the *madrasas* of the Middle Eastern world systematically excluded philosophy and the natural sciences from any 'formal' teaching conducted within their confines during this period of time. (I put 'formal' in quotation marks because there was no formal curriculum in the *madrasas*.) Evidently, the teaching of philosophy and the natural sciences ran against the religious commitments and identity of the *madrasas*, an identity that persisted into the twentieth century." Huff, "Rise of Early Modern Science," 117, emphasis in original. In this reply to Saliba, Huff (ibid. 127n4), like Grant (see note 96 above), cites Makdisi, *Rise of Colleges*.

98 Brentjes, "On the Location," 60.

99 As Jonathan Berkey has repeatedly claimed, "an education was judged not on *loci* but on *personae*." He further argues that "institutions themselves seem to have had little or no impact on the character or the processes of the transmission of knowledge." Berkey, *Transmission of Knowledge*, 23. See also Berkey, "Madrasas Medieval and Modern," 43.

100 See Bisaha, chapter 2, this volume.

101 This focus is one of the interesting aspects of Morrison, chapter 8, this volume, which addresses the role of Jewish scholars as scientific intermediaries between Islam and Europe.

102 This is A.I. Sabra's view regarding Ibn al-Shāṭir's accomplishments in theoretical astronomy, which he argues were not related to place but

rather linked to Ibn al-Shāṭir's desire to resolve specific problems posed by earlier mathematicians. According to Sabra, attention should be directed to "*situations*, as distinguished from *tradition*." Sabra, "Situating Arabic Science," 669-70, emphasis in original. Brentjes seems to echo this sentiment in seeking to understand the mathematical sciences "in the terms of their specific time and locality." Brentjes, "Mathematical Sciences," 328.

103 Within an Islamic context, the attempt to return to a purer version of Aristotelian cosmology, one free of eccentrics and epicycles, was "a most peculiar kind of Andalusian Astronomy." Samsó, "Andalusian Astronomy," 2–3. As Sabra has pointed out, this was part of a system of ideas constructed to provide an Andalusian sense of identity consciously directed against intellectual authorities in the eastern part of Islam. Sabra, "Andalusian Revolt," 143. Although this homocentric cosmology was relatively uncommon in an Islamic context, it would have a major impact on theoretical astronomy in the Latin West, a point brought out in Morrison, chapter 8, this volume; and Shank, chapter 4, this volume.

104 See note 97.

105 See Inalcık, *Ottoman Empire*, 3–34.

106 Langermann, "From My Notebooks: A Compendium"; Morrison, "Scholarly Intermediary." See also Morrison, chapter 8, this volume.

107 F.J. Ragep, "ʿAlī Qushjī and Regiomontanus."

108 Much attention has been paid to the connections between the mathematical models of postclassical Islamic astronomers and Copernicus. But it should also be emphasized that works like the *Tadhkira* and recensions of the *Almagest* represented certain types of genres of writing that existed for centuries in the Islamic world before we see their like in such works as Peurbach's *Theoricae novae planetarum* or Regiomontanus's *Epitome*.

109 Hadzibegovic, "Compendium of the Science."

110 See Saliba, "Flying Goats and Other Obsessions." In his final exchange of views with Huff, Saliba voices his frustration with those refusing to alter opinions regarding Islamic science when shown evidence to the contrary.

111 See Sylla, chapter 3, this volume. See also Barker, "Reality of Peurbach's Orbs."

112 Aiton, "Peurbach's *Theoricae novae planetarum*," 7–8.

113 Hartner, "Mercury Horoscope," 124, emphasis in original.

114 Such mathematical techniques included the Ṭūsī-couple and possibly Ibn al-Shāṭir's double epicycle. Dobrzycki and Kremer, "Peurbach and Marāgha," 189.

115 See Sylla, chapter 3, this volume; and Shank, chapter 4, this volume. As an illustration of the fluidity of European astronomy, Regiomontanus, who wrote a first-rate exposition of Ptolemaic astronomy, still entertained the possibility of a homocentric astronomy.

CHAPTER SEVEN

1 F.J. Ragep, "Copernicus." This point is made even more forcefully in my "Ibn al-Shāṭir and Copernicus: The Uppsala Notes Revisited," where I maintain that there is a stronger connection between ʿAlāʾ al-Dīn ibn al-Shāṭir (fourteenth century) and Copernicus's models and heliocentrism than has been previously claimed.

2 Here, we need to acknowledge Mario Di Bono, who, in a valuable article, insists on distinguishing the various versions of the Ṭūsī-couple. Di Bono, "Copernicus, Amico, Fracastoro." Di Bono is building on the earlier work of Noel Swerdlow, especially his "Aristotelian Planetary Theory in the Renaissance."

3 On *Risālah-i Muʿīniyya*, see F.J. Ragep, "Persian Context," and *Naṣīr al-Dīn al-Ṭūsī's Memoir*, vol. 1, 65–6. See also Kennedy, "Two Persian Astronomical Treatises."

4 The relevant parts of the Persian text discussed in this paragraph, along with translation, are in F.J. Ragep, "Persian Context," 123–5.

5 F.J. Ragep, *Naṣīr al-Dīn al-Ṭūsī's Memoir*, vol. 1, 208.

6 The name "*Dhayl-i Muʿīniyya*" is found in the only dated manuscript of Ṭūsī's text, namely Tashkent, Uzbekistan, Al-Biruni Institute of Oriental Studies, MS 8990, fols 1a, 33a, 33b.

7 Ṭūsī, *Dhayl-i Muʿīniyya*, Tashkent, Al-Biruni Institute of Oriental Studies, MS 8990, fol. 46a (original foliation):

تمت الرسالة اتفق فراغ المصنف رفع الله مراتبه في معارج القدس من تأليفه أوائل جمادى
الآخرة سنة ٦٤٣ هجرية بمقام بلدة تون بالبستان المعروف بباغ بركه

8 On Tūn as one of the residences of the local Ismāʿīlī rulers, see Daftary, "Dāʿī," 592, col. 1.

9 Ṭūsī, *Ḥall-i mushkilāt-i Muʿīniyya*, 7:

اما استقامت حركت مركز تدوير از محيط مايل بر سمت مركزش وبعد از آن رجوع او هم بر
آن سمت تا بمحيط رسيدن بی آنکه خرقی و التيامی لازم آيد يا خللی باستدارت حركات راه
يابذ بر آن وجه تواند بود كه ياد كنيم.

"The rectilinear motion of the center of the epicycle away from the circumference of the inclined [orb] in the direction of its centre and then its return on that same line until it reaches the circumference – without there being any tearing and mending, or any rupture in the circular motions – can be in the way we shall mention."

10 See F.J. Ragep, *Naṣīr al-Dīn al-Ṭūsī's Memoir*, vol. 1, 208–23, vol. 2, 448–56.

11 The relevant passages from *Risālah-i Muʿīniyya*, book 2, chs 5, 6, 8, with English translation, can be found in F.J. Ragep, "Persian Context," 123–5.

12 For details and an edition and translation of the relevant chapter from the *Ḥall*, see F.J. Ragep, "Ibn al-Haytham and Eudoxus."

13 This chronology contradicts George Saliba's contention, followed by Di Bono and others, that the two-equal-circle version in *Taḥrīr al-Majisṭī* was the first occurrence of any version of the Ṭūsī-couple. But clearly the new dating of the *Ḥall* should put to rest this earlier proposal. Compare Saliba, "Role of the *Almagest* Commentaries."

14 This comment corresponds to the *Almagest*, book 13, ch. 2; Ptolemy, *Ptolemy's Almagest*, 599-601.

15 Ṭūsī, *Taḥrīr al-Majisṭī*, fols 201a–202a:

أقول هذا كلام خارج من الصناعة {201ب} غير مقنع في هذا الموضع فإنّ من الواجب على صاحب هذه الصناعة أن يضع دوائر وأجراماً ذوات حركات متشابهة على نضد وترتيب يتركّب من جميعها هذه الحركات المحسوسة المختلفة ثم إنّ كون هذه الحركات على محيط الدوائر الصغار المذكورة كما تقتضي خروج أقطار التداوير عن سطوح الخارجة المراكز في العرض شمالاً وجنوباً كذلك تقتضي خروجها عن محاذاة مركز البروج أو موازاة أقطار على سطح البروج بأعيانها في الطول إقبالاً وإدباراً بقدر تلك العروض بأعيانها وذلك مخالف للوجود ولا يمكن أن يقال إنّ ذلك التفاوت محسوس في العرض وغير محسوس في الطول لتساويها في المقدار والبعد من مركز البروج فإن جعل قطر الدائرة الصغيرة بقدر جميع العرض في إحدى الجهتين وتوهّم أنّ مركزها يتحرّك على محيط دائرة أخرى مساوية لها مركزها في سطح الخارج المركز بقدر نصف حركة طرف قطر التدوير على محيط الدائرة الأولى وإلى خلاف جهتها حدث الانتقال إلى الشمال والجنوب بقدر العرض من غير أن يحدث في الطول تقدّم وتأخّر وليكن لبيانه اب قطعة من الخارج و جد من دائرة العرض المارّة بطرف قطر التدوير وقد تقاطعا على ه و هز هم جميع العرض في الجهتين وهح نصفه في إحديهما ونرسم على ح ببعد هح دائرة هز وعلى ه ببعد حه دائرة ح ط ك ل ونتوهّم طرف قطر التدوير على نقطة ز متحرّكاً على دائرة هز في جهة ج إلى ب ومركز ح متحرّكاً على دائرة ح ط ك ل في جهة ج إلى ا نصف تلك الحركة فظاهر أنّه إذا قطع ح ربعاً وانتهى إلى ط قطع ز نصفاً وانتهى إلى ه ثم إذا قطع ح ربعاً آخر وانتهى إلى ك قطع ز نصفاً آخر وانتهى {202آ} إلى م وإذا قطع ح ربعاً ثالثاً وانتهى إلى ل قطع ز نصفاً آخر وانتهى ثانياً إلى ه وإذا تمّ ح دورة عاد ز إلى موضعه الأول فهو دائماً يتردّد فيما بين زم على خطّ ج د غير مائل عنه إلى جهتي اب فهذا بيان هذا الوجه ولكن يلزم عليه أن يكون زمان كون القطر في الشمال مساوياً لزمان كونه في الجنوب والوجود بخلاف ذلك وأمّا القول بحركته على محيط دائرة حول نقطة غير مركزها على ما ذكر بطلميوس فمحتاج إلى نظر يحقّقه على ما مرّ ونعود إلى الكتاب

16 The unequal times in the *Almagest* occur because this motion in latitude is coordinated with the irregular motion, brought about by the equant, of the epicycle centre on the deferent. F.J. Ragep, *Naṣīr al-Dīn al-Ṭūsī's Memoir,* vol. 2, 455.

17 Ibid., vol. 1, 216–21.

18 For a fuller account of the curvilinear version, see ibid., vol. 2, 453–6. It should be noted that the curvilinear version does not in fact produce motion on a great circle arc; there is a small discrepancy resulting in a narrow, pinched figure-eight motion. This was noticed by at least one commentator on the *Tadhkira,* Shams al-Dīn al-Khafrī (fl. 1525 CE). But the maximum deviation from a great circle arc, which occurs when using the curvilinear version to deal with the problem of the Moon's prosneusis, is only 0.214°, which is about 0.87 per cent. Ibid., vol. 2, 455n55, 455n56. For an illustration of the deviation, see figure C26, in ibid., vol. 1, 361.

19 The purpose of Ibn al-Haytham's proposal was to provide a physical basis for the circular path of the epicycle apex A in Ptolemy's latitude theory; as far as is known, he was not concerned with the resultant motion of S, which traces a "hippopede" in Eudoxus of Cnidus's theory (as shown in figure 7.9). It is interesting that Regiomontanus's version of this device resulted in a curvilinear oscillation of S along a great circle arc, something that had been proposed earlier by Joseph ibn Naḥmias. For details, see Morrison, chapter 8, this volume, especially figure 8.3. For the reason that the Eudoxan-couple should produce a hippopede, not a curvilinear oscillation, see Neugebauer, "On the 'Hippopede' of Eudoxus."

20 F.J. Ragep, *Naṣīr al-Dīn al-Ṭūsī's Memoir,* vol. 1, 220–3.

21 Shīrāzī, *Al-Tuḥfa al-shāhiyya,* fol. 34a:

ويمكن أن يجعل هذا دليلاً على امتناع السكون بين حركتين صاعدة وهابطة على سمت قطر من أقطار الأرض

22 Langermann, "Quies Media," provides an excellent summary of the *quies media* question and discusses a number of Islamic thinkers, including Shīrāzī, who dealt with it.

23 The restriction of the date will exclude a discussion of the translation into Sanskrit of part of ʿAbd al-ʿAlī al-Birjandī's (d. 1525–26) commentary on Ṭūsī's *Tadhkira,* the part containing the presentation of the Ṭūsī-couple. On this translation, see Kusuba and Pingree, *Arabic Astronomy in Sanskrit.*

24 On the use of *aṣl* to translate the Greek term *hypothesis,* see Morrison, chapter 8, this volume, note 10.

25 These works are currently extant in three codices, two in the Vatican and one in the Biblioteca Medicea Laurenziana in Florence.

26 Edition and translation in Paschos and Sotiroudis, *Schemata of the Stars,* 26–53.

27 This resemblance was first recognized by Otto Neugebauer, who reproduced diagrams from Vatican Gr. 211, fol. 116r, in his *History*, part 3, 1456.
28 F.J. Ragep, "New Light on Shams."
29 This use of the earlier works can most easily be established from the list of star names found in Paschos and Sotiroudis, *Schemata of the Stars*, 30–7. For a discussion and the evidence, see F.J. Ragep, "New Light on Shams," 239, 241–2.
30 It was reported that there was great reluctance by the Persians to teach astronomy to a Byzantine because of a legend that doing so would lead to the former's demise. F.J. Ragep, "New Light on Shams," 231–2.
31 Paschos and Sotiroudis, *Schemata of the Stars*, 42–5. On the *Ḥall*, see above and F.J. Ragep, "New Light on Shams," 242.
32 David Pingree states that Vatican Gr. 211 is listed in the Vatican inventory of 1475 and that Vatican Gr. 1058 is listed in the inventory made around 1510 but may well have been in the collection earlier. Pingree, *Astronomical Works*, vol. 1, 23, 25.
33 See above and F.J. Ragep, "Ibn al-Haytham and Eudoxus."
34 Langermann, "Medieval Hebrew Texts," 34.
35 Droppers, "Questiones de Spera," 462–4; Kren, "Rolling Device."
36 Goddu, *Copernicus*, 481, 484.
37 The parts of Kren's translation in "Rolling Device," 490, that have been changed are in italics; my suggested revisions are in brackets immediately following. Droppers, "Questiones de Spera," 285, 287, 289, also provides a translation, somewhat more literal than Kren's, that I have also taken into account.
38 Here is Kren's Latin version in "Rolling Device," 491n3 (compare Droppers, "Questiones de Spera," 284, 286, 288):

Circa hanc questionem, pono 3 pulcras conclusiones. Prima est quod possibile est quod aliquis planeta secundum quodlibet sui moveatur in perpetuum motu recto composito ex pluribus motibus circularibus, ita quod iste motus proveniat a pluribus intelligentiis quarum quelibet intenderet movere motu circulari nec frustratur ab intentione sua.

Pro cuius probatione, suponatur per ymaginationem, sicut faciunt astrologi, quod A sit circulus deferens alicuius planete, vel centrum eius, et sit B circulus epiciclus eiusdem planete, et C sit corpus planete vel centrum eius; hoc habeo pro eodem. Et ymaginetur linea BC, exiens de centro epicicli ad centrum planete, et CD sit linea in planeta supra quam alia cadat perpendiculariter. Moveatur etiam A circulus supra centrum ad orientem, et B ad occidentem, et C planeta supra centrum suum volvatur ad orientem. Cum ergo linea BC semper sit equalis, quia est semidyameter, ponatur quod quantum B descendit ad motum deferentis, tantum C punctus ascendat per motum epicicli. Ex quo patet intuenti quod punctus C per aliquod certum tempus movebitur super lineam rectam. Tunc ponatur ultra quod perifora qua punctus B ascenderet motu suo tantum descendat motu planete. Et patet iterum quod punctus D continue movebit in eadem linea. Ergo totum corpus planete movebitur motu recto usque ad aliquem terminum, et iterum poterit reverti in motu consimilli.

39 Figure 7.10 is from Droppers, "Questiones de Spera," 287, reproduced by Goddu, *Copernicus*, 481. Note that despite the use of *corpus* in referring to the planet, Goddu insists that "there is no indication that Oresme was directly concerned with the physical characteristics of the bodies or the mechanisms" (481). This interpretation of Oresme may be why both Droppers and Goddu seem capable of ignoring Oresme's clear statement that it is the "entire body of the planet" that moves in a straight line. We should also note here that the title of this *questio* is "Whether any heavenly body (*corpus celeste*) is moved circularly."

40 Kren, "Rolling Device," 492.

41 In contrast, Goddu, *Copernicus*, 480, finds Kren's reconstruction "implausible," but this assessment seems to be based on the grounds that Ṭūsī's construction requires two circles whereas Oresme's requires three. He apparently is unaware of Ṭūsī's physicalization of his geometrical device and his explicit use of three spheres in the *Tadhkira*. F.J. Ragep, *Naṣīr al-Dīn al-Ṭūsī's Memoir*, vol. 1, 200–1, 350–1, vol. 2, 435–7. Kren is able to see this use of three spheres even though she was depending, as mentioned, on an earlier French translation of this passage in which Ṭūsī describes how to physicalize his device. Kren, "Rolling Device," 493n8. Goddu had access to a new translation and discussion of this passage in the *Tadhkira*, so his claim that Ṭūsī does not have a three-sphere model is odd.

42 What follows is a modified version of what is described in the *Tadhkira*, book 2, ch. 11, para. 4. F.J. Ragep, *Naṣīr al-Dīn al-Ṭūsī's Memoir*, vol. 1, 200–1; see also fig. C13, in ibid., vol. 1, 351. For a discussion of this passage, see ibid., vol. 2, 435–8.

43 Droppers, "Questiones de Spera," 291.

44 See Morrison, chapter 8, this volume.

45 Dobrzycki and Kremer, "Peurbach and Marāgha."

46 Aiton, "Peurbach's *Theoricae novae planetarum*," 36, 36n118.

47 Dobrzycki and Kremer, "Peurbach and Marāgha," 233n53.

48 Dobrzycki, "Theory of Precession," 51.

49 In addition to the previous reference, see also Dobrzycki, "Astronomical Aspects," 122; and Räumer, "Johannes Werners Abhandlung."

50 On Amico, see Swerdlow, "Aristotelian Planetary Theory"; and Di Bono, *Le sfere omocentriche*.

51 Ṭūsī refers to this third as "the enclosing sphere" (*al-kura al-muḥīṭa*). F.J. Ragep, *Naṣīr al-Dīn al-Ṭūsī's Memoir*, vol. 1, 220–1. Amico calls it a "withstanding (*obsistens*) sphere." Swerdlow, "Aristotelian Planetary Theory," 41.

52 Di Bono, "Copernicus, Amico, Fracastoro," 141. Ṭūsī does not mention this problem, but it is mentioned by at least one commentator on the *Tadhkira*. See F.J. Ragep, *Naṣīr al-Dīn al-Ṭūsī's Memoir*, vol. 2, 455; and note 18 above.

53 Di Bono, "Copernicus, Amico, Fracastoro," 143–4.

54 Swerdlow and Neugebauer, *Mathematical Astronomy*, part 1, 47.

55 Here, we follow the lead of Di Bono, "Copernicus, Amico, Fracastoro," esp. 138–41.

56 Swerdlow and Neugebauer, *Mathematical Astronomy*, part 1, 136.

57 Swerdlow, "Derivation and First Draft," 483, 497.

58 Ibid., 503.

59 See Di Bono, "Copernicus, Amico, Fracastoro," 140–1.

60 I do not deal here with all the "transmission skeptics" but focus only on the ones who have dealt specifically, using original ideas, with the transmission of the Ṭūsī-couple to medieval and early modern Europe. In particular, I do not consider here the derivative arguments of Viktor Blåsjö in "A Critique of the Arguments for Maragha Influence," 185–6, or those of Michel-Pierre Lerner and Alain-Philippe Segonds in their translation of Copernicus, *De revolutionibus (Des révolutions)*, vol. 1, 551–7. Likewise, I do not deal with André Goddu's response to criticisms by Peter Barker and Matjaž Vesel of his handling of the issue of transmission of Islamic astronomy to Copernicus since it is not germane to my own criticisms contained here. Goddu, "Response to Peter Barker," 251–4.

61 Swerdlow and Neugebauer, *Mathematical Astronomy*, part 1, 47. The emphatic way that this acceptance of late-Islamic influence is stated is most likely due more to Swerdlow than to Neugebauer, for see the latter's earlier remark that "[t]he mathematical logic of these methods is such that the purely historical problem of contact or transmission, as opposed to independent discovery, becomes a rather minor one." Neugebauer, "On the Planetary Theory," 90. Nonetheless, in a personal communication, Swerdlow assured me that Neugebauer completely endorsed the phrasing in their *Mathematical Astronomy in Copernicus's De Revolutionibus*. Edward S. Kennedy and Willy Hartner also entertain little doubt that Copernicus's work was heavily influenced by his Islamic predecessors. Kennedy, "Late Medieval Planetary Theory"; Hartner, "Copernicus, the Man, the Work." A recent rejoinder to André Goddu's skepticism regarding an Islamic influence on Copernicus has been made by Barker and Vesel, "Goddu's Copernicus," 327–32. Goddu's answer, in which he distances himself from an outright rejection of Islamic influence, can be found in his "Response to Peter Barker," 251–4.

62 Veselovsky, "Copernicus and Naṣīr al-Dīn al-Ṭūsī."

63 Copernicus, *De revolutionibus*, book 5, ch. 25.

64 F.J. Ragep, *Naṣīr al-Dīn al-Ṭūsī's Memoir*, vol. 2, 430.

65 Copernicus, *On the Revolutions*, 369, 429 (commentary by Rosen); Swerdlow, "Copernicus's Four Models," 146n5, 155n8; Prowe, *Nicolaus Coppernicus*, vol. 1, part 2, 407, cited by Rosen in Copernicus, *On the Revolutions*, 369.

66 Di Bono, "Copernicus, Amico, Fracastoro," 146.

67 Copernicus, *On the Revolutions*, 279.

68 Ibid., 126 (in book 3, ch. 4, where it was crossed out in the autograph, and in book 3, ch. 5, where it was left in).

69 This idea is also the main thrust of Blåsjö, "Critique of the Arguments."

70 Di Bono, "Copernicus, Amico, Fracastoro," 149.

71 Ibid., 133.

72 Ibid., 149 (referring to Neugebauer's statement quoted in note 61 above).

73 Note again that Goddu dismisses out of hand Kren's mostly correct reconstruction.

74 Grażyna Rosińska claims that Brudzewo owes his two-sphere model for the Moon to Sandivogius, but this is far from clear. Rosińska, "Naṣīr al-Dīn al-Ṭūsī?" Sandivogius seems to be proposing one additional orb (not two) for the Moon and for an entirely different purpose, namely to keep its single face oriented toward the observer.

75 Mancha, "Ibn al-Haytham's Homocentric Epicycles."

76 This conclusion, as part of a longer study on Brudzewo, is also reached by Barker, "Albert of Brudzewo's *Little Commentary*," 137–9. Peter Barker seems unaware of José Luis Mancha's earlier work.

77 Goddu, *Copernicus*, 157: "Experts have exaggerated the supposed identity between Copernicus's and al-Shatir's models and the Tusi couple. Di Bono explains the similarities plausibly as matters of notation and convention. Di Bono also shows that Copernicus's use of the models required an adaptation, and, we may add, if he was capable of adapting geometrical solutions, then why not the solution in Albert's [i.e., Brudzewo's] treatise? The question should be reconsidered." One hardly knows where to begin. First, Di Bono does not deal with Ibn al-Shāṭir's models. Second, the adaptation about which Di Bono is speaking (i.e., the two-equal-sphere model) already occurred with Ṭūsī, as we have seen. Third, for Goddu to think that Copernicus could have simply adapted Brudzewo's cryptic and ultimately unrelated remarks to come up with Ibn al-Shāṭir's models in the *Commentariolus*, one must assume that Goddu has never examined those models.

78 It should be noted that some of this evidence would have been available to Di Bono and even more to Goddu, whose book was published in 2010. It is unfortunate that the presumed lack of transmission that Di Bono and Goddu point to does seem to be at work in the present when we consider how slowly the work of scholars working on Islamic science seems to get transmitted to their colleagues working on the Latin West. For example, Goddu, who is mainly concerned with Copernicus's relation to the Aristotelian tradition, completely ignores the possible transmission from

Islamic sources of a number of Copernican ideas related to natural philosophy, such as the motion of the Earth, the assertion of a non-Aristotelian astronomical physics, and the heliocentric transformation itself. Summarized in F.J. Ragep, "Copernicus."

79 F.J. Ragep, "New Light on Shams," 243–5.

80 Pingree, *Astronomical Works*, 18. But there are certainly examples of Arabic works going into Greek. See Mavroudi, *Byzantine Book*; Touwaide, "Arabic Urology in Byzantium"; and Touwaide, "Arabic Medicine." Joseph Leichter believes that Chioniades may have learned or improved his Arabic at some point. Leichter, "Zīj as-Sanjarī," 11–12.

81 Mercier, "Greek 'Persian Syntaxis,'" 35–6, reproduced in Leichter, "Zīj as-Sanjarī," 3.

82 Information on the manuscripts is from Pingree, *Astronomical Works*, 23–8.

83 Swerdlow and Neugebauer, *Mathematical Astronomy*, part 1, 48n9.

84 Pingree, *Astronomical Works*, 25.

85 An excellent summary of what is known of Copernicus's life can be found in Swerdlow and Neugebauer, *Mathematical Astronomy*, part 1, 3–32.

86 Comes, "Possible Scientific Exchange." Note also that Tzvi Langermann alludes to the possibility of a link between Alfonso's court and Muḥyī al-Dīn al-Maghribī, who was of Andalusian origin but spent most of his career in Syria and Iran. Langermann, "Medieval Hebrew Texts," 35.

87 Lévy, "Gersonide, commentateur d'Euclide," 90–1, 100–15.

88 Heath, trans., *Thirteen Books*, vol. 1, 208–12. See also Jones, "Medici Oriental Press."

89 Langermann, "Medieval Hebrew Texts," 34–5. Finzi's notebook into which he copied the construction is currently preserved at the Bodleian Library in Oxford.

90 For example, Finzi states that he made a "translation with the help of a non-Jew here in the city of Mantua," and in another context he states, "I saw them in the Toledan Tables in the possession of a certain Christian." Langermann, "Scientific Writings," 26, 41.

91 These numbers are based upon the Islamic Scientific Manuscripts Initiative (ISMI) database, which is being developed collaboratively by the Institute of Islamic Studies at McGill University and the Max Planck Institute for the History of Science in Berlin. This is most definitely a conservative estimate of witnesses since the number of extant manuscripts in this database will surely increase as other libraries and private collections come to be catalogued and examined.

92 For example, in the encyclopaedic work entitled *Unmūdhaj al-ʿulūm* by Muḥammad Shāh al-Fanārī (d. 1435–36 CE), the author includes a discussion of the latitude problems of Mercury and Venus, as well as Ṭūsī's solution for them. F.J. Ragep, "Astronomy in the Fanārī-Circle," 168–9, 176.

93 George Saliba has done some interesting work on Islamic scientific manuscripts in Europe, but his examples are after 1500. Saliba, "Arabic Science," 154, 159. He points to an early copy of the *Tadhkira* (Vatican MS ar. 319), which was brought to Rome in 1623 as part of the Palatine collection, one of the spoils of the Thirty Years' War that was offered by Maximilian I of Bavaria to Pope Gregory XV. Ibid., 159–62. But it was certainly in central Europe by the mid-sixteenth century, where it was used and perhaps annotated by Jakob Christmann (1554–1613), professor of Hebrew and Arabic at the University of Heidelberg. Levi Della Vida, *Ricerche Sulla Formazione*, 329ff., esp. 332. See also Swerdlow, "Recovery of the Exact Sciences."

94 An example would be the treatise by ʿAlī Qushjī discussed in the next paragraph. Other possibilities include manuscripts held by the Biblioteca Medicea Laurenziana in Florence, such as a copy of Quṭb al-Dīn al-Shīrāzī's *Nihāyat al-idrāk* (MS Orientali 110) and two copies of his *Al-Tuḥfa al-shāhiyya fī al-hayʾa* (MS Orientali 116c; and MS Orientali 215). In addition to Ṭūsī's models, Shīrāzī in these two works deals with models of Muʾayyad al-Dīn al-ʿUrḍī as well as his own contributions to planetary theory. Unfortunately, we do not know at present when these manuscripts first appeared in Italy.

95 F.J. Ragep, "ʿAlī Qushjī and Regiomontanus." The diagrams found in the 1496 Venice printing of Regiomontanus's *Epitome* and in the manuscripts of Qushjī's treatise are quite similar.

96 Independent rediscovery now seems even less likely, given that Regiomontanus not only does not claim ownership of the proposition but also incorrectly attributes it to Ptolemy. See Shank, chapter 4, this volume.

97 Bisaha, chapter 2, this volume, discusses Bessarion's attitudes and his relationship to European humanist scholars.

98 This point is emphasized in Sabra, "Situating Arabic Science."

99 For an elaboration, see F.J. Ragep, "Review of *The Beginnings*." A more global approach is taken by Van Brummelen, *Mathematics of the Heavens*.

100 See, for example, Dursteler, *Venetians in Constantinople*. Bisaha, chapter 2, this volume, also discusses some of the complex issues involving cross-cultural transmission during this period.

101 Leo Africanus comes to mind.

102 Such travel has been noted in the case of Moses ben Judah Galeano.

103 Swerdlow and Neugebauer, *Mathematical Astronomy*, part 1, 48, emphasis added.

104 See, for example, Dannenfeldt, "Renaissance Humanists"; and Saliba, "Arabic Science."

105 This was even the case in the early seventeenth century. Feingold, "Decline and Fall."

106 Although things are changing, it is disheartening to note that Robert Westman in his recent book *The Copernican Question*, a tome of 681

double-columned pages, devotes precisely one short, off-handed endnote to the "Maragha school" (531n136). Ṭūsī and the Ṭūsī-couple are completely absent; Jews and Byzantines fare little better.

107 Di Bono, "Copernicus, Amico, Fracastoro," 149: "In conclusion, we note that this same question of transmission may be reduced in significance, in that from a mathematical point of view – as Neugebauer has already noted – it is the internal logic of the methods used that leads the Arabs and Copernicus to such similar results."

108 Ibid., 153–4n77.

109 F.J. Ragep, "Islamic Reactions."

110 These criticisms include, but certainly are not limited to, the equant. F.J. Ragep, *Naṣīr al-Dīn al-Ṭūsī's Memoir*, vol. 1, 48–51.

111 This is not to say that the equant as an issue was unknown in the Latin West; but perhaps with the limited exception of Henry of Hesse, one does not find the sustained criticism of Ptolemy's irregularities that is comparable to Ibn al-Haytham's *Al-Shukūk ʿalā Baṭlamyūs* (Doubts about Ptolemy). This criticism is of course different from criticisms of Ptolemy based upon an Aristotelian-Averroist insistence on a homocentric cosmology. The lack of sustained criticism is surprisingly still the case even in the generation before Copernicus; as Dobrzycki and Kremer put it, "We know of no extant text by Peurbach or Regiomontanus in which the Ptolemaic models are criticized explicitly on the grounds that they violate uniform, circular motion." Dobrzycki and Kremer, "Peurbach and Marāgha," 211n27.

112 Celenza, chapter 1, this volume, emphasizes the very different kind of referencing practice that was followed in the premodern world, where the need to document the source of one's ideas or scientific models was less strongly felt. However, it would be quite unusual for someone who invented as significant a device as the Ṭūsī-couple not to claim it as his own. Bisaha, chapter 2, this volume, provides another reason that early modern European thinkers may have hesitated to credit postclassical Islamic scholars with innovative ideas.

113 In "Ibn al-Shāṭir and Copernicus: The Uppsala Notes Revisited," I speculate that Copernicus's incorrect adaptation of Ibn al-Shāṭir's models in the *Commentariolus* may indicate some influence of an Aristotelian-Averroist insistence on a single centre – in this case, the Sun.

114 F.J. Ragep, *Naṣīr al-Dīn al-Ṭūsī's Memoir*, vol. 1, 208–13.

115 Barker, "Albert of Brudzewo's *Little Commentary*," 137–9, comes to a similar conclusion.

116 This is to repeat a point that I make more generally in F.J. Ragep, "Copernicus."

117 Dobrzycki and Kremer, "Peurbach and Marāgha," 211.

118 See note 68 above.

CHAPTER EIGHT

Some of the material dealt with in this chapter has appeared, but with a different emphasis, in Morrison, "Scholarly Intermediary."

1 Regarding the Ṭūsī-couple, Otto Neugebauer observes, "I do not know through what medium Copernicus knew about Ṭūsī's construction." Neugebauer, *Exact Sciences in Antiquity*, 207. The first publication to point out the connection between Copernicus's astronomy and that of ʿAlāʾ al-Dīn ibn al-Shāṭir was Roberts, "Solar and Lunar Theory," 428: "What is of most interest, however, is that his lunar theory, except for trivial differences in parameters, is identical with that of Copernicus (1473–1543)." See also Kennedy and Roberts, "Planetary Theory," 227. For a summary of the parallels between Copernicus and Islamic astronomers, parallels that cannot be explained by an assumption of independent discovery, see Saliba, *Islamic Science*, 196–209.

2 These sources were Georg Peurbach's *Theoricae novae planetarum*; Peurbach and Johannes Regiomontanus's *Epitome of the Almagest, The Alfonsine Tables*, and *The Almagest* (in the 1515 Venice edition of Gerard of Cremona's translation); and the work of Marāgha astronomers. See Swerdlow, "Derivation and First Draft," 425–6. On Renaissance citation practices, see Celenza, chapter 1, this volume.

3 On a proof of the possibility of eccentric models for Mercury and Venus, which was a possible foundation for the hypothesis of a heliocentric cosmos, see F.J. Ragep: "ʿAlī Qushjī and Regiomontanus," 359, 363. See also Gingerich, "Review of *Islamic Science*," 311. Although Owen Gingerich's review of George Saliba's *Islamic Science and the Making of the European Renaissance* sees little of value in Saliba's argument for Copernicus's roots in Islamic astronomy, Gingerich acknowledges that ʿAlī Qushjī's proof about retrograde motion may have been part of Copernicus's move to a heliocentric cosmos (see F.J. Ragep, "ʿAlī Qushjī and Regiomontanus"). For another recent explanation of the origins of Copernicus's heliocentric hypothesis, see Goldstein, "Copernicus and the Origin." Although there is no evidence of any astronomer in Islamic civilization proposing a heliocentric astronomy, discussions of a rotating Earth did exist. See F.J. Ragep, "Ṭūsī and Copernicus."

4 In their seminal book on Copernicus, Otto Neugebauer and Noel Swerdlow argue that Copernicus's immediate antecedents were the astronomers of Islamic societies. Neugebauer and Swerdlow, *Mathematical Astronomy*, part 1, 41–64. For more recent arguments that build on Swerdlow's and Neugebauer's work, see, for example, Saliba, *Islamic Science*, 193–232; F.J. Ragep: "ʿAlī Qushjī and Regiomontanus"; and F.J. Ragep, "Copernicus."

5 Neugebauer, *History*, vol. 1, 11, vol. 2, 1035, vol. 3, 1456 (for a reproduction of the relevant manuscript folio); Neugebauer, *Exact Sciences in Antiquity*, 214, 217–26.

6 Morrison, "Scientific Intermediary"; Saliba, "Arabic Science."

7 André Goddu has written that without a clear path of transmission to explain the parallels between the astronomy of the Islamic world and the work of Renaissance astronomers, including Copernicus, we should entertain other explanations. Goddu, *Copernicus.* I have commented further on Goddu's argument in Morrison, "Scholarly Intermediary," 57. See, more recently, Blåsjö, "Critique of the Arguments"; and Lerner and Segonds in Copernicus, *De revolutionibus (Des révolutions)*, vol. 1, 545–62. Peter Barker and Tofigh Heidarzadeh have recently addressed Viktor Blåsjö's arguments about the Ṭūsī-couple. Barker and Heidarzadeh, "Copernicus, the Ṭūsī Couple."

8 F.J. Ragep, "Copernicus," 70: "What seems to be overlooked by those who advocate a reinvention by Copernicus and/or his contemporaries of the mathematical models previously used by Islamic astronomers is the lack of an historical context for those models within European astronomy."

9 For a sample of this research, see Contadini and Norton, eds, *Renaissance*; Hess, ed., *Arts of Fire*; Saliba, "Arabic Science"; Russell, *"Arabick" Interest*; Brentjes, *Travellers from Europe*; Brentjes, "Early Modern"; and Brentjes, "Republic of Letters," esp. 439–48, 451, where Brentjes calls attention to the interest of members of the Republic of Letters in manuscripts in Middle Eastern languages on topics such as religion, science, history, and the occult. According to Brentjes, European travellers' negative appraisals of the state of the sciences in the Ottoman Empire were "less an expression of an evolution than an expression of the genre."

10 Dimitri Gutas's and Gerhard Endress's research has found that the word *aṣl* was used to translate the Greek word ὑπόθεσις. Gutas and Gerhard, eds, *Greek and Arabic Lexicon*, vol. 1, fascicle 2, 224–31. There were other possible Arabic translations of the word ὑπόθεσις; see ibid., vol. 1, fascicle 2, note 23. G.J. Toomer, in his translation of the *Almagest*, rendered that Greek word as "hypothesis." See Ptolemy, *Ptolemy's Almagest*, 23–4, 141ff., where the word "hypothesis" is applied to the epicycle and eccentric. Olaf Pedersen sometimes refers to the epicycle and eccentric as "hypotheses" but more frequently as "models." Pedersen, *Survey of the Almagest*, 134 ("hypotheses"), 137–9 ("models"). I also use "hypothesis" to translate the Arabic word *aṣl* in Morrison, *Light of the World*. In my translation, hypotheses are propositions that are the building blocks for models, the complex set of orbs necessary to explain a single planet's motions. The Arabic *aṣl* lacks the speculative connotations of the English "hypothesis"; many of the astronomers described in this chapter attached a probable physical reality to their hypotheses. See Morrison, "Quṭb al-Dīn al-Shīrāzī's Hypotheses," 23–4. Further justification for my choice of the English word "hypothesis" to translate *aṣl* exists in Morrison, *Light of the World*, 269–71 (§0.1). Page references to my study of *The Light of the World* are to the relevant commentary. The parenthetical

references (e.g., §0.1) are to the relevant section(s) of the Judeo-Arabic original and Hebrew recension of the text.

11 On the various varieties of the Ṭūsī-couple, see F.J. Ragep, *Naṣīr al-Dīn al-Ṭūsī's Memoir*, vol. 2, 427–57. For an update, see F.J. Ragep, chapter 7, this volume.

12 Zonta, "Jewish Mediation," 90. Mauro Zonta's comments here note the difficulty of defining Jewish culture in the fifteenth and sixteenth centuries in Europe; Zonta's paper itself treats a slightly earlier period. See also Shulvass, *Jews in the World*, 309–23, for his discussion of science, medicine, and philosophy.

13 Bisaha, *Creating East and West*, 172: "[I]t is important to remember that many humanists transferred their disdain for the Turks to other Islamic peoples – the conflict of West versus East was not limited to the Turks alone."

14 Setton, *Papacy and the Levant*, vol. 2, 296. In a covert attempt at negotiations to stall further Ottoman military advances, the Venetian Senate sent David Mavrogonato, a Jew, to Istanbul in 1469.

15 Modena, *Medici e chirurghi*, 49: "David di Guiseppe Namias (ebreo veneto), dottorato in fil. e med … Nella prima metà del '600 una famiglia Namias si trova stabilita a Venezia. Il medico Abram Namias fu scolaro dello Studio nel 1645, ma pare si sia laureato altrove … Nel 1645 furono pure scolari dell'Università artista Joseph ed Isach Namias."

16 Offenberg, "First Printed Book."

17 On the Qusṭanṭīnī family as copyists of the *Almagest* manuscript, the copying of which was completed in 1475, see Kunitzsch, "Role of al-Andalus," 148. On the flight of the Qusṭanṭīnī family to Candia, see Modena, *Medici e chirurghi*, 22.

18 Ibn Naḥmias's dates and location make his work part of the astronomy of Islamic societies in the fourteenth and fifteenth centuries. On *The Light of the World*'s location in an Islamic context, see Morrison, *Light of the World*, 6–12. See also S.P. Ragep, chapter 6, this volume.

19 On Biṭrūjī, see Goldstein, ed. and trans., *Al-Biṭrūjī*. Goldstein's book includes the Hebrew translation of Biṭrūjī's *Kitāb fī al-hay'a*. See also Robinson, "First References in Hebrew." For the Latin translation, see Biṭrūjī, *De motibus celorum*. On Copernicus's reference to Biṭrūjī in *De revolutionibus*, book 1, ch. 10, see Copernicus, *Gesamtausgabe*, vol. 2, 17; Goldstein, ed. and trans., *Al-Biṭrūjī*, 44; and Rosen, "Copernicus and al-Biṭrūjī," 152.

20 Pierre Duhem criticized Biṭrūjī and other astronomers of Islamic civilization who did not incorporate epicycles and eccentrics into their astronomical models for their inability to account for the planets' observed motions. Duhem, *To Save the Phenomena*, 29–35.

21 On Ibn Naḥmias's improvements on Biṭrūjī, see Morrison, *Light of the World*, 300 (§B.1.II.13), 309 (§B.1.II.19). On the divergences between the

positions predicted by the lunar model and observations, see ibid., 328–32 (§B.2.II.5–6), 344–6 (§B.2.V.4–9).

22 Profiat Duran (d. ca. 1415) criticized Ibn Naḥmias "for trying to force reality into conformity with theory." See Kozodoy, *Secret Faith*, 88–90, quotation at 89. See also Freudenthal, "Towards a Distinction." Gad Freudenthal's assessment of Ibn Naḥmias seems to depend on Duran's criticisms of Ibn Naḥmias, which cite Levi ben Gerson's observational proof of the need for eccentric orbs.

23 Shank, "'Notes on al-Biṭrūjī,'" 15: "And yet, paradoxical though it may seem, Regiomontanus was very interested in the homocentric tradition, in spite of the fact that he was an exceptionally competent mathematical astronomer." Shank, chapter 4, this volume, discusses Regiomontanus and homocentric astronomy in Europe.

24 Shank, "Regiomontanus as a Physical Astronomer," 327, where he further comments, "Indeed, if one is concerned with apparent planetary areas or even brightnesses, a homocentric model more nearly approximates the observations of the Moon than do the variations in area that follow from Ptolemaic models." See also Shank, "Regiomontanus and Homocentric Astronomy," 158–9; and Shank, "'Notes on al-Biṭrūjī,'" 15.

25 Barker, "Copernicus and the Critics," 346–52. In those pages, Barker explains Copernicus's recognition of the Averroist critique of Ptolemy and Copernicus's own exposure to Averroism.

26 Goldstein, "Copernicus and the Origin," 219–21.

27 Sabra, "Andalusian Revolt."

28 Ibid., 134.

29 For a summary of the debate about whether there could be apparently opposite motions in the heavens, see Morrison, *Light of the World*, 272–4.

30 On Ibn Naḥmias's criticism of Biṭrūjī's position that any appearances of opposite motions should be avoided, see ibid., 25, 272–6 (§0.2).

31 Ibid., 46–7. In ibid., chs 3, 4, and 6, I examine the Hebrew recension of *The Light of the World*. See also Freudenthal, "Towards a Distinction," 918, who both raised and questioned the possibility that Ibn Naḥmias was responsible for the Hebrew version of *The Light of the World*.

32 In addition to nonuniform motions in planetary longitude, astronomers also had to account for the planets' retrograde motion, in which the planets appear to slow down, stop, and reverse direction before resuming their direct motion. Biṭrūjī did attempt to address the planets' retrograde motion in *On the Principles of Astronomy*, but in Ibn Naḥmias's *The Light of the World* the section on the planets ends abruptly, before any models are presented, so a comparison is impossible. The most *The Light of the World* says about the motions of the five planets is the beginning of the explanation for the variations in Mars, Jupiter, and Saturn's motions in longitude. See Morrison,

Light of the World, 356 (§B.4.II.1); and Goldstein, ed. and trans., *Al-Biṭrūjī*, vol. 1, 112–13. Reconstructions of Eudoxus of Cnidus's (fourth century BCE) homocentric astronomy explain how Eudoxus might have been able to account for retrograde motion. Yavetz, "On the Homocentric Spheres." The older, classic reconstruction of Eudoxus's astronomy exists in Schiaparelli, "Le Sfere Omocentriche." Whether Biṭrūjī (and, in turn, Ibn Naḥmias) was trying to reconstruct Eudoxus is a matter of scholarly debate. Goldstein, ed. and trans., *Al-Biṭrūjī*, vol. 1, 45. Goldstein disagrees with Edward S. Kennedy's view that Biṭrūjī was influenced directly by Eudoxus's models. Kennedy, "Review of *Al-Biṭrūjī*." See also Yavetz, "On the Homocentric Spheres." On the relationship of Ibn Naḥmias's *The Light of the World* to Eudoxus's homocentric astronomy, see Morrison, *Light of the World*, 19–23. The Hebrew recension of Ibn Naḥmias's *The Light of the World* (ibid., 29) contains a key hypothesis found in Giovanni Virginio Schiaparelli's reconstruction of Eudoxus's astronomy.

33 Shank, "Regiomontanus and Homocentric Astronomy," 158. See also Shank, "'Notes on al-Biṭrūjī,'" 22.

34 Swerdlow, "Regiomontanus's Concentric-Sphere Models," 14–15. Here, Swerdlow comments that the slider-crank mechanism is the key element of Regiomontanus's homocentric models, the element responsible for producing a linear oscillation through the motion of orbs.

35 Ibid., 17. Swerdlow wonders how the point that was supposed to oscillate in the ecliptic (on arc LBM in figure 8.2) would indeed know to oscillate rather than move outside of the ecliptic.

36 Morrison, *Light of the World*, 365–8 (§B.1.II.26/X.7–9).

37 Swerdlow, "Regiomontanus's Concentric-Sphere Models," 4. Still, there are similarities between Quṭb al-Dīn al-Shīrāzī's model for trepidation and Regiomontanus's model. Note the similarities between the figure in the Paris manuscript of *Al-Tuḥfa al-shāhiyya fī al-hay'a* (Paris, BNF, MS Arabe 2516, fol. 14b) and those in Regiomontanus's *Notes on al-Biṭrūjī*. Shank, "'Notes on al-Biṭrūjī,'" 20–1.

38 The preceding sentence and much of the rest of the paragraph are from Morrison, *Light of the World*, 368.

39 On the Hebrew recension's acknowledgment that there will be a slight deviation, see Morrison, *Light of the World*, 248 (§B.1.II.26/X.12).

40 The question of how Ibn Naḥmias could have learned of the Ṭūsī-couple is complex. See Morrison, *Light of the World*, 23. On connections between France and the Iberian Peninsula, and figures who worked at Marāgha, see Lévy, "Gersonide, commentateur d'Euclide," 90–1. On the exchange of instruments, see Comes, "Possible Scientific Exchange." On the curvilinear Ṭūsī-couple in the *Tadhkira*, see F.J. Ragep, *Naṣīr al-Dīn al-Ṭūsī's Memoir*, vol. 2, 448–9.

41 Morrison, *Light of the World*, 361 (§B.1.II.20/X).

42 The preceding sentence and the rest of the paragraph are from ibid., 306.

43 Ibid., 297–300 (§B.1.II.13).

44 Ibid., 308.

45 Ibid., 313–14, 361–2 (for the Hebrew recension). Ibn Naḥmias preferred the solar model that combined the double-circle hypothesis and the circle of the path of the centre of the Sun, pictured in figure 8.4, as he claimed that the inclusion of the circle of the path of the centre would enable the model to reflect the asymmetries of the Sun's motion. My analysis has shown this not to be the case. Ibid., 313–15.

46 Di Bono, "Copernicus, Amico, Fracastoro." See also Goldstein, *Astronomy of Levi ben Gerson,* 14. Goldstein writes that Amico's criticism of the epicycle may have been inspired by Levi ben Gerson.

47 Swerdlow, "Aristotelian Planetary Theory," 41.

48 On Fracastoro, see Di Bono, "Copernicus, Amico, Fracastoro," 143. This is the only scholarship I could find on this technical point of Fracastoro's astronomy. I searched unsuccessfully Enrico Peruzzi's *La nave di Ermete* for any discussion of a double-circle hypothesis.

49 In that sense, the double-circle hypothesis found in Ibn Naḥmias's *The Light of the World* resembles the earliest versions of the Ṭūsī-couple. On the earliest appearance of the Ṭūsī-couple, see F.J. Ragep, "Ibn al-Haytham and Eudoxus," 787–8. See also Saliba, "Role of the *Almagest* Commentaries," 17, 19. The double-circle hypothesis in *The Light of the World,* being on the surface of an orb, also resembled the later, curvilinear version of the Ṭūsī-couple, but its geometrical foundation bore a closer similarity to the version of the Ṭūsī-couple in Ṭūsī's *Taḥrīr al-Majisṭī.* On the different versions of the Ṭūsī-couple, see F.J. Ragep, chapter 7, this volume.

50 See Di Bono, "Copernicus, Amico, Fracastoro," 146–9. Indeed, Swerdlow and Neugebauer, *Mathematical Astronomy,* part 1, 48, use Amico's theory as evidence that Ṭūsī's hypothesis was known in Italy.

51 On Avner de Burgos and the Ṭūsī-couple, see Langermann, "Medieval Hebrew Texts," 33–5. Tzvi Langermann has discovered that the philosopher and polemicist Avner de Burgos (widely thought to be the apostate known as Alfonso de Valladolid) proved a theorem identical to a planar Ṭūsī-couple. Langermann is confident that Avner learned of the theorem from an astronomy text and did not discover the theorem on his own. (Ibn Naḥmias did not cite a source for the double-circle hypothesis that he used.) Gita Mendelevna Gluskina, the editor and translator of Avner's treatise, identifies Alfonso with Avner of Burgos. Gluskina, ed. and trans., *Alfonso,* 17–24, 134–6. For the double-circle hypothesis, see folios 126b–127a of the manuscript and pages 196–7 of Gluskina's translation. Gad Freudenthal has more recently provided better evidence for the same identification of Alfonso with Avner. Freudenthal, "Two Notes."

52 Di Bono, "Copernicus, Amico, Fracastoro," 148. *Planetarum theorica* was the Latin translation of Biṭrūjī's *On the Principles of Astronomy*. On this title, see Barker, "Reality of Peurbach's Orbs," 14, 18.

53 Langermann, "From My Notebooks: A Compendium"; Langermann, "From My Notebooks: Medicine." On the identity of Galeano (Jālīnūs), see Langermann, "From My Notebooks: A Compendium," 288; and Morrison, "Astronomical Treatise," 386. See also Şeşen and İzgi, eds, *Osmanlı astronomi*, vol. 1, 224.

54 Langermann, "From My Notebooks: A Compendium," 290–1. On the connections between Ibn al-Shāṭir and Copernicus, see Saliba, *Islamic Science*, 193–232, esp. 204–9. See also Morrison, "Scholarly Intermediary."

55 Langermann, "From My Notebooks: A Compendium," 290–1.

56 I present the relevant evidence in Morrison, "Astronomical Treatise."

57 Langermann, "From My Notebooks: A Compendium," 295–6.

58 I thank F. Jamil Ragep and İhsan Fazlıoğlu for bringing Ilyās al-Yahūdī to my attention. My information about Ilyās al-Yahūdī, unless otherwise noted, comes from Şeşen and İzgi, eds, *Osmanlı astronomi*, vol. 1, 71–3.

59 Langermann, "From My Notebooks: A Compendium," 312–14. See also Langermann, "From My Notebooks: Medicine," 356–7.

60 Morrison, "Role of Oral Transmission."

61 Morrison, "Astronomical Treatise," 388.

62 Bartolocci, *Bibliotheca magna rabbinica*, vol. 4, 501. The report came via a certain Petrus Rivier, who was associated with the Collegium Neophytorum, where Bartolocci had also been a professor, and who was, not surprisingly, a convert from Judaism. Ibid., vol. 4, 228. Rivier was likely a contemporary of Bartolocci. Ibid., vol. 4, 229. The Collegium Neophytorum was founded in 1543. Wilkinson, *Orientalism, Aramaic, and Kabbalah*, 42. The date of the publication of Bartolocci's book is the *terminus ante quem* for the appearance of Ibn Naḥmias's *The Light of the World* at Padua.

63 Ivry, "Remnants," 243–4. Most scholarship on Elijah Delmedigo has focused on his *Beḥinat ha-dat*, an examination of the proper relationship between philosophy and religion. Ibid., 250ff.

64 Geffen, "Insights," 71–2.

65 Bartòla, "Eliyahu Del Medigo," 255. See Ben Abdeljelil, "Drei jüdische Averroisten," 971–2, for more on Delmedigo's locations in Italy.

66 Delmedigo was often critical of the Qabbala, finding the Averroism of the university preferable to the Neoplatonic cosmology that underpinned the Qabbala. Ben Abdeljelil, "Drei jüdische Averroisten," 972.

67 On Pico and Gersonides, see Goldstein, *Astronomy of Levi ben Gerson*, 12. Goldstein hypothesized that Pico learned of Gersonides from a Jewish informant, whom perhaps was Mordechai Finzi. An awareness of Hebrew astronomy texts was not necessarily confined to Renaissance humanists. See

Goldstein, "Astronomy in the Medieval," 231–2: "Nevertheless, the fact that John of Lignères, one of the main collaborators in the Parisian Latin version of the Alfonsine tables, praised a Hebrew text that he had not read, is surely evidence of the high esteem of Jewish astronomers at the time." For more on Pico and Hebrew, see Novak, "Giovanni Pico della Mirandola," 128–9.

68 Barzilay, *Yoseph Shlomo Delmedigo*, 35. In ibid., 35n2, one finds that there were Jewish students studying at the University of Padua in 1501 under assumed names. Jews did not attend the University of Padua in large numbers until the end of the sixteenth century. Ruderman, "Medicine and Scientific Thought," 191.

69 Barzilay, *Yoseph Shlomo Delmedigo*, 35n2.

70 Goddu, "Copernicus, Nicholas," 177.

71 Barzilay, *Yoseph Shlomo Delmedigo*, 36.

72 Note, however, that on page 74b of the same manuscript, Delmedigo refers to Averroes (Ibn Rushd) as one of the *aḥaronim* (which can mean any post-antique figure) in philosophy.

73 Paris, Bibliothèque nationale de France (BNF), MS Hébreu 968, fol. 49a.

74 On the *Epitome*, see Lay, "*L'Abrégé de l'Almageste*," 47–8 (for the *Shukūk*) and, for example, 46–8 (for discussion of eccentrics and epicycles).

75 Paris, BNF, MS Hébreu 968, fol. 74a. For more on the dating of the Hebrew and Latin versions of Delmedigo's commentary on *On the Substance of the Celestial Orb*, see Kieszkowski, "Les Rapports," 45.

76 On Delmedigo's return to Crete, see Levinger, "Delmedigo, Elijah ben Moses Abba."

77 There is evidence that Peurbach knew of at least some version of the Ṭūsī-couple. Dobrzycki and Kremer, "Peurbach and Marāgha," 210. Jerzy Dobrzycki and Richard Kremer argue that since Ibn al-Haytham's *Al-Maqāla fī ḥarakat al-iltifāf* is neither extant in Arabic nor known to have been translated into Latin, Ṭūsī's *Tadhkira*, which describes the double-epicycle model, might be a plausible source for Europeans' knowledge of the model. See also Mancha, "Ibn al-Haytham's Homocentric Epicycles."

78 Langermann, "Scientific Writings," 15–20.

79 Ibid., 8–11, *passim.*

80 Lacerenza, "Rediscovered Autograph Manuscript."

81 Swerdlow, "Derivation and First Draft," 425–6.

82 Goldstein and Chabás, "Ptolemy, Bianchini, and Copernicus," 470: "Copernicus depended on Bianchini's tables for planetary latitude which, in turn, are based on Ptolemy's models in the *Almagest.*" See also ibid., 453, which cites a copy of Bianchini's latitude tables in Copernicus's own hand.

83 Finzi cited Bianchini's tables as a source, and Bianchini's tables existed in Hebrew. Chabás and Goldstein, *Astronomical Tables*, 21. On Finzi's citation of Bianchini's tables, see Langermann, "Scientific Writings," 20.

84 Bianchini received a letter from Regiomontanus in the 1460s criticizing the implications of the Ptolemaic models for planetary distances. See Swerdlow, "Regiomontanus's Concentric-Sphere Models," 5.
85 Curtze, "Der Briefwechsel Regiomontans," 220.
86 Swerdlow, "Regiomontanus on the Critical Problems," 166.
87 Ibid.
88 Solon, "*Six Wings*," 1. There was also a Latin translation from 1406 and, eventually, a Russian version.
89 Tihon, "L'astronomie byzantine," 253. Bonjorn was baptized in 1391. Roth, "Bonjorn, Bonet Davi(d)."
90 On Solomon ben Elijah Sharbiṭ ha-Zahab, see Gardette, "Judaeo-Provençal Astronomy," 196. See also Tihon, "L'astronomie byzantine," 253. On George Chrysococces, see Tihon and Mercier, *Georges Gémiste Pléthon*, 253. Chrysococces acknowledged the role of Shams al-Dīn al-Bukhārī in the exchange of the *Persian Tables*. Mercier, "Shams al-Dīn al-Bukhārī." On the differences between the different manuscripts of the *Persian Tables*, see Kunitzsch, "Das Fixsternverzeichnis," 386. Based on an analysis of the star indices, Paul Kunitzsch has found two different dates of composition, 1346 and before 1308.
91 Paris, BNF, MS Hébreu 1085, fols 1a–33b.
92 Paris, BNF, MS Hébreu 1085, fol. 31a and esp. fol. 33a–b.
93 Hacker, "Mizraḥi, Elijah," 393–4. See also Hacker, *Those Banished from Spain*, 144–7. Mizrahi's astronomy included a commentary on Abraham Bar Ḥiyya's *Ṣurat ha-areṣ* (The Form of the Earth); one of his Talmudic responsa alluded to his having composed a commentary on Ptolemy's *Almagest*.
94 Tihon and Mercier, *Georges Gémiste Pléthon*, 255–60. Other possible Hebrew sources are mentioned in ibid., 261, but Mercier does not believe they shaped Pletho's work. For Tihon's discussion of Pletho's Jewish sources, see ibid., 10–13. See also Gardette, "Judaeo-Provençal Astronomy."
95 Tihon and Mercier, *Georges Gémiste Pléthon*, 6–7. It is possible that the Jew in question could be a certain Elisha, author of a text called *The Key of Medicine*. Anne Tihon has commented that Jewish philosophers were sought out at this time. Ibid., 8: "En ce début du XVe siècle, les philosophes juifs, et non plus seulement les Latins, exerçaient une grande attraction sur les penseurs byzantins: dans leur polémique à propos d'Aristote, aussi bien Scholarios que Pléthon se réclament des exégètes latins et juifs dans leurs commentaires d'Averroès." For more information on Elisha, see Langermann, "Science in the Jewish Communities." See also Wust, "Elisha the Greek," 50. Elisha was Pletho's teacher, but there is no evidence in Wust's article that Elisha was a scholar of astronomy.
96 Paris, BNF, MS Hébreu 1085, fol. 33a. He criticized his predecessors and opponents for their insufficient attention to how their tables were

connected to astronomy's physical models and to mathematical astronomy's connection to natural philosophy.

97 Paris, BNF, MS Hébreu 1085, fols 56a–84b.

98 Paschos and Sotiroudis, *Schemata of the Stars*, 6.

99 But see Pingree, "Some Fourteenth-Century Byzantine," 105. David Pingree suggests that some of the texts in the same codices that contain *The Schemata of the Stars* are from the early fifteenth century. Given certain similarities between the *Schemata* and Ibn al-Shāṭir's astronomy, such as the epicyclic solar model (Paschos and Sotiroudis, *Schemata of the Stars*, 39–43), the later date perhaps deserves consideration. See F.J. Ragep, "New Light on Shams," 242, who argues that virtually the entire contents of the *Schemata*, with the exception of the epicyclic solar model, were taken from Ṭūsī's Persian works entitled *Risālah-i Muʿīniyya* and *Ḥall-i mushkilāt-i Muʿīniyya*.

100 Paschos and Sotiroudis, *Schemata of the Stars*, 16. The attribution to Chioniades is originally due to Pingree, "Gregory Chioniades."

101 See Pingree, "Gregory Chioniades," 141. See also Neugebauer, *History*, vol. 3, 1109. On Chioniades's knowledge of the Ṭūsī-couple and Ṭūsī's lunar model, see Swerdlow and Neugebauer, *Mathematical Astronomy*, part 1, 47–8. To be clear, the Marāgha models are *not* reflected in the *Persian Tables*. On the Hebrew translation of the *Persian Tables*, see Goldstein, "Survival of Arabic Astronomy," 36–7. On other Hebrew translations of Greek texts and on Greek translations of Hebrew texts, see Tihon, "L'astronomie byzantine," 252–4.

102 Tihon, "Astronomy of George," 116. A few sentences earlier, Tihon writes: "Sometimes the scientific activities of the Karaite schools of Constantinople are invoked, but one must equally consider the possibility of transmission through the intermediary of Latin, and via Italy: the *Cycles* of Bonjorn are called 'Italian tables' in the Greek version of Mark Eugenikos."

Bibliography

Abbud, Fuad. "The Planetary Theory of Ibn al-Shāṭir: Reduction of the Geometric Models to Numerical Tables." *Isis* 53, no. 4 (1962): 492–9.

Abdukhalimov, Bahrom. "Aḥmad al-Farghānī and His Compendium of Astronomy." *Journal of Islamic Studies* 10, no. 2 (1999): 142–58.

Aeneas Silvius Piccolomini (Enea Silvio). *Commentaries.* 5 vols. Ed. Leona C. Gabel. Trans. Florence A. Gragg. Northampton, MA: Smith College Studies in History, 1937–57.

– *Commentaries.* 2 vols. Ed. Margaret Meserve and Marcello Simonetta. Cambridge, MA: Harvard University Press, 2003.

– *De Bohemorum origine et gestorum historia.* Solingen: n.p., 1538.

– *De Europa.* Ed. Adrian Van Heck. Vatican City: Biblioteca Apostolica Vaticana, 2001.

– *Der Briefwechsel des Aeneas Silvius Piccolomini.* In *Fontes rerum austriacarum,* vols 61, 62, 67, 68, ed. Rudolf Wolkan. Vienna: Alfred Hölder, 1909–18.

– *Epistola ad Mahomatem II (Epistle to Mohammed II).* Ed. and trans. Albert R. Baca. New York: Peter Lang, 1990.

– *Europe (c. 1400–1458).* Trans. Robert Brown. Annot. Nancy Bisaha. Washington, DC: Catholic University of America Press, 2013.

– "On the Origin and Authority of the Roman Empire." In *Three Tracts on Empire,* ed. and trans. Thomas M. Izbicki and Cary J. Nederman, 95–112. Bristol: Thoemmes, 2000.

– *Opera quae extant omnia.* 1571. Photostat reprint, Frankfurt am Main: Minerva, 1967.

– *Reject Aeneas, Accept Pius: Selected Letters of Aeneas Sylvius Piccolomini (Pope Pius II).* Ed. and trans. Thomas M. Izbicki, Gerald Christianson, and Philip Krey. Washington DC: Catholic University Press of America, 2006.

Aiton, E.J. "Celestial Spheres and Circles." *History of Science* 19, no. 2 (1981): 75–114.

– "Peurbach's *Theoricae novae planetarum*: A Translation with Commentary." *Osiris*, 2nd series, 3 (1987): 4–43.

Akbari, Suzanne Conklin. "From Due East to True North: Orientalism and Orientation." In *The Postcolonial Middle Ages*, ed. Jeffrey Jerome Cohen, 19–34. New York: St Martin's, 2000.

Albert of Saxony. *Questiones et decisiones physicales insignium virorum.* Ed. George Lokert. Paris: Jodocus Badius Ascensius u. Conradus Resch, 1518.

– *Quaestiones subtilissime in libros de caelo et mundo.* 1492. Reprint, Hildesheim: Georg Olms Verlag, 1986.

Alberti, Leon Battista. *On Painting: A New Translation and Critical Edition.* Ed. and trans. Rocco Sinisgalli. Cambridge, UK: Cambridge University Press, 2011.

– *On Painting and On Sculpture: The Latin Texts of De Pictura and De Statua.* Ed. and trans. Cecil Grayson. London: Phaidon, 1972.

Albertus Magnus. *Metaphysica.* In *Alberti Magni Opera Omnia*, vol. 16, part 2. Monasterium Westfalorum: Aschendorff, 1964.

Andrews, Walter K., and Mehmet Kalpaklı. *The Age of the Beloveds: Love and the Beloved in Early-Modern Ottoman and European Culture and Society.* Durham, NC: Duke University Press, 2005.

Ansari, S.M. Razaullah. "On the Transmission of Arabic-Islamic Science to Medieval India." *Archives internationales d'histoire des sciences* 45 (1995): 273–97.

– "Transmission of Islamic Exact Science to India and Its Neighbours and Repercussions Thereof." *Studies in the History of Natural Sciences* 24 (2005): 31–5.

Aristotle. *Metaphysics.* Trans. W.D. Ross. In *Complete Works of Aristotle: The Revised Oxford Translation*, vol. 2, ed. Jonathan Barnes. Princeton, NJ: Princeton University Press, 1984.

– *Meteorologica.* Trans. H.D.P. Lee. Cambridge, MA: Harvard University Press, 1952.

– *On the Parts of Animals.* Trans. W. Ogle. In *Complete Works of Aristotle: The Revised Oxford Translation*, vol. 1, ed. Jonathan Barnes. Princeton, NJ: Princeton University Press, 1984.

– *Posterior Analytics.* Trans. Jonathan Barnes. In *Complete Works of Aristotle: The Revised Oxford Translation*, vol. 1, ed. Jonathan Barnes. Princeton, NJ: Princeton University Press, 1984.

– *Prior Analytics.* Trans. A.J. Jenkinson. In *Complete Works of Aristotle: The Revised Oxford Translation*, vol. 1, ed. Jonathan Barnes. Princeton, NJ: Princeton University Press, 1984.

Aschbach, Joseph. *Geschichte der Wiener Universität.* 3 vols. Vienna: K.K. Universität, 1865.

Averroes (Ibn Rushd). *Aristotelis De Caelo, de Generatione et corruptione … cum Averrois Cordubensis variis in eosdem commentariis.* Vol. 5, *Aristotelis Opera cum Averrois commentariis.* 1562. Reprint, Frankfurt am Main: Minerva, 1962.

– *Large Commentary on Aristotle's De Caelo et Mundo.* Ed. Francis J. Carmody and Rüdiger Arnzen. Leuven: Peeters, 2003.

Avi-Yonah, Reuven. "Ptolemy vs. al-Biṭrūjī: A Study of Scientific Decision-Making in the Middle Ages." *Archives internationales d'histoire des sciences* 35 (1985): 124–47.

Babicz, Józef. "Die exakten Wissenschaften an der Universität zu Krakau und der Einfluss Regiomontanus auf ihre Entwicklung." In *Regiomontanus-Studien,* ed. Günther Hamann, 301–14. Vienna: Verlag der Österreichischen Akademie der Wissenschaften, 1980.

Bacon, Roger. *De celestibus.* In *Opera hactenus inedita Rogeri Baconi,* fasc. 4, book 2, *Communium naturalium Fratris Rogeri de celestibus,* ed. Robert Steele. Oxford: Clarendon, 1913.

Bagheri, Mohammad. "A Newly Found Letter of al-Kāshī on Scientific Life in Samarkand." *Historia Mathematica* 24, no. 3 (1997): 241–56.

Baldi, Barbara. "Enea Silvio Piccolomini e il *De Europa*: Umanesimo, religion, e politica." *Archivio storico italiano* 161, no. 4 (2003): 619–83.

Barker, Peter. "Albert of Brudzewo's *Little Commentary on George Peurbach's 'Theoricae Novae Planetarum.'*" *Journal for the History of Astronomy* 44, no. 2 (2013): 125–48.

– "Copernicus and the Critics of Ptolemy." *Journal for the History of Astronomy* 30, no. 4 (1999): 343–58.

– "The Reality of Peurbach's Orbs: Cosmological Continuity in Fifteenth and Sixteenth Century Astronomy." In *Change and Continuity in Early Modern Cosmology,* ed. Patrick J. Boner, 7–32. Dordrecht: Springer-Verlag, 2011.

– "Why Was Copernicus a Copernican?" *Metascience* 23, no. 2 (2014): 203–8.

Barker, Peter, and Matjaž Vesel. "Goddu's Copernicus: An Essay Review of André Goddu's *Copernicus* and the Aristotelian Tradition." *Aestimatio* 9 (2012): 304–36.

Barker, Peter, and Tofigh Heidarzadeh. "Copernicus, the Ṭūsī Couple and East-West Exchange in the Fifteenth Century." In *Unifying Heaven and Earth: Essays in the History of Early Modern Cosmology,* ed. Miguel Á. Granada, Patrick J. Boner, and Dario Tessicini, 19–57. Barcelona: Edicions de la Universitat de Barcelona, 2016.

Bartòla, Alberto. "Eliyahu Del Medigo e Giovanni Pico della Mirandola." *Rinascimento: Rivista dell'Istituto nazionale di studi sul Rinascimento,* n.s., 33 (1993): 253–78.

Bartolocci, Giulio. *Bibliotheca magna rabbinica de scriptoribus, & scriptis hebraicis, ordine alphabetico Hebraicè, & Latinè digestis … In qua complures identidem interseruntur dissertationes, et digressiones.* 5 vols. Rome: Sacrae Congregationis de Propaganda Fide, 1675–94.

Barzilay, Isaac. *Yoseph Shlomo Delmedigo, Yashar of Candia: His Life, Works and Times.* Leiden: Brill, 1974.

Beck, Christopher I. *Warrior of the Cloisters: The Central Asian Origins of Science in the Medieval World.* Princeton, NJ: Princeton University Press, 2012.

Ben Abdeljelil, Jameleddine. "Drei jüdische Averroisten: Höhepunkt und Niedergang des jüdischen Averroismus im Mittelalter." *Asiatische Studien* 62, no. 4 (2008): 933–86.

Benjamin, Francis S., Jr, and G.J. Toomer, eds. *Campanus of Novara and Medieval Planetary Theory: Theorica planetarum.* Madison: University of Wisconsin Press, 1971.

Benson, Larry, ed. *The Riverside Chaucer.* 3rd ed. Boston: Houghton Mifflin, 1987.

Berkey, Jonathan. "Madrasas Medieval and Modern: Politics, Education, and the Problem of Muslim Identity." In *Schooling Islam: The Culture and Politics of Modern Muslim Education,* ed. Robert W. Hefner and Muhammad Qasim Zaman, 40–61. Princeton, NJ: Princeton University Press, 2007.

– *The Transmission of Knowledge in Medieval Cairo.* Princeton, NJ: Princeton University Press, 1992.

Bernardus de Virduno (Bernard of Verdun). *Tractatus super totam astrologiam.* Ed. Polykarp Hartmann. Werl: Dietrich-Coelde-Verlag, 1961.

Biard, Joël. *Science et nature: La théorie buridanienne du savoir.* Paris: Vrin, 2012.

Birkenmajer, Aleksander. "Marcin Bylica." In *Studia Copernicana,* vol. 4, *Études d'histoire des sciences en Pologne,* 533–40. Wrocław: Ossolineum, 1972.

Birkenmajer, Ludwik Antoni. *Mikolaj Kopernik.* 1900. Translated and catalogued as Jerzy Dobrzycki, *Nicholas Copernicus: Studies on the Works of Copernicus and Biographical Materials (Parts 1 and 2).* Ann Arbor, MI: University Microfilms, 1985.

al-Bīrūnī, Abū Rayḥān Muḥammad ibn Aḥmad. *Kitāb al-Tafhīm li-awāʾil ṣināʿat al-tanjīm* [The book of instruction in the elements of the art of astrology]. Trans. R. Ramsay Wright from the facsimile reproduction of an Arabic manuscript. London: Luzac and Company, 1934. Persian text edited by Jalāl al-Dīn Humāʾī (Tehran: Intishārāt-i Bābak, 1983–84).

– *Al-Qānūn al-Masʿūdī.* 3 vols. Hyderabad: Dāʾirat al-maʿārif al-ʿUthmāniyya, 1954–56.

Bisaha, Nancy. "Barbarians or Intellectual Peers? Byzantine Perceptions of Islamic Learning." Paper presented at the Renaissance Society of American Annual Conference, Montreal, March 2011.

– *Creating East and West: Renaissance Humanists and the Ottoman Turks.* Philadelphia: University of Pennsylvania Press, 2004.

– "Discourses of Power and Desire": The Letters of Aeneas Silvius Piccolomini (1453)." In *Florence and Beyond: Culture, Society, and Politics in Renaissance Italy: Essays in Honour of John M. Najemy*, ed. David S. Peterson with Daniel E. Bornstein, 121–34. Toronto: Center for Reformation and Renaissance Studies, 2008.

– "Petrarch's Vision of the Muslim and Byzantine East." *Speculum* 76, no. 2 (2001): 284–314.

– "Pope Pius II's Letter to Sultan Mehmed II: A Reexamination." *Crusades* 1 (2002): 183–200.

al-Biṭrūjī, Nūr al-Dīn. *Al-Biṭrūjī: On the Principles of Astronomy*. 2 vols. Ed. and trans. Bernard R. Goldstein. New Haven, CT: Yale University Press, 1971.

– *De motibus celorum*. Ed. Francis Carmody. Berkeley: University of California Press, 1952.

Bjørnstad, Hall, ed. *Borrowed Feathers: Plagiarism and the Limits of Imitation in Early Modern Europe*. Oslo: Unipub, 2008.

Blair, Ann. "Note-Taking as an Art of Transmission." *Critical Inquiry* 31, no. 1 (2004): 85–107.

– "Reading Strategies for Coping with Information Overload ca. 1550–1700." *Journal of the History of Ideas* 64, no. 1 (2003): 11–28.

– *Too Much to Know: Managing Scholarly Information before the Information Age*. New Haven, CT: Yale University Press, 2010.

Blåsjö, Viktor. "A Critique of the Arguments for Maragha Influence on Copernicus." *Journal for the History of Astronomy* 45, no. 2 (2014): 183–95.

Bolens, Guillemette, and Lukas Erne, eds. *Medieval and Early Modern Authorship*. Tübingen: Narr Verlag, 2011.

Bolzoni, Lina. *The Gallery of Memory: Literary and Iconographic Models in the Age of the Printing Press*. Trans. J. Parzen. Toronto: University of Toronto Press, 2001.

– *La stanza della memoria: Modelli letterari e iconografici nell' età della stampa*. Turin: Einaudi, 1995.

Bonaventure. *Opera Omnia*. 10 vols. Florence: Quaracchi, 1882–1901.

Boncompagni, Baldassare. "Catalogo dei Lavori di Andalò di Negro." In *Bullettino di bibliografia e di storia delle scienze matematiche e fisiche*, ed. Baldassare Boncompagni, vol. 7, 339–76. Rome: Scienze matematiche e fisiche, 1874.

Bònoli, Fabrizio, C. Colavita, and C. Mataix. "L'ambiente culturale bolognese del Quattrocento attraverso Domenico Maria Novara e la sua influenza in Nicolò Copernico." *Memorie della Società astronomica italiana* 66, no. 4 (1995): 871–80.

Brann, Noel L. "Humanism in Germany." In *Renaissance Humanism: Foundations, Forms, and Legacy*, vol. 2, *Humanism beyond Italy*, ed. Albert Rabil Jr, 123–55. Philadelphia: University of Pennsylvania Press, 1988.

Brentjes, Sonja. "Courtly Patronage of the Ancient Sciences in Post-Classical Islamic Societies." *Al-Qanṭara* 29 (2008): 403–36.

– "Early Modern Western European Travellers in the Middle East and Their Reports about the Sciences." In *Sciences, techniques et instruments dans le monde iranien (Xe–XIXe siècle)*, ed. N. Pourjavady and Ž. Vesel, 379–420. Tehran: Presses Universitaires d'Iran and Institut Français de Recherche en Iran, 2004.

– "The Mathematical Sciences in Safavid Iran: Questions and Perspective." In *Muslim Cultures in the Indo-Iranian World during the Early-Modern and Modern Periods*, ed. Denis Hermann and Fabrizio Speziale, 325–402. Berlin: Klaus Schwarz Verlag, 2010.

– "On the Location of the Ancient or 'Rational' Sciences in Muslim Education Landscapes (AH 500–1100)." *Bulletin of the Royal Institute for Inter-Faith Studies* 4, no. 1 (2002): 47–71.

– "Reflections on the Role of the Exact Sciences in Islamic Culture and Education between the Twelfth and the Fifteenth Centuries." In *Études des sciences arabes*, ed. Mohammed Abattouy, 15–33. Casablanca: Fondation du Roi Abdul-Aziz Al Saoud, 2007.

– "The Republic of Letters in the Middle East." *Science in Context* 12, no. 3 (1999): 435–68.

– *Travellers from Europe in the Ottoman and Safavid Empires, 16th–17th Centuries: Seeking, Transforming, Discarding Knowledge.* Farnham, UK: Ashgate-Variorum, 2010.

Brudzewo, Albert of. *Commentaria utilissima in theoricis planetarum.* Milan: n.p., 1495.

– *Commentariolum super Theoricas novas planetarum Georgii Purbachii.* Ed. Ludovicus Antonius Birkenmajer. Cracow: Typis et Sumptibus Universitatis Jagellonicae, 1900.

Brummett, Palmira. "The Lepanto Paradigm Revisited: Knowing the Ottomans in the Sixteenth Century." In *The Renaissance and the Ottoman World*, ed. Anna Contadini and Claire Norton, 63–93. Farnham, UK: Ashgate, 2013.

– *Ottoman Seapower and Levantine Diplomacy in the Age of Discovery.* Albany, NY: SUNY Press, 1994.

Buridan, John. *In Metaphysicen Aristotelis quaestiones argutissimae.* 1518. Reprint, Frankfurt am Main: Minerva, 1964.

Burley, Walter. *Quaestiones super librum posteriorum.* Ed. Mary Catherine Sommers. Toronto: Pontifical Institute of Mediaeval Studies, 2000.

Burnett, Charles. "The Certitude of Astrology: The Scientific Methodology of al-Qabīṣī and Abū Maʿshar." *Early Science and Medicine* 7, no. 3 (2002): 198–213.

– "The Second Revelation of Arabic Philosophy and Science: 1492–1562." In *Islam and the Italian Renaissance*, ed. Charles Burnett and Anna Contadini, 185–98. London: Warburg Institute, 1999.

Byrne, James Steven. "A Humanist History of Mathematics? Regiomontanus's Padua Oration in Context." *Journal of the History of Ideas* 67, no. 1 (2006): 41–61.

– "The Mean Distances of the Sun and Commentaries on the *Theorica Planetarum*." *Journal for the History of Astronomy* 42, no. 2 (2011): 205–21.

– "The Stars, the Moon, and the Shadowed Earth: Viennese Astronomy in the Fifteenth Century." PhD diss., Princeton University, 2007.

Campanus of Novara. *Theorica planetarum*. In *Campanus of Novara and Medieval Planetary Theory: Theorica planetarum*, ed. Francis S. Benjamin Jr and G.J. Toomer. Madison: University of Wisconsin Press, 1971.

Cardini, Franco. *Europa e Islam: Storia di un malinteso*. Rome: Laterza, 1999.

Carlino, Andrea. "*Kunstbüchlein* and *Imagines Contrafactae*: A Challenge to the Notion of Plagiarism." In *Borrowed Feathers: Plagiarism and the Limits of Imitation in Early Modern Europe*, ed. Hall Bjørnstad, 87–108. Oslo: Unipub, 2008.

Carman, Charles H. *Leon Battista Alberti and Nicholas Cusanus: Towards an Epistemology of Vision for Italian Art and Culture*. Farnham, UK: Ashgate, 2014.

Carmody, Francis J., ed. *Al-Biṭrūjī: De Motibus Celorum: Critical Edition of the Latin Translation of Michael Scot*. Berkeley: University of California Press, 1952.

Carruthers, Mary. *The Craft of Thought: Meditation, Rhetoric, and the Making of Images, 400–1200*. Cambridge, UK: Cambridge University Press, 1998.

Casamassima, Emanuele. *L'autografo Riccardiano della seconda lettera del Petrarca a Urbino V (Senile IX 1)*. Rome: Valerio Levi Ediotre, 1986.

Casella, Nicola. "Pio II tra geografia e storia: La 'Cosmographia.'" *Archivio della Società romana di storia patria* 95 (1972): 35–112.

Celenza, Christopher S. "End Game: Humanist Latin in the Late Fifteenth Century." In *Latinitas Perennis II: Appropriation and Latin Literature*, ed. Yanick Maes, Jan Papy, and Wim Verbaal, 201–42. Leiden: Brill, 2009.

– "Humanism and the Classical Tradition." *Annali d'Italianistica* 26 (2008): 25–49.

– "Late Antiquity and Florentine Platonism: The 'Post-Plotinian' Ficino." In *Marsilio Ficino: His Theology, His Philosophy, His Legacy*, ed. Michael J.B. Allen and Valery R. Rees, 71–97. Leiden: Brill, 2002.

– *The Lost Italian Renaissance: Humanists, Historians, and Latin's Legacy*. Baltimore: Johns Hopkins University Press, 2004.

– "Petrarch, Latin, and Italian Renaissance Latinity." *Journal of Medieval and Early Modern Studies* 35, no. 3 (2005): 509–36.

– "What Counted as Philosophy in the Italian Renaissance? The History of Philosophy, the History of Science, and Styles of Life." *Critical Inquiry* 39, no. 2 (2013): 367–401.

– ed. *Angelo Poliziano's Lamia in Context: Text, Translation, and Introductory Studies*. Leiden: Brill, 2010.

Chabás, José, and Bernard R. Goldstein. *The Astronomical Tables of Giovanni Bianchini.* Leiden: Brill, 2009.

Charette, François. "The Locales of Islamic Astronomical Instrumentation." *History of Science* 44, no. 2 (2006): 123–38.

Chenu, Marie-Dominique. "Auctor, actor, autor." *Bulletin du Cange* 3 (1927): 81–6.

Clagett, Marshall. *Archimedes in the Middle Ages.* Vol. 3. Madison: University of Wisconsin Press, 1978.

Clark, Stuart. *Thinking with Demons: The Idea of Witchcraft in Early Modern Europe.* Oxford: Oxford University Press, 1997.

Colish, Marsha. *Peter Lombard.* 2 vols. Leiden: Brill, 1994.

Comes, Mercè. "The Possible Scientific Exchange between the Courts of Hūlāgū and Alfonso X." In *Sciences, techniques et instruments dans le monde iranien (Xe–XIXe siècle)*, ed. N. Pourjavady and Ž. Vesel, 29–50. Tehran: Presses Universitaires d'Iran and Institut Français de Recherche en Iran, 2004.

Connell, William. "The Republican Idea." In *Renaissance Civic Humanism: Reappraisals and Reflections*, ed. James Hankins, 14–29. Cambridge, UK: Cambridge University Press, 2004.

Contadini, Anna. "Sharing a Taste? Material Culture and Intellectual Curiosity around the Mediterranean, from the Eleventh to the Sixteenth Century." In *The Renaissance and the Ottoman World*, ed. Anna Contadini and Claire Norton, 23–61. Farnham, UK: Ashgate, 2013.

Copernicus, Nicholas. *Commentariolus.* http://copernicus.torun.pl/en/archives/astronomical/1/?view=transkrypcja&lang=latina.

– *Complete Works.* 3 vols. Ed. Paweł Czartoryski. Trans. Edward Rosen with Erna Hilfstein. London: Macmillan, 1992.

– *Das neue Weltbild: Drei Texte Commentariolus, Brief gegen Werner, De revolutionibus I, Im Anhang eine Auswahl aus der Narratio prima des G.J. Rheticus.* Ed. and trans. Hans Günter Zekl. Hamburg: Felix Meiner Verlag, 1990.

– *De revolutionibus orbium coelestium.* Nuremberg: Johannes Petreius, 1543.

– *De revolutionibus orbium coelestium (Des révolutions des orbes célestes).* 3 vols. Trans. Michel-Pierre Lerner and Alain-Philippe Segonds with the collaboration of Concetta Luna, Isabelle Pantin, and Denis Savoie. Paris: Les Belle Lettres, 2015.

– *Erster Entwurf seines Weltsystems: Sowie eine Auseinandersetzung Johannes Keplers mit Aristoteles über die Bewegung der Erde.* Ed. and trans. Fritz Rossmann. München: Verlag Hermann Rinn, 1948.

– *Gesamtausgabe.* 9 vols. Ed. Heribert M. Nobis and Menso Folkerts. Hildesheim: H.A. Gerstenberg, 1974.

– *Letter against Werner.* In *Three Copernican Treatises.* 2nd ed. Trans. Edward Rosen. New York: Dover, 1959.

– *On the Revolutions.* Trans. Edward Rosen. Baltimore, MD: Johns Hopkins University Press, 1978.
– *On the Revolutions of the Heavenly Spheres: A New Translation with an Introduction and Notes by A.M. Duncan.* New York: Barnes and Noble, 1976.
– *Three Copernican Treatises.* 2nd ed. Trans. Edward Rosen. New York: Dover, 1959.

Coseriu, Eugenio, and Reinhard Meisterfeld. *Geschichte der romanischen Sprachwissenschaft.* Vol. 1, *Von den Anfängen bis 1492.* Tübingen: G. Narr, 2003.

Coxon, Sebastian. "Introduction." In *The Presentation of Authorship in Medieval German Literature, 1220–1290,* 1–34. Oxford: Oxford University Press, 2001.

Curtze, Maximilian. "Der Briefwechsel Regiomontans mit Giovanni Bianchini, Jacob von Speier, und Christian Roder." *Abhandlungen zur Geschichte der mathematischen Wissenschaften* 12 (1902): 185–336. Reprinted in Maximilian Curtze, *Urkunden zur geschichte der mathematik im mittelalter und der renaissance,* 185–336 (New York: Johnson Reprint Corporation, 1968).

Cusa, Nicholas of. *Complete Philosophical and Theological Treatises of Nicholas of Cusa.* 2 vols. Minneapolis, MN: Arthur J. Banning, 2001.
– *De docta ignorantia.* In *Werke: Neuausgegeben des Strassburger Drucks von 1488,* vol. 1, ed. Paul Wilpert. Berlin: De Gruyter, 1967.
– *Nicolai Cusae Cardinalis: Opera.* 1514. Reprint, Frankfurt am Main: Minerva, 1962.
– *Nikolaus von Kues: Die mathematischen Schriften.* Trans. Joseph Ehrenfried Hofmann. 1952. Reprint, Hamburg: F. Meiner, 1980.
– *On Learned Ignorance: A Translation and an Appraisal of De Docta Ignorantia.* 2nd ed. Trans. Jasper Hopkins. Minneapolis, MN: Arthur J. Banning, 1990.
– *The Vision of God.* Trans. Emma Gurney Salter. London: J.M. Dent and Sons, 1928.

Czartoryski, Paweł. "The Library of Copernicus." In *Science and History: Studies in Honor of Edward Rosen,* ed. Paweł Czartoryski and Erna Hilfstein, 355–401. Wrocław: Ossolineum and Polish Academy of Sciences Press, 1978.

Daftary, Farhad. "Dāʿī." In *Encyclopaedia Iranica,* vol. 6, fasc. 6, ed. Ehsan Yarshater, 590–3. Costa Mesa, CA: Mazda, 1993. http://www.iranicaonline.org/articles/dai-propagandists.

d'Ailly, Pierre. *Questions on the Sphere.* In *Sphera cum commentis.* Venice: Octavianus Scotus, 1518.

d'Alessandro, Paolo, and Pier Daniele Napolitani, eds. *Archimede Latino: Iacopo da San Cassciano et il corpus archimedeo all metà del quattrocento con edizione della Circuli dimensio et della Quadratura parabola.* Paris: Les Belles Lettres, 2012.

Dallal, Ahmad. *Islam, Science, and the Challenge of History.* New Haven, CT: Yale University Press, 2010.

Danielson, Dennis. *The First Copernican: Georg Joachim Rheticus and the Rise of the Copernican Revolution*. New York: Walker and Company, 2006.

Dannenfeldt, Karl H. "The Renaissance Humanists and the Knowledge of Arabic." *Studies in the Renaissance* 2 (1955): 96–117.

Daston, Lorraine. "Preternatural Philosophy." In *Biographies of Scientific Objects*, ed. Lorraine Daston, 15–41. Chicago: University of Chicago Press, 2000.

Daston, Lorraine, and Elizabeth Lunbeck, eds. *Histories of Scientific Observation*. Chicago: University of Chicago Press, 2011.

Daston, Lorraine, and Katherine Park. *Wonders and the Order of Nature: 1150–1750*. New York: Zone, 1998.

De Groot, Jean. *Aristotle's Empiricism: Experience and Mechanics in the 4th Century* BC. Las Vegas: Parmenides, 2014.

– "Modes of Explanation in the Aristotelian *Mechanical Problems*." In *Evidence and Interpretation in Studies on Early Science and Medicine: Essays in Honor of John E. Murdoch*, ed. Edith Dudley Sylla and William R. Newman, 22–42. Leiden: Brill, 2009.

De Simoni, Cornelio. "Intorno alla vita ed ai lavori de Andalò di Negro matematico ed astronomo genovese del secolo decimoquarto et …" In *Bullettino di bibliografia e di storia delle scienze matematiche e fisiche*, ed. Baldassare Boncompagni, vol. 7, 313–36. Rome: Scienze matematiche e fisiche, 1874.

Di Bono, Mario. "Copernicus, Amico, Fracastoro, and Ṭūsī's Device: Observations on the Use and Transmission of a Model." *Journal for the History of Astronomy* 26, no. 2 (1995): 133–54.

– *Le sfere omocentriche di Giovan Battista Amico nell'astronomia del Cinquecento: Con il testo del "De motibus corporum coelestium …"* Genoa: Centro di Studio sulla Storia della Tecnica, 1990.

Dicks, D.R. *Early Greek Astronomy to Aristotle*. Ithaca, NY: Cornell University Press, 1970.

Diels, H., ed. *Commentaria in Aristotelem Graeca*. Vol. 10, *Simplicii in Aristotelis Physicorum libros quattuor priores commentaria*. Berlin: G. Reiner, 1982.

Dijksterhuis, E.J. *Mechanization of the World Picture*. Trans. C. Dikshoorn. Oxford: Clarendon, 1961.

Diogenes Laertius. *The Lives, Opinions, and Remarkable Sayings of the Most Famous Ancient Philosophers*. Trans. T. Fetherstone. London: London: Printed for R. Bentley in Covent Garden, W. Hensman in Westminster Hall, J. Taylor in St Paul's Church-Yard, and T. Chapman in the Pall Mall, 1696.

Dobrzycki, Jerzy. "Astronomical Aspects of the Calendar Reform." In *Gregorian Reform of the Calendar: Proceedings of the Vatican Conference to Commemorate Its 400th Anniversary*, ed. G.V. Coyne, M.A. Hoskin, and O. Pedersen, 117–26. Vatican City: Pontificia Academia Scientiarum, Specola Vaticana, 1983.

– "Notes on Copernicus's Early Heliocentrism." *Journal for the History of Astronomy* 32, no. 3 (2001): 223–5.

– "The Theory of Precession in Medieval Astronomy." In *Selected Papers on Medieval and Renaissance Astronomy by Jerzy Dobrzycki*, ed. Jacek Włodarczyk and Richard L. Kremer, trans. Aleksandra Niemirycz, 15–60. Warsaw: Institute for the History of Science, Polish Academy of Sciences, Copernicus Center for Interdisciplinary Studies, 2010. Originally published in Polish as Jerzy Dobrzycki, "Teoria precesji w astronomii średniowiecznej," *Studia i materiały z dziejów nauki polskiej*, series c, 11 (1965): 3–47.

Dobrzycki, Jerzy, and Richard L. Kremer. "Peurbach and Marāgha Astronomy? The Ephemerides of Johannes Angelus and Their Implications." *Journal for the History of Astronomy* 27, no. 3 (1996): 187–237.

Doucet, Victorin. *Supplément au répertoire de M. Frédéric Stegmüller*. Florence: Collegii S. Bonaventurae, Ad Claras Aquas, 1954.

Droppers, Garrett. "The Questiones de Spera of Nicole Oresme: Latin Text with English Translation, Commentary and Variants." PhD diss., University of Wisconsin, 1966.

Duhem, Pierre. *Le Système du Monde: Histoire des doctrines cosmologiques de Platon à Copernic*. Vol. 4, *L'Astronomie latine au Moyen Age (suite): La crue de l'aristotélisme*. Paris: Hermann et Cie, 1954.

– *To Save the Phenomena: An Essay on the Idea of Physical Theory from Plato to Galileo*. Trans. Edmund Doland and Chaninah Maschler. 1969. Reprint, Chicago: University of Chicago Press, 1985.

Dupuis, J. *Théon de Smyrne, philosophe platonicien: Exposition des connaissances mathématiques utiles pour la lecture de Platon traduites pour la première fois du grec en français*. 1892. Reprint, Brussels: Culture et Civilisation, 1966.

Dursteler, Eric. *Venetians in Constantinople: Nation, Identity, and Coexistence in the Early Modern Mediterranean*. Baltimore, MD: Johns Hopkins University Press, 2006.

Eagleton, Catherine. *Monks, Manuscripts, and Sundials: The Navicula in Medieval England*. Leiden: Brill, 2010.

Eisenstein, Elizabeth. *The Printing Press as an Agent of Change*. 2 vols. Cambridge, UK: Cambridge University Press, 1979.

Eisenstein, Elizabeth, Anthony T. Grafton, and Adrian Johns. "Forum: How Revolutionary Was the Print Revolution?" *American Historical Review* 107, no. 1 (2002): 84–128.

Endress, Gerhard, and Dimitri Gutas, eds. *A Greek and Arabic Lexicon (GALex): Materials for a Dictionary of the Mediaeval Translations from Greek into Arabic*. Leiden: Brill, 1992.

Eshenkulova, Kishimjan. "Timurlular Devri Medrese Eğitimi ve Ulum el-Evail (Matematik-Astronomi ve Tib)." MA thesis, Istanbul University, 2001.

al-Farghānī, Aḥmad ibn Muḥammad ibn Kathīr (Alfraganus). *Jawāmiʿ ʿilm al-nujūm wa-uṣūl al-ḥarakāt al-samāwīyah*. Ed. Jacobus Golius. 1669. Reprint by Fuat Sezgin, Frankfurt am Main: Institut für Geschichte der

Arabisch-Islamischen Wissenschaften an der Johann Wolfgang Goethe Universität, 1986.

Fazlıoğlu, İhsan. "Kamāl al-Dīn al-Turkmānī: Kamāl al-Dīn Muḥammad ibn Aḥmad ibn ʿUthmān ibn Ibrāhīm ibn Muṣṭafā al-Māridīnī." In *The Biographical Encyclopedia of Astronomers*, vol. 1, ed. Thomas Hockey et al., 609. New York: Springer-Verlag, 2007.

– "Qushjī: Abū al-Qāsim ʿAlāʾ al-Dīn ʿAlī ibn Muḥammad Qushči-zāde." In *The Biographical Encyclopedia of Astronomers*, vol. 1, ed. Thomas Hockey et al., 946–8. New York: Springer-Verlag, 2007.

– "The Samarqand Mathematical-Astronomical School: A Basis for Ottoman Philosophy and Science." *Journal for the History of Arabic Science* 14 (2008): 3–68.

Feingold, Mordechai. "Decline and Fall: Arabic Science in Seventeenth-Century England." In *Tradition, Transmission, Transformation: Proceedings of Two Conferences on Premodern Science Held at the University of Oklahoma*, ed. F. Jamil Ragep and Sally Ragep, 441–69. Leiden: Brill, 1996.

Ficino, Marsilio. *De vita libri tres*. Ed. and trans. Carol. V. Kaske and J.R. Clark. Binghamton, NY: MRTS, 1989.

Field, J.V. *The Invention of Infinity: Mathematics and Art in the Renaissance*. Oxford: Oxford University Press, 1997.

Folkerts, Menso. "Conrad Landvogt, ein bisher unbekannter Algebraiker um 1500." In *Amphora: Festschrift für Hans Wussing zu seinem 65 Geburtstag*, ed. Sergei Demidov, Menso Folkerts, David Rowe, and Christoph Scriba, 229–59. Basel: Birkhäuser, 1992.

Freudenthal, Gad. "'Instrumentalism' and 'Realism' as Categories in the History of Astronomy: Duhem vs. Popper, Maimonides vs. Gersonides." *Centaurus* 45, nos 1–4 (2003): 227–48.

– "Towards a Distinction between the Two Rabbis Joseph ibn Joseph ibn Naḥmias." *Qiryat Sefer* 62 (1988–89): 917–19 (in Hebrew).

– "Two Notes on *Sefer Meyashsher ʿaqov* by Alfonso, alias Abner of Burgos." Hebrew. *Qiryat Sefer* 63 (1990–91): 984–6 (in Hebrew).

Gal, Ofer, and Raz Chen-Morris. *Baroque Science*. Chicago: University of Chicago Press, 2013.

Galileo Galilei. *Tractatio de praecognitionibus et praecognitis and Tractatio de demonstratione*. Transcribed from the Latin authography by William F. Edwards. Padua: Editrice Antenore, 1988.

Gardette, Philippe. "Judaeo-Provençal Astronomy in Byzantium and Russia." *Byzantinoslavica* 63 (2005): 195–209.

Garin, Eugenio. *Portraits from the Quattrocento*. New York: Harper and Row, 1963.

– ed. *Prosatori latini del Quattrocento*. Milan: Ricciardi, 1952.

Geffen, David. "Insights into the Life and Thought of Elijah Medigo Based on His Published and Unpublished Works." *Proceedings of the American Academy for Jewish Research* 41–42 (1973–74): 69–86.

George of Trebizond. *Commentary on the Almagest.* 1451. Vienna, Österreichische Nationalbibliothek (ÖNB), cod. 3106.
Ghalandari, Hanif. "A Survey of the Works of '*Hayʾa*' in the Islamic Period with a Critical Edition, Translation and Commentary of the Treatise *Muntaha al-Idrāk fī Taqāsīm al-Aflāk* written by Bahāʾ al-Dīn al-Kharaqī (d. 553 AH/1158 AD)." PhD diss., Islamic Azad University, 2012.
Gilbert, Joan Elizabeth. "Institutionalization of Muslim Scholarship and Professionalization of the ʿUlamāʾ in Medieval Damascus." *Studia Islamica* 52 (1980): 105–34.
– "The Ulama of Medieval Damascus and the International World of Islamic Scholarship." PhD diss., University of California, Berkeley, 1977.
Gingerich, Owen. "Review of *Islamic Science and the Making of the European Renaissance*, by George Saliba." *Journal of Interdisciplinary History* 39, no. 2 (2008): 310–11.
Glasner, Ruth. *Averroes' Physics: A Turning Point in Medieval Natural Philosophy.* Oxford: Oxford University Press, 2009.
Głogów, John of (Jan Głogowczyk). *Introductorium compendiosum in tractatum sphere materialis magistri Joannis de Sacrobusto.* Strasbourg: n.p., 1518. Digital reproduction in Bayerische Staatsbibliothek, Münchener Digitalisierungs Zentrum.
– *Quaestiones super libros Analyticorum posteriorum Aristotelis cum textu.* Leipzig: Wolfgang de Monaco, 1499. Digital reproduction in Bayerische Staatsbibliothek, Münchener Digitalisierungs Zentrum.
Gluskina, Gita Mendelevna, ed. and trans. *Alfonso: Meyashsher ʿaqov.* Moscow: Izdatelsvo Nauka, 1983.
Goddu, André. *Copernicus and the Aristotelian Tradition: Education, Reading, and Philosophy in Copernicus's Path to Heliocentrism.* Leiden: Brill, 2010.
– "Copernicus's Annotations: Revisions of Czartoryski's 'Copernicana.'" *Scriptorium* 58, no. 2 (2004): 202–26.
– "Copernicus, Nicholas." In *Complete Dictionary of Scientific Biography*, vol. 20, ed. Charles Coulston Gillispie, Frederic Lawrence Holmes, Noretta Koertge, and Gale Thomson, 176–82. Detroit: Charles Scribner's Sons, 2008.
– "The Logic of Copernicus's Arguments and His Education in Logic at Cracow." *Early Modern Science and Medicine* 1, no. 1 (1996): 26–68.
– "Reflections on the Origins of Copernicus's Cosmology." *Journal for the History of Astronomy* 37, no. 1 (2006): 37–53.
– "A Response to Peter Barker and Matjaž Vesel, 'Goddu's Copernicus.'" *Aestimatio* 10 (2013): 248–76.
Goldstein, Bernard R. "Astronomy in the Medieval Spanish Jewish Community." In *Between Demonstration and Imagination: Essays in the History of Science and Philosophy Presented to John D. North*, ed. Lodi Nauta and Arjo Vanderjagt, 225–41. Leiden: Brill, 1999.
– *The Astronomy of Levi ben Gerson.* New York: Springer-Verlag, 1985.

– "Copernicus and the Origin of His Heliocentric System." *Journal for the History of Astronomy* 33, no. 3 (2002): 219–35.
– "The Survival of Arabic Astronomy in Hebrew." *Journal for the History of Arabic Science* 3 (1979): 31–9.
– ed. and trans. *Al-Biṭrūjī: On the Principles of Astronomy.* 2 vols. New Haven, CT: Yale University Press, 1971.
Goldstein Bernard R., and José Chabás. "Ptolemy, Bianchini, and Copernicus: Tables for Planetary Latitudes." *Archive for History of Exact Sciences* 58, no. 5 (2004): 453–73.
Gottlieb, Theodor. *Mittelalterliche Bibliothekskataloge Österreichs.* Vol. 1. Vienna: A. Holzhausen, 1915.
Grafton, Anthony. *The Footnote: A Curious History.* Cambridge, MA: Harvard University Press, 1997.
– "The Importance of Being Printed." *Journal of Interdisciplinary History* 11, no. 2 (1980): 265–86.
– *Leon Battista Alberti: Master Builder of the Italian Renaissance.* New York: Hill and Wang, 2000.
Grant, Edward. "Celestial Motions in the Late Middle Ages." *Early Science and Medicine* 2, no. 2 (1997): 129–48.
– "Celestial Orbs in the Latin Middle Ages." *Isis* 78, no. 2 (1987): 152–73.
– "The Fate of Ancient Greek Natural Philosophy in the Middle Ages: Islam and Western Christianity." *Review of Metaphysics* 61, no. 3 (2008): 503–26.
– "Late Medieval Thought, Copernicus, and the Scientific Revolution." *Journal of the History of Ideas* 23, no. 2 (1962): 197–220.
– *Planets, Stars, and Orbs: The Medieval Cosmos, 1200–1687.* Cambridge, UK: Cambridge University Press, 1994.
Grellard, Christophe. *Croire et savoir: Les principes de la connaissance selon Nicolas d'Autrécourt.* Paris: Vrin, 2005.
Griffel, Frank. *Al-Ghazālī's Philosophical Theology.* Oxford: Oxford University Press, 2009.
Grosseteste, Robert. *Commentarius in posteriorum analyticorum libros.* Ed. Pietro Rossi. Florence: L.S. Olschki, 1981.
Grössing, Helmuth. *Humanistische Naturwissenschaft: Zur Geschichte der Wiener mathematischen Schulen des 15. und 16. Jahrhunderts.* Baden-Baden: Verlag Valentin Koerner, 1983.
Grössing, Helmuth, and Franz Stuhlhofer. "Versuch einer Deutung der Rolle der Astrologie in den persönlichen Entscheidungen einiger Habsburger des Spätmittelalters." *Anzeiger der phil.-hist. Klasse der Österreichischen Akademie der Wissenschaften* 117 (1980): 267–83.
Gruen, Eric. *Culture and National Identity in Imperial Rome.* Ithaca, NY: Cornell University Press, 1992.
Guicciardini, Francesco. *Ricordi.* Milan: Mursia, 1994.

Habermas, Jürgen. *Strukturwandel der Öffentichkeit: Untersuchungen zu einer Kategorie der bürgerlichen Gesellschaft.* Berlin: Luchterhand, 1962.

Hacker, Joseph. "Mizraḥi, Elijah." In *Encyclopaedia Judaica*, vol. 14, 2nd ed., ed. Michael Berenbaum and Fred Skolnik, 393–5. Detroit: Macmillan Reference USA, 2007.

– *Those Banished from Spain and Their Descendants in the Ottoman Empire.* Jerusalem: Hebrew University, 1966.

Hadot, Pierre. *Philosophy as a Way of Life.* Ed. Arnold I. Davidson. Trans. Michael Chase. Oxford: Blackwell, 1995.

– *What Is Ancient Philosophy?* Cambridge, MA: Harvard University Press, 2002.

Hadzibegovic, Zalkida. "Compendium of the Science of Astronomy by al-Jaghmīnī Used in Bosnia for Teaching and Learning Planetary Motions." Paper presented at the GIREP-EPEC Conference, Opatija, Croatia, 2007.

Hamann, Günther, ed. *Regiomontanus-Studien.* Vienna: Verlag der Österreichischen Akademie der Wissenschaften, 1980.

Hankins, James. *Plato in the Italian Renaissance.* 2 vols. Leiden: Brill, 1990.

– "Renaissance Crusaders: Humanist Crusade Literature in the Age of Mehmed II." *Dumbarton Oaks Papers* 49 (1995): 111–207.

Harries, Karsten. "On the Power and Poverty of Perspective: Cusanus and Alberti." In *Cusanus: The Legacy of Learned Ignorance*, ed. Peter J. Casarella, 105–26. Washington, DC: Catholic University of America Press, 2006.

Hartner, Willy. "Copernicus, the Man, the Work, and Its History." *Proceedings of the American Philosophical Society* 117, no. 6 (1973): 413–22.

– "The Mercury Horoscope of Marcantonio Michiel of Venice: A Study in the History of Renaissance Astrology and Astronomy." In *Vistas in Astronomy*, vol. 1, ed. Arthur Beer, 84–138. London: Pergamon, 1955. Reprinted in Willy Hartner, *Oriens-Occidens: Ausgewählte Schriften zur Wissenschafts- und Kulturgeschichte: Festschrift zum 60. Geburtstag*, vol. 1, 440–95 (Hildesheim: George Olms, 1968).

Hashemipour, Behnaz. "Gayhānshenākht: A Cosmological Treatise." In *Sciences, techniques et instruments dans le monde iranien (Xe–XIXe siècle)*, ed. N. Pourjavady and Ž. Vesel, 77–84. Tehran: Presses Universitaires d'Iran and Institut Français de Recherche en Iran, 2004.

– "Qaṭṭān al-Marwazī: ʿAyn al-Zamān Abū ʿAlī Ḥasan ibn ʿAlī Qaṭṭān al-Marwazī." In *The Biographical Encyclopedia of Astronomers*, vol. 2, ed. Thomas Hockey et al., 943–4. New York: Springer-Verlag, 2007.

Hasse, Dag Nikolaus. "The Soul's Faculties." In *The Cambridge History of Medieval Philosophy*, ed. Robert Pasnau and Christina van Dyke, 305–19. Cambridge, UK: Cambridge University Press, 2010.

Hay, Denys. *Europe: The Emergence of an Idea.* 1957. Reprint, Edinburgh: Edinburgh University Press, 1968.

Hayton, Darin. *The Crown and the Cosmos: Astrology and the Politics of Maximilian I.* Pittsburgh: University of Pittsburgh Press, 2015.

Heath, Thomas. *Mathematics in Aristotle.* Oxford: Clarendon, 1949.

– trans. *The Thirteen Books of Euclid's Elements.* 3 vols. 2nd ed. New York: Dover, 1956.

Hess, Catherine, ed., with contributions by Linda Komaroff and George Saliba. *The Arts of Fire: Islamic Influences on Glass and Ceramics of the Italian Renaissance.* Los Angeles, CA: J. Paul Getty Museum, 2004.

Hildebrandt, Thomas. "Waren Ǧamāl ad-Dīn al-Afġānī und Muḥammad ʿAbduh Neo-Muʿtaziliten?" *Die Welt des Islams* 42, no. 2 (2002): 207–62.

Hockey, Thomas, et al., eds. *The Biographical Encyclopedia of Astronomers.* 2 vols. New York: Springer-Verlag, 2007.

Hoenen, Maarten J.F.M. "*Via Antiqua* and *Via Moderna* in the Fifteenth Century: Doctrinal, Institutional, and Church Political Factors in the *Wegestreit.*" In *The Medieval Heritage in Early Modern Metaphysics and Modal Theory, 1400–1700*, ed. Russell L. Friedman and Lauge O. Nielsen, 9–36. Dordrecht: Kluwer, 2003.

Hopkins, Jasper, trans. *Nicholas of Cusa: Metaphysical Speculations.* 2 vols. Minneapolis, MN: Arthur J. Banning, 1998–2000.

Housley, Norman. *Later Crusades: From Lyons to Alcazar, 1274–1580.* Oxford: Oxford University Press, 1992.

Houzel, Christian. "The New Astronomy of Ibn al-Haytham." *Arabic Sciences and Philosophy* 19, no. 1 (2009): 1–41.

Huff, Toby E. "The Rise of Early Modern Science: A Reply to George Saliba." *Bulletin of the Royal Institute for Inter-Faith Studies* 4, no. 2 (2002): 115–28.

Hugonnard-Roche, Henri. "Problèmes méthodologiques dans l'astronomie au début du XIVe siècle." In *Studies on Gersonides: A Fourteenth-Century Jewish Philosopher-Scientist*, ed. Gad Freudenthal, 55–70. Leiden: Brill, 1992.

Hunt, Lynn. *Inventing Human Rights: A History.* New York: St Martin's, 2007.

Huxley, G.L. "Theon of Smyrna." In *Dictionary of Scientific Biography*, vol. 13, ed. Charles Coulston Gillispie, 326. New York: Charles Scribner's Sons, 1976.

Ibn Khaldūn. *The Muqaddimah: An Introduction to History.* 3 vols. Trans. Franz Rosenthal. Princeton, NJ: Princeton University Press, 1967.

İhsanoğlu, Ekmeleddin. "Institutionalisation of Science in the *Medreses* of Pre-Ottoman and Ottoman History." In *Turkish Studies in the History and Philosophy of Science*, ed. Gürol Irzik and Güven Güzeldere, 265–83. Dordrecht: Springer-Verlag, 2005.

– ed. *History of the Ottoman State, Society and Civilisation.* 2 vols. Istanbul: IRCICA, 2001–02.

IJsewijn, Jozef. *Companion to Neo-Latin Studies.* Vol. 1. Louvain: Louvain University Press, 1990.

Inalcık, Halil. *The Ottoman Empire: The Classical Age, 1300–1600.* Trans. Norman Itzkowitz and Colin Imber. London: Weidenfeld and Nicolson, 1973.

Iskander, A.Z. *A Catalogue of Arabic Manuscripts on Medicine and Science in the Wellcome Historical Medical Library*. London: Wellcome Historical Medical Library, 1967.

Isom-Verhaaren, Christine. *Allies with the Infidel: The Ottoman and French Alliance in the Sixteenth Century*. London: I.B. Tauris, 2011.

Ivry, Alfred. "Remnants of Jewish Averroism in the Renaissance." In *Jewish Thought in the Sixteenth Century*, ed. Bernard D. Cooperman, 243–65. Cambridge, MA: Harvard University Press, 1983.

Izbicki, Thomas M., and Cary J. Nederman, eds and trans. *Three Tracts on Empire*. Bristol: Thoemmes, 2000.

İzgi, Cevat. *Osmanlı Medreselerinde İlim: Riyazî ilimler*. 2 vols. Istanbul: İz, 1997.

Jacob, Margaret. *Strangers Nowhere in the World: The Rise of Cosmopolitanism in Early Modern Europe*. Philadelphia: University of Pennsylvania Press, 2006.

Jardine, Lisa, and Jerry Brotton. *Global Interests: Renaissance Art between East and West*. Ithaca, NY: Cornell University Press, 2000.

Jardine, Nicholas. *The Birth of History and Philosophy of Science: Kepler's* A Defence of Tycho against Ursus *with Essays on Its Provenance and Significance*. Cambridge, UK: Cambridge University Press, 1984.

– "The Significance of the Copernican Orbs." *Journal for the History of Astronomy* 13, no. 3 (1982): 168–94.

Jarzombek, Mark. *On Leon Battista Alberti: His Literary and Aesthetic Theories*. Cambridge, MA: MIT Press, 1989.

John of Jandun. *In Duodecim Libros Metaphysicae*. Venice: Girolamo Scoto, 1553.

John Marsilius Inguen. *Quaestiones subtillissime super octo libros physicorum secundum nominalium viam*. 1518. Reprint, Frankfurt am Main: Minerva, 1964.

John of Salisbury. *Metalogicon*. Ed. John B. Hall and K.S.B. Keats-Rohan. Turnhout: Brepols, 1991.

John of Sicily. *Scriptum super canones Azarchelis de tabulis Toletanis*. Ed. Fritz S. Pedersen. *Cahiers de l'Institut du Moyen-Âge Grec et Latin* (Copenhagen) 51–52 (1986).

Johns, Adrian. *The Nature of the Book: Print and Knowledge in the Making*. Chicago: University of Chicago Press, 1998.

Jones, Robert. "The Medici Oriental Press (Rome 1584–1614) and the Impact of Its Arabic Publications on Northern Europe." In *The "Arabick" Interest of the Natural Philosophers in Seventeenth-Century England*, ed. Gül Russell, 88–108. Leiden: Brill, 1993.

Jordan, William C. "'Europe' in the Middle Ages." In *The Idea of Europe: From Antiquity to the European Union*, ed. Anthony Pagden, 72–90. Cambridge, UK: Cambridge University Press, 2002.

Kafadar, Cemal. *Between Two Worlds: The Construction of the Ottoman State*. Berkeley: University of California Press, 1995.

Kemp, Martin. *Behind the Picture: Art and Evidence in the Italian Renaissance*. New Haven, CT: Yale University Press, 1997.

Kennedy, Edward S. "Al-Bīrūnī (or Bērūnī), Abū Rayḥān (or Abu'l-Rayḥān) Muḥammad Ibn Aḥmad." In *Dictionary of Scientific Biography*, vol. 2, ed. Charles Coulston Gillispie, 147–58. New York: Charles Scribner's Sons, 1970.

– "The Exact Sciences in Timurid Iran." In *The Cambridge History of Iran*, vol. 6, *The Timurid and Safavid Periods*, ed. P. Jackson and L. Lockhart, 568–80. Cambridge, UK: Cambridge University Press, 1986.

– "Late Medieval Planetary Theory." *Isis* 57, no. 3 (1966): 365–78.

– "A Letter of Jamshīd al-Kāshī to His Father: Scientific Research and Personalities at a Fifteenth Century Court." *Orientalia* 29, no. 2 (1960): 191–213. Reprinted in Edward S. Kennedy, Colleagues, and Former Students, *Studies in the Islamic Exact Sciences*, ed. David A. King and Mary Helen Kennedy, 722–44 (Beirut: American University of Beirut, 1983).

– *The Planetary Equatorium of Jamshid Ghiyāth al-Dīn al-Kāshī*. Princeton, NJ: Princeton University Press, 1960.

– "Review of *Al-Biṭrūjī: De Motibus Coelorum*, by Carmody." *Speculum* 29, no. 2, part 1 (1954): 246–51.

– "A Survey of Islamic Astronomical Tables." *Transactions of the American Philosophical Society*, n.s., 46, part 2 (1956): 121–77.

– "Two Persian Astronomical Treatises by Naṣīr al-Dīn al-Ṭūsī." *Centaurus* 21, no. 2 (1984): 109–20.

Kennedy, Edward S., and Victor Roberts. "The Planetary Theory of Ibn al-Shāṭir." *Isis* 50, no. 3 (1959): 227–35.

Kepler, Johannes. *Gesammelte Werke*. 21 vols. Munich: C.H. Beck'sche Verlag, 1937–.

Kieszkowski, Bohdan. "Les Rapports entre Elie Del Medigo et Pic de la Mirandole (d'après le ms. lat. 6508 de le Bibliothèque Nationale)." *Rinascimento* 15 (1964): 41–91.

King, David A. *Astronomy in the Service of Islam*. Adlershot, UK: Ashgate Variorum Reprints, 1993.

– "The Astronomy of the Mamluks." *Isis* 74, no. 4 (1983): 531–55. Reprinted in David A. King, *Islamic Mathematical Astronomy* (London: Variorum Reprints, 1986).

– *Islamic Mathematical Astronomy*. London: Variorum Reprints, 1986.

– "On the Role of the Muezzin and the Muwaqqit in Medieval Islamic Society." In *Tradition, Transmission, Transformation: Proceedings of Two Conferences on Premodern Science Held at the University of Oklahoma*, ed. F. Jamil Ragep and Sally P. Ragep with Steven Livesey, 285–346. Leiden: Brill, 1996.

– *A Survey of the Scientific Manuscripts in the Egyptian National Library*. Winona Lake, IN: Eisenbrauns, 1986.

King, David A., and Julio Samsó. "Astronomical Handbooks and Tables from the Islamic World (750–1900): An Interim Report." *Suhayl* 2 (2001): 9–105.

Kircher, Timothy. *Living Well in Renaissance Italy: The Virtues of Humanism and the Irony of Leon Battista Alberti.* Tempe: Arizona Center for Medieval and Renaissance Studies, 2012.

Klausner, Carla L. *The Seljuk Vezirate: A Study of Civil Administration, 1055–1194.* Cambridge, MA: Harvard University Press, 1973.

Koenigsberger, Dorothy. *Renaissance Man and Creative Thinking: A History of Concepts of Harmony, 1400–1700.* Atlantic Highlands, NJ: Humanities, 1979.

Koerner, Joseph Leo. *The Moment of Self-Portraiture in German Renaissance Art.* Chicago: University of Chicago Press, 1993.

Koestler, Arthur. *The Sleepwalkers: A History of Man's Changing Vision of the Universe.* London: Hutchinson, 1959.

Kozodoy, Maud. *The Secret Faith of Maestre Honoratus: Profayt Duran and Jewish Identity in Late Medieval Iberia.* Philadelphia: University of Pennsylvania Press, 2015.

– "A Study of the Life and Works of Profiat Duran." PhD diss., Jewish Theological Seminary of America, 2006.

Krause, Max. "Al-Biruni: Ein iranischer Forscher des Mittelalters." *Der Islam* 26, no. 1 (1942): 1–15.

Kremer, Richard L. "Bernard Walther's Astronomical Observations." *Journal for the History of Astronomy* 11, no. 3 (1980): 174–91.

– "Text to Trophy: Shifting Representations of Regiomontanus's Library." In *Lost Libraries: The Destruction of Great Book Collections since Antiquity*, ed. James Raven, 75–90. London: Palgrave Macmillan, 2004.

– "The Use of Bernard Walther's Astronomical Observations: Theory and Observation in Early Modern Astronomy." *Journal for the History of Astronomy* 12, no. 2 (1981): 124–32.

– "War Bernard Walther, Nürnberger astronomischer Beobachter des 15. Jahrhunderts, auch ein Theoretiker?" In *Astronomie in Nürnberg*, ed. Gudrun Wolfschmidt, 156–83. Hamburg: Tredition Science, 2010.

Kren, Claudia. "Homocentric Astronomy in the Latin West: The *De reprobatione ecentricorum et epiciclorum* of Henry of Hesse." *Isis* 59, no. 3 (1968): 269–81.

– "A Medieval Objection to 'Ptolemy.'" *British Journal for the History of Science* 4, no. 4 (1969): 378–93.

– "The Rolling Device of Naṣīr al-Dīn al-Ṭūsī in the *De spera* of Nicole Oresme?" *Isis* 62, no. 4 (1971): 490–8.

Kristeller, Paul Oskar. *Iter Italicum: A Finding List of Uncatalogued or Incompletely Catalogued Humanistic Manuscripts of the Renaissance in Italian and Other Libraries.* 7 vols. London: Warburg Institute; Leiden: Brill, 1963–97.

Kuhn, Thomas S. *The Copernican Revolution: Planetary Astronomy in the Development of Western Thought.* Cambridge, MA: Harvard University Press, 1957.

– *The Structure of Scientific Revolutions.* Chicago: University of Chicago Press, 1962.

Kunitzsch, Paul. "Das Fixsternverzeichnis in der 'Persichen Syntaxis' des Georgios Chrysokokkes." *Byzantinische Zeitschrift* 57, no. 2 (1964): 382–411.

– "The Role of al-Andalus in the Transmission of Ptolemy's *Planisphaerum* and *Almagest*." *Zeitschrift für Geschichte der arabisch-islamischen Wissenschaften* 10 (1995–96): 147–55.

Kusuba, Takanori, and David Pingree. *Arabic Astronomy in Sanskrit: Al-Birjandī on Tadhkira II, Chapter 11, and Its Sanskrit Translation.* Leiden: Brill, 2002.

Labowski, Lotte. *Bessarion's Library and the Biblioteca Marciana: Six Early Inventories.* Rome: Edizioni di Storia e Letteratura, 1979.

Lacerenza, Giancarlo. "A Rediscovered Autograph Manuscript by Mordekay Finzi." *Aleph* 3 (2003): 301–25.

Laird, Walter R. "The *Scientiae Mediae* in Medieval Commentaries on Aristotle's *Posterior Analytics*." PhD diss., University of Toronto, 1983.

Langermann, Y. Tzvi "From My Notebooks: A Compendium of Renaissance Science: *Taʿalumot ḥokmah* by Moses Galeano." *Aleph* 7 (2007): 285–318.

– "From My Notebooks: Medicine, Mechanics, and Magic from Moses ben Judah Galeano's *Taʿalumot ḥokmah*." *Aleph* 9, no. 2 (2009): 353–77.

– *Ibn al-Haytham's "On the Configuration of the World."* New York: Garland, 1990.

– "Kharaqī: Shams al-Dīn Abū Bakr Muḥammad ibn Aḥmad al-Kharaqī [al-Khiraqī]." In *The Biographical Encyclopedia of Astronomers*, vol. 1, ed. Thomas Hockey et al., 627. New York: Springer-Verlag, 2007.

– "Medieval Hebrew Texts on the Quadrature of the Lune." *Historia Mathematica* 23, no. 1 (1996): 31–53.

– "Quies Media: A Lively Issue in Medieval Physics." In *International Ibn Sina Symposium Papers*, vol. 2, ed. Nevzat Bayhan, Mehmet Mazak, Nevzat Özkaya, and Raşit Küçük, 53–67. Istanbul: İstanbul Büyükşehir Belediyesi Kültür A.S. Yayınları, ca. 2009.

– "Science in the Jewish Communities of the Byzantine Cultural Orbit: New Perspectives." In *Science in the Medieval Jewish Communities*, ed. Gad Freudenthal, 435–53. Cambridge, UK: Cambridge University Press, 2011.

– "The Scientific Writings of Mordekhai Finzi." *Italia: Studi e ricerche sulla storia, la cultura e la letteratura degli ebrei d'Italia* 7 (1988): 7–44.

Lay, Juliane. "*L'Abrégé de l'Almageste*: Un inédit d'Averroès en version hébraïque." *Arabic Sciences and Philosophy* 6, no. 1 (1996): 23–61.

Leichter, Joseph. "The Zīj as-Sanjarī of Gregory Chioniades: Text, Translation and Greek to Arabic Glossary." PhD diss., Brown University, 2004.

Lerner, Michel-Pierre. *Le Monde des sphères.* 2 vols. 2nd ed. Paris: Les Belles Lettres, 1996.

Levao, Ronald. *Renaissance Minds and Their Fictions: Cusanus, Sidney, Shakespeare.* Berkeley: University of California Press, 1985.

Levi Della Vida, Giorgio. *Ricerche sulla formazione del più antico fondo dei manoscritti orientali della Biblioteca Vaticana.* Vatican City: Biblioteca Apostolica Vaticana, 1939.

Levinger, Jacob S. "Delmedigo, Elijah ben Moses Abba." In *Encyclopaedia Judaica*, vol. 5, 2nd ed., ed. Michael Berenbaum and Fred Skolnik, 542–3. Detroit: Macmillan Reference USA, 2007.

Lévy, Tony. "Gersonide, commentateur d'Euclide: Traduction annotée de ses gloses sur les *Éléments*." In *Studies on Gersonides: A Fourteenth-Century Jewish Philosopher-Scientist*, ed. Gad Freudenthal, 83–147. Leiden: Brill, 1992.

Lines, David. "Humanism and the Italian Universities." In *Humanism and Creativity: Essays in Honor of Ronald G. Witt*, ed. Christopher S. Celenza and Kenneth Gouwens, 327–46. Leiden: Brill, 2006.

– "Natural Philosophy and Mathematics in Sixteenth-Century Bologna." *Science and Education* 15, no. 2 (2006): 131–50.

Livesey, Steven J. "*Metabasis*: The Interrelationship of the Sciences in Antiquity and the Middle Ages." PhD diss., University of California, Los Angeles, 1982.

Lloyd, G.E.R. "Saving the Appearances." *Classical Quarterly* 28, no. 1 (1978): 202–22.

Lombardus, Petrus. *Sententiae in IV libris distinctae.* 2 vols. Ed. V. Doucet. Grottaferrata: Collegium S. Bonaventurae ad Claras Aquas, 1971–81.

Lorch, Richard. "The Astronomy of Jābir ibn Aflaḥ." *Centaurus* 19, no. 2 (1975): 85–107.

– "Some Remarks on the *Almagestum parvum*." In *Amphora: Festschrift für Hans Wussing zu seinem 65. Geburtstag*, ed. Sergei Demidov, Menso Folkerts, David Rowe, and Christoph Scriba, 407–37. Basel: Birkhäuser, 1992.

Maffei, Domenico, ed. *Enea Silvio Piccolomini Papa Pio II.* Siena: Varese, 1968.

Maggi, Armando. *In the Company of Demons: Unnatural Beings, Love, and Identity in the Italian Renaissance.* Chicago: University of Chicago Press, 2006.

Maier, Anneliese. *Metaphysische Hintergründe der Spätscholastischen Naturphilosophie.* Rome: Edizioni de Storia e Letteratura, 1955.

Makdisi, George. "Baghdad, Bologna, and Scholasticism." In *Centres of Learning: Learning and Location in Pre-modern Europe and the Near East*, ed. Jan Willem Drijvers and Alasdair A. MacDonald, 141–57. Leiden: Brill, 1995.

– "Madrasa and University in the Middle Ages." *Studia Islamica* 32 (1970): 255–64.

– "Muslim Institutions of Learning in Eleventh-Century Baghdad." *Bulletin of the School of Oriental and African Studies, University of London* 24, no. 1 (1961): 1–56.

– *The Rise of Colleges: Institutions of Learning in Islam and the West.* Edinburgh: Edinburgh University Press, 1981.

Malpangotto, Michela. "L'univers auquel s'est confronté Copernic: La sphère de Mercure dans les *Theoricae novae planetarum* de Georg Peurbach." *Historia Mathematica* 40, no. 3 (2013): 262–308.

– "The Original Motivation for Copernicus's Research: Albert of Brudzewo's *Commentariolum super Theoricas novas Georgii Purbachii*." *Archive for History of Exact Sciences* 70, no. 4 (2016): 361–411.

– *Regiomontano e il rinnovamento del sapere matematico e astronomico nel Quattrocento.* Bari: Caccucci Editore, 2008.

Mancha, José Luis. "Ibn al-Haytham's Homocentric Epicycles in Latin Astronomical Texts of the XIVth and XVth Centuries." *Centaurus* 33, no. 1 (1990): 70–89.

– "The Latin Translation of Levi ben Gerson's *Astronomy*." In *Studies on Gersonides: A Fourteenth-Century Jewish Philosopher-Scientist*, ed. Gad Freudenthal, 21–46. Leiden: Brill, 1992.

Markowski, Mieczsław. "Die kosmologische Anschauungen des Prosdocimo de' Beldomandi." In *Studi sul XIV secolo in memoria di Anneliese Maier*, ed. Alfonso Maierù and Agostino Paravicini Bagliani, 263–73. Rome: Edizioni di Storia e Letteratura, 1981.

Martin, Craig. "Conjecture, Probabilism, and Provisional Knowledge in Renaissance Meteorology." *Early Science and Medicine* 14, nos 1–3 (2009): 265–89.

Matar, Nabil. *Europe through Arab Eyes, 1578–1727*. New York: Columbia University Press, 2009.

Mavroudi, Maria. *A Byzantine Book on Dream Interpretation: The Oneirocriticon of Achmet and Its Arabic Sources.* Leiden: Brill, 2002.

– "Exchanges with Arab Writers during the Late Byzantine Period." In *Byzantium: Faith and Power (1261–1557)*, ed. Sarah Brooks, 62–75. New York: Metropolitan Museum of Art, 2006.

McKitterick, David. *Print, Manuscript, and the Search for Order, 1450–1830.* Cambridge, UK: Cambridge University Press, 2003.

McLaughlin, Martin. *Literary Imitation in the Italian Renaissance.* Oxford: Oxford University Press, 1995.

Mercati, Angelo. "Le due lettere di Giorgio da Trebizonda a Maometto II." *Orientalia Christiana Periodica* 9, nos 1–2 (1943): 65–99.

Mercier, Raymond. "The Greek 'Persian Syntaxis' and the *Zīj-i Īlkhānī*." *Archives internationales d'histoire des sciences* 34 (1984): 35–60.

– "Shams al-Dīn al-Bukhārī." In *The Biographical Encyclopedia of Astronomers*, vol. 2, ed. Thomas Hockey et al., 1047–8. New York: Springer-Verlag, 2007.

Meserve, Margaret. *Empires of Islam in Renaissance Historical Thought.* Cambridge, MA: Harvard University Press, 2008.

Migne, Jacques-Paul. *Patrologia Latina.* 221 vols. Paris: Migne 1844–1903.

Mitchell, R.J. *The Laurels and the Tiara: Pope Pius II, 1458–1464.* Garden City, NY: Doubleday and Company, 1962.

Modena, Abdelkader. *Medici e chirurghi ebrei dottorati e licenziati nell'Università di Padova dal 1617 al 1816.* Bologna: Forni, 1967.

Mohler, Ludwig. *Kardinal Bessarion als Theologe, Humanist und Staatsmann: Funde und Forschungen.* 3 vols. 1923–42. Reprint, Aalen: Scientia Verlag, 1967.

Monfasani, John. *Byzantine Scholars in Renaissance Italy: Cardinal Bessarion and Other Emigrés: Selected Essays.* Aldershot, UK: Variorum, 1995.

– *Collectanea Trapezuntiana: Texts, Documents, and Bibliographies of George of Trebizond.* Binghamton, NY: Medieval and Renaissance Texts and Studies, 1984.
– *George of Trebizond: A Biography and a Study of His Rhetoric and Logic.* Leiden: Brill, 1976.
Morrison, Robert. "Andalusian Responses to Ptolemy in Hebrew." In *Studies in Arabic and Hebrew Letters: In Honor of Raymond P. Scheindlin*, ed. Jonathan P. Decter and Michael Rand, 69–87. Piscataway, NJ: Gorgias, 2007.
– "An Astronomical Treatise by Mūsā Jālīnūs alias Moses Galeano." *Aleph* 11, no. 2 (2011): 385–413.
– *The Light of the World: Astronomy in al-Andalus.* Berkeley and Los Angeles: University of California Press, 2016.
– "Quṭb al-Dīn al-Shīrāzī's Hypotheses for Celestial Motions." *Journal for the History of Arabic Science* 13 (2005): 21–140.
– "The Reception of Early Modern European Astronomy by Ottoman Religious Scholars." *Archivum Ottomanicum* 21 (2003): 187–95.
– "The Role of Oral Transmission for Astronomy among Romaniot Jews." In *Texts in Transit in the Medieval Mediterranean*, ed. Y. Tzvi Langermann and Robert Morrison, 10–28. University Park: Pennsylvania State University Press, 2016.
– "A Scholarly Intermediary between the Ottoman Empire and Renaissance Europe." *Isis* 105, no. 1 (2014): 32–57.) ck
– "The Solar Model in Joseph ibn Joseph ibn Naḥmias' *Light of the World*." *Arabic Sciences and Philosophy* 15, no. 1 (2005): 57–108.
Mulsow, Martin, and Marcelo Stamm, eds. *Konstellationsforschung.* Frankfurt am Main: Suhrkamp, 2005.
Murdoch, John E. "Thomas Bradwardine: Mathematics and Continuity in the Fourteenth Century." In *Mathematics and Its Application to Science and Natural Philosophy in the Middle Ages: Essays in Honour of Marshall Clagett*, ed. Edward Grant and John E. Murdoch, 103–37. Cambridge, UK: Cambridge University Press, 1987.
Murphy, Jane H. "Improving the Mind and Delighting the Spirit: Jabarti and the Sciences in Eighteenth-Century Ottoman Cairo." PhD diss., Princeton University, 2006.
Nagy, Zoltan. "Ricerche cosmologiche nella corte umanistica de Giovanni Vitéz ad Esztergom." In *Rapporti veneto-ungheresi all' epoca del Rinascimento*, ed. Tibor Klaniczay, 65–93. Budapest: Akadémiai Kiadó, 1975.
Nardi, Bruno. *Saggi sul pensiero inedito di Pietro Pomponazzi.* Padua: Antenore, 1970.
– *Studi su Pietro Pomponazzi.* Florence: Le Monnier, 1965.
Nederman, Cary J. "Aeneas Sylvius Piccolomini, Cicero, and the Imperial Ideal." *Historical Journal* 36, no. 3 (1993): 499–515.

Neuber, Wolfgang, Thomas Rahn, and Claus Zittel, eds. *The Making of Copernicus: Early Modern Transformations of the Scientist and His Science.* Leiden: Brill, 2015.

Neugebauer, Otto. *The Exact Sciences in Antiquity.* 1957. Reprint, New York: Dover, 1969.

– *A History of Ancient Mathematical Astronomy.* 3 parts. Berlin: Springer-Verlag, 1975.

– "On the 'Hippopede' of Eudoxus." *Scripta Mathematica* 19, no. 4 (1953): 225–9. Reprinted in Otto Neugebauer, *Astronomy and History: Selected Essays,* 305–9 (New York: Springer-Verlag, 1983).

– "On the Planetary Theory of Copernicus." *Vistas in Astronomy* 10 (1968): 89–103. Reprinted in Otto Neugebauer, *Astronomy and History: Selected Essays,* 491–505 (New York: Springer-Verlag, 1983).

Nicolle, Jean-Marie. *Nicolas de Cues: Les écrits mathématiques.* Paris: H. Champion, 2007.

Nikfahm-Khubravan, Sajjad, and F. Jamil Ragep. "Ibn al-Shāṭir and Copernicus on Mercury." *Journal for the History of Astronomy* (forthcoming).

North, John. *Richard of Wallingford: An Edition of His Writings with Introductions, English Translation and Commentary.* 3 vols. Oxford: Clarendon, 1976.

Novak, B.C. "Giovanni Pico della Mirandola and Jochanan Allemano." *Journal of the Warburg and Courtauld Institutes* 45 (1982): 125–47.

Offenberg, A.K. "The First Printed Book Produced at Constantinople." *Studia Rosenthaliana* 3, no. 1 (1969): 96–112.

Omodeo, Pietro Daniel. *Copernicus in the Cultural Debates of the Renaissance: Reception, Legacy, Transformation.* Leiden: Brill, 2014.

Pagden, Anthony, ed. *The Idea of Europe from Antiquity to the European Union.* Cambridge, UK: Cambridge University Press, 2002.

Panofsky, Erwin. *Abbot Suger on the Abbey Church of St. Denis and Its Art Treasures.* Princeton, NJ: Princeton University Press, 1946.

Pantin, Isabelle. "The First Phases of the *Theoricae planetarum* Printed Tradition (1474–1535): The Evolution of a Genre Observed through Its Images." *Journal for the History of Astronomy* 43, no. 1 (2012): 3–26.

ch 17 → Park, Katherine. "Observation in the Margins, 500–1500." In *Histories of Scientific Observation,* ed. Lorraine Daston and Elizabeth Lunbeck, 15–44. Chicago: University of Chicago Press, 2011.

Paschos, E.A., and P. Sotiroudis. *The Schemata of the Stars: Byzantine Astronomy from A.D. 1300.* Singapore and River Edge, NJ: World Scientific, 1998.

Pasnau, Robert. "Science and Certainty." In *The Cambridge History of Medieval Philosophy,* vol. 1, ed. Robert Pasnau and Christina van Dyke, 357–68. Cambridge, UK: Cambridge University Press, 2010.

Pasnau, Robert, and Christina van Dyke, eds. *The Cambridge History of Medieval Philosophy.* 2 vols. Cambridge, UK: Cambridge University Press, 2010.

Pastore-Stocchi, Manlio. "G.B. Abioso e l'umanesimo a Treviso." In *La letteratura, la rappresentazione, la musica al tempo e nei luoghi di Giorgione*, ed. Michelangelo Muraro, 17–34. Rome: Jouvence, 1987.

Pedersen, Olaf. "The Decline and Fall of the Theorica Planetarum: Renaissance Astronomy and the Art of Printing." In *Science and History: Studies in Honor of Edward Rosen*, ed. Paweł Czartoryski and Erna Hilfstein, 157–85. Wrocław: Ossolineum and Polish Academy of Sciences Press, 1978.

– *A Survey of the Almagest.* Odense: Odense University Press, 1974. Reprint with annotation and new commentary by Alexander Jones, New York: Springer Verlag, 2011.

– "The 'Theorica Planetarum' and Its Progeny." In *Filosofia, scienza e astrologia nel Trecento europeo: Biagio Pelacani Parmense*, ed. Graziella Federici-Vescovini and Francesco Barocelli, 53–78. Padua: Il Poligrafo, 1992.

Pelttari, Aaron. *The Space That Remains: Reading Latin Poetry in Late Antiquity.* Ithaca, NY: Cornell University Press, 2014.

Pertusi, Agostino, ed. *La caduta di Costantinopoli.* 2 vols. Milan: Arnoldo Mondadori Editore, 1976.

Peruzzi, Enrico. *La nave di Ermete: La cosmologia di Girolamo Fracastoro.* Florence: Olschki, 1995.

Petrucci, Armando. *Writers and Readers in Medieval and Renaissance Italy.* Ed. and trans. Charles M. Radding. New Haven, CT: Yale University Press, 1995.

Peurbach, Georg. *Theoricae novae planetarum.* Basel: n.p., 1569.

– *Theoricae novae planetarum.* In Johannes Regiomontanus, *Joannis Regiomontani Opera collectanea: Faksimiledrucke von neun Schriften Regiomontans und einer von ihm gedruckten Schrift seines Lehres Purbach*, ed. Felix Schmeidler. Osnabrück: Otto Zeller Verlag, 1972.

Peurbach, Georg, and Johannes Regiomontanus. *Tabulae eclypsium magistri Georgii Peurbachii. Tabula primi mobilis Joannis de Monteregio. Indices præterea monumentorum.* Ed. Georg Tanstetter. Vienna: n.p., 1514.

Pine, Martin. *Pietro Pomponazzi: Radical Philosopher of the Italian Renaissance.* Padua: Antenore, 1986.

Pines, Shlomo. "The Semantic Distinction between the Terms Astronomy and Astrology According to al-Bīrūnī." *Isis* 55, no. 3 (1964): 343–9.

Pingree, David. *The Astronomical Works of Gregory Chioniades.* Vol. 1, *The Zīj al-ʿAlāʾī.* Part 1, *Text, Translation, Commentary.* Amsterdam: J.C. Grieben, 1985.

– "Gregory Chioniades and Paleologan Astronomy." *Dumbarton Oaks Papers* 18 (1964): 135–60.

– "Some Fourteenth-Century Byzantine Astronomical Texts." *Journal for the History of Astronomy* 29, no. 2 (1998): 103–8.

– "The Teaching of the *Almagest* in Late Antiquity." *Apeiron* 27, no. 4 (1994): 75–98.

Plato. "The Republic." In *The Collected Dialogues of Plato*, ed. Edith Hamilton and Huntington Cairns, trans. Paul Shorey, 576–844. Princeton, NJ: Princeton University Press, 1961.

Pocock, J.G.A. *The Machiavellian Moment: Florentine Political Thought and the Atlantic Republican Tradition*. Rev. ed. Princeton, NJ: Princeton University Press, 2003.

Poliziano, Angelo. *Lamia: Praelectio in priora Aristotelis analytica*. Ed. A. Wesseling. Leiden: Brill, 1986.

Pomata, Gianna. "Observation Rising: Birth of an Epistemic Genre, 1500–1650." In *Histories of Scientific Observation*, ed. Lorraine Daston and Elizabeth Lunbeck, 45–80. Chicago: University of Chicago Press, 2011.

Pomponazzi, Pietro. *De naturalium effectuum causis sive de incantationibus*. 1567. Reprint, Hildesheim: Georg Olms, 1970.

Possevino, Antonio. *Bibliotheca selecta de ratione studiorum*. Vatican City: Typographia Apostolica Vaticana, 1593.

Poulle, Emmanuel. "L'horloge planétaire de Regiomontanus." In *Regiomontanus-Studien*, ed. Günther Hamann, 335–41. Vienna: Verlag der Österreichischen Akademie der Wissenschaften, 1980.

Price, Derek de Solla. *The Equatorie of the Planetis*. Cambridge, UK: Cambridge University Press, 1955.

Proclus. *A Commentary on the First Book of Euclid's Elements*. Trans. Glenn R. Morrow. Princeton, NJ: Princeton University Press, 1970.

– *Procli Diadochi Hypotyposis astronomicarum positionum*. Ed. Carolus Manitius. 1909. Reprint, Stuttgart: B.G. Teubner, 1974.

Prowe, Leopold. *Nicolaus Coppernicus*. 2 vols. 1883–84. Reprint, Osnabrück: Zeller, 1967.

Ptolemy, Claudius. *Ptolemy's Almagest*. Trans. G.J. Toomer. London: Duckworth, 1984.

Quintilian. *Institutio oratoria*. 4 vols. Ed. and trans. Donald A. Russell. Cambridge, MA: Harvard University Press, 2001.

Ragep, F. Jamil. "ʿAlī Qushjī and Regiomontanus: Eccentric Transformations and Copernican Revolutions." *Journal for the History of Astronomy* 36, no. 4 (2005): 359–71.

– "Astronomy." In *Encyclopaedia of Islam*, 3rd ed., part 1, ed. Kate Fleet, Gudrun Krämer, Denis Matringe, John Nawas, and Everett Rowson, 120–50. Leiden: Brill, 2009. http://dx.doi.org/10.1163/1573-3912_ei3_COM_22652.

– "Astronomy in the Fanārī-Circle: The Critical Background for Qāḍīzāde al-Rūmī and the Samarqand School." In *Uluslararası Molla Fenârî Sempozyumu (4-6 Aralık 2009 Bursa)* [International symposium on Molla Fanārī (4–6 December 2009 Bursa)], ed. Tevfik Yücedoğru, Orhan Koloğlu, U. Murat Kılavuz, and Kadir Gömbeyaz, 165–76. Bursa: Bursa Büyükşehir Belediyesi, 2010.

– "Copernicus and His Islamic Predecessors: Some Historical Remarks." *History of Science* 45, no. 1 (2007): 65–81.
– "Freeing Astronomy from Philosophy: An Aspect of Islamic Influence on Science." *Osiris*, 2nd series, 16 (2001): 49–71.
– "*Hay'a*." In *Encyclopaedia of the History of Science, Technology, and Medicine in Non-Western Cultures*, 2nd ed., ed. Helaine Selin, 1061–2. Berlin: Springer-Verlag, 2008.
– "Ibn al-Haytham and Eudoxus: The Revival of Homocentric Modeling in Islam." In *Studies in the History of the Exact Sciences in Honour of David Pingree*, ed. Charles Burnett, Jan P. Hogendijk, Kim Plofker, and Michio Yano, 786–809. Leiden: Brill, 2004.
– "Ibn al-Shāṭir and Copernicus: The Uppsala Notes Revisited." *Journal for the History of Astronomy* 47, no. 4 (2016).
– "Islamic Reactions to Ptolemy's Imprecisions." In *Ptolemy in Perspective: Use and Criticism of His Work from Antiquity to the Nineteenth Century*, ed. Alexander Jones, 121–34. Dordrecht: Springer-Verlag, 2010.
– "Ḳāḍīzāde Rūmī." In *Encyclopaedia of Islam*, vol. 12, 2nd ed., ed. P.J. Bearman, Th. Bianquis, C.E. Bosworth, E. van Donzel, and W.P. Heinrichs, 502. Leiden: Brill, 2004.
– "Naṣīr al-Dīn al-Ṭūsī." In *Encyclopaedia of the History of Science, Technology, and Medicine in Non-Western Cultures*, ed. Helaine Selin, 757–8. Dordrecht: Kluwer Academic, 1997.
– *Naṣīr al-Dīn al-Ṭūsī's Memoir on Astronomy (al-Tadhkira fī ʿilm al-hayʾa)*. 2 vols. New York: Springer-Verlag, 1993.
– "New Light on Shams: The Islamic Side of Σὰμψ Πουχάρης." In *Politics, Patronage, and the Transmission of Knowledge in 13th–15th Century Tabriz*, ed. Judith Pfeiffer, 231–47. Leiden: Brill, 2014.
– "The Persian Context of the Ṭūsī Couple." In *Naṣīr al-Dīn al-Ṭūsī: Philosophe et Savant du XIIIe Siècle*, ed. N. Pourjavady and Ž. Vesel, 113–30. Tehran: Institut français de recherche en Iran and Presses universitaires d'Iran, 2000.
– "Qāḍīzāde al-Rūmī: Ṣalāḥ al-Dīn Mūsā ibn Muḥammad ibn Maḥmūd al-Rūmī." In *The Biographical Encyclopedia of Astronomers*, vol. 2, ed. Thomas Hockey et al., 942. New York: Springer-Verlag, 2007.
– "Review of *The Beginnings of Western Science*, 2nd edition, by David C. Lindberg. Chicago: University of Chicago Press 2007." *Isis* 100, no. 2 (2009): 383–5.
– "Shīrāzī's *Nihāyat al-idrāk*: Introduction and Conclusion." *Tarikh-e Elm* (Tehran) 11 (2013): 41–57.
– "Ṭūsī and Copernicus: The Earth's Motion in Context." *Science in Context* 14, nos 1–2 (2001): 145–63.
– "The Two Versions of the Ṭūsī Couple." In *From Deferent to Equant: Studies in Honor of E.S. Kennedy*, ed. David King and George Saliba, 329–56. New York: New York Academy of Sciences, 1987.

– "When Did Islamic Science Die (and Who Cares)?" *Newsletter of the British Society for the History of Science* 85 (2008): 1–3.

Ragep, F. Jamil, and Sally P. Ragep. "The Astronomical and Cosmological Works of Ibn Sīnā: Some Preliminary Remarks." In *Sciences, techniques et instruments dans le monde iranien (Xe–XIXe siècle)*, ed. N. Pourjavady and Ž. Vesel, 3–15. Tehran: Presses Universitaires d'Iran and Institut Français de Recherche en Iran, 2004.

Ragep, F. Jamil, and Sally P. Ragep, eds, with Steven J. Livesey. *Tradition, Transmission, Transformation: Proceedings of Two Conferences on Premodern Science Held at the University of Oklahoma*. Leiden: Brill, 1996.

Ragep, Sally P. *Jaghmīnī's Mulakhkhaṣ: An Islamic Introduction to Ptolemaic Astronomy*. New York: Springer-Verlag, 2016.

– "Maḥmūd ibn Muḥammad ibn ʿUmar al-Jaghmīnī's *al-Mulakhkhaṣ fī al-hayʾa al-basīṭa*: An Edition, Translation, and Study." PhD diss., McGill University, 2015.

– "The Teaching of Theoretical Astronomy in Pre-modern Islam: Looking beyond Individual Initiatives." In *Schüler und Meister*, ed. Andreas Speer and Thomas Jeschke, 557–68. Berlin: De Gruyter, 2016.

Ramminger, Johann. "Humanists and the Vernacular: Creating the Terminology for a Bilingual Universe." In *Humanists and the Vernacular*, ed. Trine Arlund Hass and Johann Ramminger, special issue of *Renaessanceforum* 6 (2010): 1–22. http://www.renaessanceforum.dk.

Ranković, Slavica, ed. *Modes of Authorship in the Middle Ages*. Toronto: Pontifical Institute of Medieval Studies, 2012.

Rashed, Roshdi. "The Celestial Kinematics of Ibn al-Haytham." *Arabic Sciences and Philosophy* 17, no. 1 (2007): 7–55.

– "The Configuration of the Universe: A Book by al-Ḥasan ibn al-Haytham?" *Revue d'histoire des sciences* 60, no. 1 (2007): 47–63.

– *Les mathématiques infinitésimales du IXe au XIe siècle*. Vol. 5, *Ibn al-Haytham, astronomie, géométrie, sphérique et trigonométrie*. London: Al-Furqan Islamic Heritage Foundation, 2006.

Räumer, Anne. "Johannes Werners Abhandlung 'Über die Bewegung der achten Sphäre' (De motu octavae sphaerae, Nürnberg 1522)." *Wolfenbütteler Renaissance Mitteilungen* 12, no. 2 (1988): 49–61.

Regiomontanus, Johannes. *Defensio Theonis contra Trapezuntium* [Defence of Theon against George of Trebizond]. Facsimile and diplomatic transcription by Michael H. Shank. Archive of the Russian Academy of Sciences, St Petersburg, MS IV-1-935. http://regio.dartmouth.edu.

– *Epytoma in Almagestum Ptolemaei* (Venice: Johannes Hamann, 1496). Facsimile in *Joannis Regiomontani Opera collectanea: Faksimiledrucke von neun Schriften Regiomontans und einer von ihm gedruckten Schrift seines Lehrers Purbach*, ed. Felix Schmeidler. Osnabrück: Otto Zeller Verlag, 1972.

– *Joannis Regiomontani Opera collectanea: Faksimiledrucke von neun Schriften Regiomontans und einer von ihm gedruckten Schrift seines Lehrers Purbach.* Ed. Felix Schmeidler. Osnabrück: Otto Zeller Verlag, 1972.

– *Oratio Iohannis de Monteregio, habita in Patavii in praelectione Alfragani.* In *Joannis Regiomontani Opera collectanea: Faksimiledrucke von neun Schriften Regiomontans und einer von ihm gedruckten Schrift seines Lehrers Purbach,* ed. Felix Schmeidler, 43–53. Osnabrück: Otto Zeller Verlag, 1972.

– *De triangulis omnimodis.* Trans. Barnabas Hughes. Madison: University of Wisconsin Press. 1967.

Rheticus, Georg Joachim. *Narratio prima.* Danzig: n.p., 1540.

Rigo, Antonio. "Bessarione, Giovanni Regiomontano e i loro studi su Tolomeo a Venezia e Roma (1462–1464)." *Studi Veneziani* 21 (1991): 48–110.

Rizzo, Silvia. *Il lessico filologico degli umanisti.* Rome: Edizioni di storia e letteratura, 1984.

– *Ricerche sul latino umanistico.* Rome: Edizioni di storia e letteratura, 2002.

Roberts, Victor. "The Planetary Theory of Ibn al-Shāṭir: Latitudes of the Planets." *Isis* 57, no. 2 (1966): 208–19.

– "The Solar and Lunar Theory of Ibn ash-Shāṭir: A Pre-Copernican Copernican Model." *Isis* 48, no. 4 (1957): 428–32.

Robinson, James T. "The First References in Hebrew to al-Biṭrūjī's 'On the Principles of Astronomy.'" *Aleph* 3 (2003): 145–63.

Rome, Adolphe. *Commentaires de Pappus et Théon d'Alexandrie sur l'Almageste.* 3 vols. Rome: Biblioteca Apostolica Vaticana, 1931–43.

Rose, Paul L. *The Italian Renaissance of Mathematics.* Geneva: Droz, 1975.

Rosen, Edward. "Copernicus and al-Biṭrūjī." *Centaurus* 7, no. 2 (1960): 152–6.

– "Copernicus' Spheres and Epicycles." *Archives internationales d'histoire des sciences* 25 (1975): 82–92.

– "Reply to N. Swerdlow." *Archives internationales d'histoire des sciences* 26 (1976): 301–4.

Rosenfeld, Boris A., and Ekmeleddin Ihsanoğlu. *Mathematicians, Astronomers and Other Scholars of Islamic Civilization and Their Works (7th–19th c.).* Istanbul: IRCICA, 2003.

Rosińska, Grażyna. "Naṣīr al-Dīn al-Ṭūsī and Ibn al-Shāṭir in Cracow?" *Isis* 65, no. 2 (1974): 239–43.

Roth, Cecil. "Bonjorn, Bonet Davi(d)." In *Encyclopaedia Judaica,* vol. 4, 2nd ed., ed. Michael Berenbaum and Fred Skolnik, 63. Detroit: Macmillan Reference USA, 2007.

Ruderman David B. "Medicine and Scientific Thought in the Ghetto: The Cultural World of Tobias Cohen." In *The Jews of Venice: A Unique Renaissance Community,* ed. R.C. Davis and B. Ravid, 191–210. Baltimore, MD: Johns Hopkins University Press, 2001.

Russell, Gül, ed. *The "Arabick" Interest of the Natural Philosophers in Seventeenth-Century England.* Leiden: Brill, 1993.

Ruthven, Kenneth Knowles. *Critical Assumptions.* Cambridge, UK: Cambridge University Press, 1979.

Sabra, A.I. "Al-Farghānī, Abu'l-ʿAbbās Aḥmad ibn Muḥammad ibn Kathīr." In *Dictionary of Scientific Biography,* vol. 4, ed. Charles Coulston Gillispie, 541–5. New York: Charles Scribner's Sons, 1971.

– "The Andalusian Revolt against Ptolemaic Astronomy: Averroes and al-Biṭrūjī." In *Transformation and Tradition in the Sciences: Essays in Honor of I. Bernard Cohen,* ed. Everett Mendelsohn, 133–53. Cambridge, UK: Cambridge University Press, 1984. Reprinted in A.I. Sabra, *Optics, Astronomy and Logic: Studies in Arabic Science and Philosophy* (Aldershot, UK: Ashgate Variorum Reprints, 1994).

– "The Appropriation and Subsequent Naturalization of Greek Science in Medieval Islam: A Preliminary Statement." *History of Science* 25, no. 3 (1987): 223–43. Reprinted in A.I. Sabra, *Optics, Astronomy and Logic: Studies in Arabic Science and Philosophy* (Adlershot, UK: Ashgate Variorum Reprints, 1994); and in *Tradition, Transmission, Transformation: Proceedings of Two Conferences on Premodern Science Held at the University of Oklahoma,* ed. F. Jamil Ragep and Sally P. Ragep, 3–27 (Leiden: Brill, 1996).

– "Configuring the Universe: Aporetic, Problem Solving, and Kinematic Modeling as Themes of Arabic Astronomy." *Perspectives on Science* 6, no. 3 (1998): 288–330.

– "An Eleventh-Century Refutation of Ptolemy's Planetary Theory." In *Science and History: Studies in Honor of Edward Rosen,* ed. Paweł Czartoryski and Erna Hilfstein, 117–31. Wrocław: Ossolineum and Polish Academy of Sciences Press, 1978.

– "Ibn al-Haytham, Abū ʿAlī al-Ḥasan ibn al-Ḥasan." In *Dictionary of Scientific Biography,* vol. 6, ed. Charles Coulston Gillispie, 189–210. New York: Charles Scribner's Sons, 1972.

– "One Ibn al-Haytham or Two? Conclusion." *Zeitschrift für Geschichte der Arabisch-Islamischen Wissenschaften* 15 (2002–03): 95–108.

– "One Ibn al-Haytham or Two? An Exercise in Reading the Bio-bibliographical Sources." *Zeitschrift für Geschichte der Arabisch-Islamischen Wissenschaften* 12 (1998): 1–50.

– *The Optics of Ibn al-Haytham: Books I–III on Direct Vision.* 2 vols. Trans. by A.I. Sabra. London: Warburg Institute, 1989.

– "Reply to Saliba." *Perspectives on Science* 8, no. 4 (2000): 342–5.

– "Science, Islamic." In *Dictionary of the Middle Ages,* vol. 11, ed. Joseph Strayer, 81–9. New York: Charles Scribner's Sons, 1982.

– "Situating Arabic Science: Locality versus Essence." *Isis* 87, no. 4 (1996): 654–70.

Said, Edward W. *Orientalism.* New York: Vintage, 1978.

Saliba, George. "Al-Qushjī's Reform of the Ptolemaic Model for Mercury." *Arabic Sciences and Philosophy* 3, no. 2 (1993): 161–203.

– "Arabic Planetary Theories after the Eleventh Century AD." In *Encyclopedia of the History of Arabic Science,* vol. 1, ed. Roshdi Rashed, 58–127. London: Routledge, 1996.

– "Arabic Science in Sixteenth-Century Europe: Guillaume Postel (1510–1581) and Arabic Astronomy." *Suhayl* 7 (2007): 115–64.

– "Arabic versus Greek Astronomy: A Debate over the Foundations of Science." *Perspectives on Science* 8, no. 4 (2000): 328–41.

– "The Astronomical Tradition of Maragha: A Historical Survey and Prospects for Future Research." *Arabic Sciences and Philosophy* 1, no. 1 (1991): 67–99.

– "The Development of Astronomy in Medieval Islamic Society." *Arabic Studies Quarterly* 4, no. 3 (1982): 211–25. Reprinted in George Saliba, *A History of Arabic Astronomy: Planetary Theories during the Golden Age of Islam,* 51–65 (New York: New York University Press, 1994).

– "The First Non-Ptolemaic Astronomy at the Maraghah School." *Isis* 70, no. 4 (1979): 571–76. Reprinted in George Saliba, *A History of Arabic Astronomy: Planetary Theories during the Golden Age of Islam,* 113–18 (New York: New York University Press, 1994).

– "Flying Goats and Other Obsessions: A Response to Toby Huff's 'Reply.'" *Bulletin of the Royal Institute for Inter-Faith Studies* 4, no. 2 (2002): 129–41.

– "Islamic Astronomy in Context: Attacks on Astrology and the Rise of the *Hay'a* Tradition." *Bulletin of the Royal Institute for Inter-Faith Studies* 2, no. 1 (2002): 25–46.

– *Islamic Science and the Making of the European Renaissance.* Cambridge, MA: MIT Press, 2007.

– "Reform of Ptolemaic Astronomy at the Court of Ulugh Beg." In *Studies in the History of the Exact Sciences in Honour of David Pingree,* ed. Charles Burnett, Jan P. Hogendijk, Kim Plofker, and Michio Yano, 810–24. Leiden: Brill, 2004.

– "The Role of the *Almagest* Commentaries in Medieval Arabic Astronomy: A Preliminary Survey of Ṭūsī's Redaction of Ptolemy's *Almagest.*" *Archives internationales d'histoire des sciences* 37 (1987): 3–20. Reprinted in George Saliba, *A History of Arabic Astronomy: Planetary Theories during the Golden Age of Islam,* 143–60 (New York: New York University Press, 1994).

– "The Role of the Astrologer in Medieval Islamic Society." *Bulletin d'Études Orientales* 44 (1992): 45–67.

– "Theory and Observation in Islamic Astronomy: The Work of Ibn al-Shāṭir of Damascus." *Journal for the History of Astronomy* 28, no. 1 (1987): 35–43.

Saliba, George, and Edward S. Kennedy. "The Spherical Case of the Ṭūsī Couple." *Arabic Sciences and Philosophy* 1, no. 2 (1991): 285–91.

Samsó, Julio. "Andalusian Astronomy: Its Main Characteristics and Influence in the Latin West." In *Islamic Astronomy and Medieval Spain*. Aldershot, UK: Variorum, 1994.

Santinello, Giovanni. *Leon Battista Alberti: Una visione estetica del mondo e della vita*. Florence: G.C. Sansoni, 1962.

Sayılı, Aydın. "The Institutions of Science and Learning in the Moslem World." PhD diss., Harvard University, 1941.

– *The Observatory in Islam and Its Place in the General History of the Observatory*. Ankara: Türk Tarih Kurumu Basımevi, 1960.

– *Uluğ Bey ve Semerkanddeki ilim faaliyeti hakkinda Giyasüddin-i Kâşî'nin mektubu* [Ghiyâth al Dîn al Kâshî's letter on Ulugh Bey and the scientific activity in Samarqand]. Ankara: Türk Tarih Kurumu Basımevi, 1960.

Schiaparelli, Giovanni Virginio. "Le Sfere omocentriche di Eudosso, di Callippo e di Aristotele." In *Scritti Sulla Storia della Astronomia Antica*, vol. 2. Bologna: Nicola Zanichelli, 1925–27. http://www.liberliber.it/mediateca/libri/s/schiaparelli/scritti_sulla_storia_2/pdf/schiaparelli_scritti_2.pdf.

Schmidl, Petra G. "Kāshī: Ghiyāth (al-Milla wa-) al-Dīn Jamshīd ibn Masʿūd ibn Maḥmūd al-Kāshī [al-Kāshānī]." In *The Biographical Encyclopedia of Astronomers*, vol. 1, ed. Thomas Hockey et al., 613–15. New York: Springer-Verlag, 2007.

– "ʿUrḍī: Muʾayyad (al-Milla wa-) al-Dīn (Muʾayyad ibn Barīk [Burayk]) al-ʿUrḍī (al-ʿĀmirī al-Dimashqī)." In *The Biographical Encyclopedia of Astronomers*, vol. 2, ed. Thomas Hockey et al., 1161–2. New York: Springer-Verlag, 2007.

Schoener, Johannes. *Scripta clarissimi mathematici M. Joannis Regiomontani*. Nuremberg: Joannes Montanus and Ulrich Neuber, 1544.

Schwoebel, Robert. *The Shadow of the Crescent: The Renaissance Image of the Turk (1453–1517)*. New York: St Martin's, 1967.

Sensi, Mario. "Niccolò Tignosi da Foligno opera e il pensiero." *Annali della facoltà di lettere e filosofia della Università degli Studi di Perugia* 9 (1971–72): 361–495.

Şeşen, Ramazan, and Cevat İzgi, eds. *Osmanlı astronomi literatürü tarihi* [History of astronomy literature during the Ottoman period]. 2 vols. General editor Ekmeleddin İhsanoğlu. Istanbul: IRCICA, 1997.

– eds. *Osmanlı matematik literatürü tarihi* [History of mathematical literature during the Ottoman period]. 2 vols. General editor Ekmeleddin İhsanoğlu. Istanbul: IRCICA, 1999.

Setton, Kenneth M. *The Papacy and the Levant, 1204–1571*. 4 vols. Philadelphia: American Philosophical Society, 1976–84.

Shank, Michael H. "Academic Consulting in Fifteenth-Century Vienna: The Case of Astrology." In *Texts and Contexts in Ancient and Medieval Science: Studies on the Occasion of John E. Murdoch's Seventieth Birthday*, ed. Edith Sylla and Michael McVaugh, 245–70. Leiden: Brill, 1997.

– "The Classical Scientific Tradition in Fifteenth-Century Vienna." In *Tradition, Transmission, Transformation: Proceedings of Two Conferences on Premodern Science Held at the University of Oklahoma*, ed. F. Jamil Ragep and Sally Ragep, 115–36. Leiden: Brill, 1996.
– "The Geometrical Diagrams in Regiomontanus's Edition of His Own *Disputationes* (c. 1475): Background, Production, and Diffusion." *Journal for the History of Astronomy* 43, no. 1 (2012): 27–55.
– "Made to Order." *Isis* 105, no. 1 (2014): 167–76.
– "Mechanical Thinking in European Astronomy (13th–15th Centuries)." In *Mechanics and Cosmology in the Medieval and Early Modern Period*, ed. Massimo Bucciantini, Michele Camerota, and Sophie Roux, 3–27. Florence: Leo Olschki, 2007.
– "The 'Notes on al-Biṭrūjī' Attributed to Regiomontanus: Second Thoughts." *Journal for the History of Astronomy* 23, no. 1 (1992): 15–30.
– "Regiomontanus and Homocentric Astronomy." *Journal for the History of Astronomy* 29, no. 2 (1998): 157–66.
– "Regiomontanus as a Physical Astronomer: Samplings from *The Defence of Theon against George of Trebizond*." *Journal for the History of Astronomy* 38, no. 3 (2007): 325–49.
– "Regiomontanus on Ptolemy, Physical Orbs, and Astronomical Fictionalism: Goldsteinian Themes in the 'Defense of Theon against George of Trebizond.'" *Perspectives on Science* 10, no. 2 (2002): 179–207.
– "Rejoinder." *Isis* 105, no. 1 (2014): 185–7.
– "Rings in a Fluid Heaven: The Equatorium-Driven Physical Astronomy of Guido de Marchia (fl. 1292–1310)." *Centaurus* 45, nos 1–4 (2003): 175–203.
– "Zwischen Berechnung und Experiment: Wissenschaftsgeschichte an der frühen Universität Wien: Ein Überblick." In *Wien 1365: Eine Universität entsteht*, ed. Heidrun Rosenberg and Michael Viktor Schwarz, 184–215. Vienna: Brandstätter, 2015.
Shapin, Stephen. *A Social History of Truth: Civility and Science in Seventeenth-Century England*. Chicago: University of Chicago Press, 1994.
al-Shīrāzī, Quṭb al-Dīn. *Al-Tuḥfa al-shāhiyya fī al-hay'a*. Istanbul, Süleymaniye Library, Turhan H. Sultan MS 220.
Shulvass, Moses A. *The Jews in the World of the Renaissance*. Leiden: Brill, 1973.
Sinisgalli, Rocco. *Il nuovo De Pictura di Leon Battista Alberti* [The new *De Pictura* of Leon Battista Alberti]. Rome: Edizioni Kappa, 2006.
Siorvanes, Lucas. *Proclus: Neo-Platonic Philosophy and Science*. New Haven, CT: Yale University Press, 1996.
Skinner, Quentin. *The Foundations of Modern Political Thought*. 2 vols. Cambridge, UK: Cambridge University Press, 1978.
– *Liberty before Liberalism*. Cambridge, UK: Cambridge University Press, 1998.

Solon, Peter. "The *Six Wings* of Immanuel Bonfils and Michael Chrysokokkes." *Centaurus* 15, no. 1 (1971): 1–20.

Sønnesyn, Sigbjørn Olsen. "Obedient Creativity and Idiosyncratic Copying: Tradition and Individuality in the Works of William of Malmesbury and John of Salisbury." In *Modes of Authorship in the Middle Ages*, ed. Slavica Ranković, 113–32. Toronto: Pontifical Institute of Medieval Studies, 2012.

Steele, John M., and F. Richard Stephenson. "Eclipse Observations Made by Regiomontanus and Walther." *Journal for the History of Astronomy* 29, no. 4 (1998): 331–44.

Stegmüller, Friedrich. *Repertorium commentariorum in Sententias Petri Lombardi.* 2 vols. Würzburg: Schöning, 1947.

Stephens, Walter. *Demon Lovers: Witchcraft, Sex, and the Crisis of Belief.* Chicago: University of Chicago Press, 2002.

Stock, Brian. "Minds, Bodies, Readers." *New Literary History* 37, no. 3 (2006): 489–525.

Streete, Thomas. *Astronomia Carolina, or A New Theorie of the Coelestial Motions, Composed According to the Best Observations and Most Rational Grounds of Art, yet Farre More Easie, Expedite and Perspicuous Than Any before Extant, with Exact and Most Easie Tables Thereunto, and Precepts for the Calculation of Eclipses &C.* London: Printed for Lodowick Lloyd, 1661.

Summers, David. *Vision, Reflection, and Desire in Western Painting.* Chapel Hill: University of North Carolina Press, 2007.

Swerdlow, Noel. "The Annals of Scientific Publishing: Johannes Petreius's Letter to Rheticus." *Isis* 83, no. 2 (1992): 270–4.

– "Aristotelian Planetary Theory in the Renaissance: Giovanni Battista Amico's Homocentric Spheres." *Journal for the History of Astronomy* 3, no. 1 (1972): 36–48.

– "Astronomy in the Renaissance." In *Astronomy before the Telescope*, ed. Christopher Walker, 197–200. London: British Library, 1996.

– "Copernicus and Astrology, with an Appendix of Translations of Primary Sources." *Perspectives on Science* 20, no. 3 (2012): 353–78 (plus online appendix).

– "Copernicus's Four Models of Mercury." In *Studia Copernicana XIII (Colloquia Copernicana, III): Astronomy of Copernicus and Its Background: Proceedings of the Joint Symposium of the IAU and IUHPS, Co-sponsored by the IAHS, Torun, 1973*, ed. Owen Gingerich and Jerzy Dobrzycki, 141–55. Warsaw: Ossolineum, 1975.

– "The Derivation and First Draft of Copernicus's Planetary Theory: A Translation of the Commentariolus with Commentary." *Proceedings of the American Philosophical Society* 117, no. 6 (1973): 423–512.

– "An Essay on Thomas Kuhn's First Scientific Revolution, *The Copernican Revolution.*" *Proceedings of the American Philosophical Society* 148, no. 1 (2004): 64–120.

– "Pseudodoxia Copernicana, or Enquiries into Very Many Received Tenets and Commonly Presumed Truths, Mostly Concerning Spheres." *Archives internationales d'histoire des sciences* 26 (1976): 108–58.
– "The Recovery of the Exact Sciences of Antiquity: Mathematics, Astronomy, Geography." In *Rome Reborn: The Vatican Library and Renaissance Culture*, ed. Anthony Grafton, 125–67. New Haven, CT: Yale University Press, 1993.
– "Regiomontanus on the Critical Problems of Astronomy." In *Nature, Experiment and the Sciences: Essays on Galileo and the History of Science*, ed. Trevor H. Levere and William R. Shea, 165–95. Dordrecht: Kluwer Academic, 1990.
– "Regiomontanus's Concentric-Sphere Models for the Sun and the Moon." *Journal for the History of Astronomy* 30, no. 1 (1999): 1–23.
– "Science and Humanism in the Renaissance: Regiomontanus's Oration on the Dignity and Utility of the Mathematical Sciences." In *World Changes: Thomas Kuhn and the Nature of Science*, ed. Paul Horwich, 131–68. Cambridge, MA: MIT Press. 1993.
Swerdlow, Noel, and Otto Neugebauer. *Mathematical Astronomy in Copernicus's De Revolutionibus*. 2 parts. New York: Springer-Verlag, 1984.
Sylla, Edith Dudley. "The A Posteriori Foundations of Natural Science: Some Medieval Commentaries on Aristotle's *Physics*, Book I, Chapters 1 and 2." *Synthese* 40 (1979): 147–87.
– "Astronomy at Cracow University in the Late Fifteenth Century: Albert of Brudzewo and John of Głogów." In *What Is New in the New Universities? Learning in Central Europe in Later Middle Ages (1348–1500)*, ed. Elżbieta Jung. Turnhout: Brepols, forthcoming.
– "Averroes and Fourteenth-Century Theories of Alteration: Minima Naturalia and the Distinction between Mathematics and Physics." In *Averroes' Natural Philosophy and Its Reception in the Latin West*, ed. Paul J.J.M. Bakker, 141–92. Leuven: Leuven University Press, 2015.
– "Averroism and the Assertiveness of the Separate Sciences." In *Knowledge and the Sciences in Medieval Philosophy*, vol. 3, ed. Reijo Työrinoja, Anja Inkeri Lehtinen, and Dagfinn Føllesdal, 171–80. Helsinki: Annals of the Finnish Society for Missiology and Ecumenics, 1990.
– "Galileo and Probable Arguments." In *Nature and Scientific Method*, ed. Daniel O. Dahlstrom, 211–34. Washington, DC: Catholic University of America Press, 1991.
– "John Buridan and Critical Realism." *Early Science and Medicine* 14, nos 1–3 (2009): 211–47.
– "The Oxford Calculators' Middle Degree Theorem in Context." *Early Science and Medicine* 15, nos 4–5 (2010): 338–70.
– "Political, Moral, and Economic Decisions and the Origins of the Mathematical Theory of Probability: The Case of Jacob Bernoulli's *The Art of Conjecturing*." In *Acting under Uncertainty: Multidisciplinary Conceptions*, ed. George von Furstenburg, 19–44. Boston: Kluwer, 1990.

– "The Status of Astronomy between Experience and Demonstration in the Commentaries on Aristotle's *Posterior Analytics* of Robert Grosseteste and Walter Burley." In *Erfahrung und Beweis: Die Wissenschaften von der Natur im 13. und 14. Jahrhundert* [Experience and demonstration: The sciences of nature in the 13th and 14th centuries], ed. Alexander Fidora and Matthias Lutz-Bachmann, 265–91. Berlin: Akademie Verlag, 2007.

Tarugi, Luisa Rotondi Secchi, ed. *Pio II e la cultura del suo tempo.* Milan: Guerini e Associati, 1991.

Ṭāshkubrīzāde, Aḥmad ibn Muṣṭafā. *Miftāḥ al-sa'āda wa-miṣbāḥ al-siyāda.* 3 vols. Beirut: Dār al-kutub al-'ilmiyya, 1985.

Taub, Liba C. *Ptolemy's Universe: The Natural Philosophical and Ethical Foundations of Ptolemy's Astronomy.* Chicago: Open Court, 1993.

Tavoni, Mirko. *Latino, grammatica, volgare: Storia di una questione umanistica.* Padua: Antenore, 1984.

Theon of Smyrna. *Mathematics Useful for Understanding Plato.* Trans. Robert Lawlor and Deborah Lawlor. San Diego: Wizards Bookshelf, 1979.

Thorndike, Lynn. *A History of Magic and Experimental Science.* 8 vols. New York: Columbia University Press, 1958.

Tihon, Anne. "L'astronomie byzantine à l'aube de la Renaissance." *Byzantion* 66 (1996): 244–80.

– "The Astronomy of George Gemistus Plethon." *Journal for the History of Astronomy* 29, no. 2 (1998): 109–16.

Tihon, Anne, and Raymond Mercier. *Georges Gémiste Pléthon: Manuel d'Astronomie.* Louvain-la-Neuve: Bruylant-Academia, 1998.

Topdemir, Hüseyin. "'Abd al-Wājid: Badr al-Dīn 'Abd al-Wājid [Wāḥid] ibn Muḥammad ibn Muḥammad al-Ḥanafī." In *The Biographical Encyclopedia of Astronomers*, vol. 1, ed. Thomas Hockey et al., 5–6. New York: Springer-Verlag, 2007.

Touwaide, Alain. "Arabic Medicine in Greek Translation: A Preliminary Report." *Journal of the International Society for the History of Islamic Medicine* 1 (2002): 45–53.

– "Arabic Urology in Byzantium." *Journal of Nephrology* 17 (2004): 583–9.

Trivellato, Francesca. *The Familiarity of Strangers: The Sephardic Diaspora, Livorno, and Cross-Cultural Trade in the Early Modern Period.* New Haven, CT: Yale University Press, 2009.

Tucci, Roberta. "Giorgio Valla e i libri matematici del 'De expetendis et fugiendis rebus': Contenuto, fonti, fortuna." PhD diss., University of Pisa, 2008.

al-Ṭūsī, Naṣīr al-Dīn. *Dhayl-i Mu'īniyya.* Tashkent, Al-Biruni Institute of Oriental Studies, MS 8990, fols 33b–46a (original foliation).

– *Ḥall-i mushkilāt-i Mu'īniyya.* Facsimile of Tehran, Malik 3503, with an introduction by Muḥammad Taqī Dānish-Pizhūh. Tehran: Intishārāt Dānishgāh Tahrān, 1956–57.

– *Taḥrīr al-Majisṭī*. Istanbul, Feyzullah MS 1360.

Uiblein, Paul. "Die Wiener Universität, ihre Magister und Studenten zur Zeit Regiomontans." In *Regiomontanus-Studien*, ed. Günther Hamann, 395–432. Vienna: Verlag der Österreichischen Akademie der Wissenschaften, 1980.

Uzielli, Gustavo, and Giovanni Celoria. *La vita e i tempi di Paolo dal Pozzo Toscanelli*. Rome: Ministero della pubblica istruzione, 1894.

Van Brummelen, Glen. *The Mathematics of the Heavens and the Earth: The Early History of Trigonometry*. Princeton, NJ: Princeton University Press, 2009.

van Dalen, Benno. "Ulugh Beg." In *The Biographical Encyclopedia of Astronomers*, vol. 2, ed. Thomas Hockey et al., 1157–9. New York: Springer-Verlag, 2007.

Vargha, Magda, and Elöd Both. "Astronomy in Renaissance Hungary." *Journal for the History of Astronomy* 18, no. 4 (1987): 279–83.

Vast, Henri. *Le cardinal Bessarion (1403–1472) Étude sur le chrétienté et la renaissance vers le milieu du XVe siècle*. Paris: Librairie Hachette et Companie, 1878.

Veselovsky, Ivan Nikolayevich. "Copernicus and Naṣīr al-Dīn al-Ṭūsī." *Journal for the History of Astronomy* 4, no. 2 (1973): 128–30.

Vollmann, Benedikt Konrad. "Aeneas Silvius Piccolomini as Historiographer: *Asia*." In *Pius II 'El Più Expeditivo Pontefice': Selected Studies on Aeneas Silvius Piccolomini (1405–1464)*, ed. Zweder Von Martels and Arjo Vanderjagt, 41–54. Leiden: Brill, 2003.

Von Martels, Zweder, and Arjo Vanderjagt, eds. *Pius II 'El Più Expeditivo Pontefice': Selected Studies on Aeneas Silvius Piccolomini (1405–1464)*. Leiden: Brill, 2003.

von Staden, Heinrich. "Liminal Perils: Early Roman Receptions of Greek Medicine." In *Tradition, Transmission, Transformation: Proceedings of Two Conferences on Premodern Science Held at the University of Oklahoma*, ed. F. Jamil Ragep and Sally Ragep, 369–418. Leiden: Brill, 1996.

Wallerstein, Immanuel. *The Modern World System*. 3 vols. New York: Academic Press, 1974–89.

Wattenberg, Dietrich. "Johannes Regiomontanus und die astronomische Instrumente seiner Zeit." In *Regiomontanus-Studien*, ed. Günther Hamann, 343–62. Vienna: Verlag der Österreichischen Akademie der Wissenschaften, 1980.

Weinberg, Steven. "A Deadly Certitude." *Times Literary Supplement*, 17 January 2007.

Weinig, Paul. *Aeneam suscipite, Pium recipite: Aeneas Silvius Piccolomini: Studien zur Rezeption eines humanistischen Schriftstellers im Deutschland des 15. Jahrhunderts*. Wiesbaden: Harrassowitz, 1998.

Wells, Peter. *The Barbarians Speak: How the Conquered Peoples Shaped Roman Europe*. Princeton, NJ: Princeton University Press, 1999.

Westman, Robert S. "The Astronomer's Role in the Sixteenth Century: A Preliminary Study." *History of Science* 18, no. 2 (1980): 105–47.

– *The Copernican Question: Prognostication, Skepticism, and Celestial Order*. Berkeley: University of California Press, 2011.

– "The Copernican Question Revisited: A Reply to Noel Swerdlow and John Heilbron." *Perspectives on Science* 21, no. 1 (2013): 100–36.
– "The Melanchthon Circle, Rheticus, and the Wittenberg Interpretation of the Copernican Theory." *Isis* 66, no. 2 (1975): 165–93.
– "Reply to Michael Shank." *Isis* 105, no. 1 (2014): 177–84.
– "Two Cultures or One? A Second Look at Kuhn's *The Copernican Revolution*." *Isis* 85, no. 1 (1994): 79–115.

Wilkinson, Robert J. *Orientalism, Aramaic, and Kabbalah in the Catholic Reformation: The First Printing of the Syriac New Testament*. Leiden: Brill, 2007.

The Wisdom of Solomon. 11:20. King James Version. http://www.kingjamesbibleonline.org/book.php?book=Wisdom+of+Solomon&chapter=11&verse=20.

Wisnovsky, Robert. "The Nature and Scope of Arabic Philosophical Commentary in Post-Classical (ca. 1100–1900 AD) Islamic Intellectual History: Some Preliminary Observations." In *Philosophy, Science and Exegesis in Greek, Arabic and Latin Commentaries*, ed. Peter Adamson, Han Baltussen, and Martin William Francis Stone, special issue of *Bulletin of the Institute of Classical Studies* (University of London) 83, nos 1–2 (2004): 149–91.

Witkam, Jan Just. *De Egyptische Arts Ibn al-Akfānī*. Leiden: Ter Lugt Pers, 1989.

Wussing, Hans. "Regiomontanus als Student in Leipzig." In *Regiomontanus-Studien*, ed. Günther Hamann, 167–74. Vienna: Verlag der Österreichischen Akademie der Wissenschaften, 1980.

Wust, Ephraim. "Elisha the Greek: A Physician and Philosopher at the Beginning of the Ottoman Period." *Pe'amim* 41 (1989): 49–57.

Yates, Frances. *The Art of Memory*. Chicago: University of Chicago Press, 1966.

Yavetz, Ido. "On the Homocentric Spheres of Eudoxus." *Archive for History of Exact Sciences* 52, no. 3 (1998): 221–78.

Zanier, Giancarlo. *Ricerche sulla diffusione e fortuna del De incantationibus di Pomponazzi*. Florence: La nuova Italia, 1975.

Zepeda, Henry. "The Medieval Latin Transmission of the Menelaus Theorem." PhD diss., University of Oklahoma, 2013.

Zinner, Ernst. *Der deutsche Kalender des Johannes Regiomontan, Nürnberg, um 1474: Faksimiledruck nach dem Exemplar der Preussischen Staatsbibliothek, mit einer Einleitung*. Leipzig: O. Harrassowitz, 1937.
– "Die wissenschaftlichen Bestrebungen Regiomontans." *Beiträge zur Inkunablekunde*, n.s., 2 (1938): 89–103.
– *Leben und Wirken des Joh. Müller von Königsberg, genannt Regiomontanus*. 2nd ed. Osnabrück: O. Zeller, 1968.

Zonta, Mauro. "The Jewish Mediation in the Transmission of Arabo-Islamic Science and Philosophy to the Latin Middle Ages: Historical Overview and Perspectives of Research." In *Wissen über Grenzen: Arabisches Wissen und lateinisches Mittelalter*, ed. Andreas Speer and Lydia Wegener, 89–105. Berlin: De Gruyter, 2006.

Contributors

NANCY BISAHA is a professor of history and the director of Medieval and Renaissance Studies at Vassar College. She is currently working on an intellectual biography of Pope Pius II and his conception of Europe and Asia.

CHRISTOPHER S. CELENZA is the vice dean of Humanities and Social Sciences and the Charles Homer Haskins Professor at Johns Hopkins University. His most recent book is *Petrarch: Everywhere a Wanderer* (2017).

RAZ CHEN-MORRIS is a senior lecturer in the Department of History at the Hebrew University of Jerusalem and a research fellow at the Minerva Humanities Center, Tel Aviv University. He is the author of *Measuring Shadows: Kepler's Optics of Invisibility* (2016) and, with Ofer Gal, *Baroque Science* (2013). His current research project, "Optics, Lenses, and the Formation of Knowledge in Early Modern Europe," examines disputes over visual experience in the early stages of the New Science, concentrating on Kepler's *Dioptrice* and examining its political implications for the formation of early modern notions of sovereignty.

RIVKA FELDHAY is a professor emerita of the history of science at the Cohn Institute for History and Philosophy of Science and Ideas and the director of the Minerva Humanities Center, Tel Aviv University. She recently published "The Global and the Local in the Study of the Humanities," in T. Arabatzis et al., eds, *Relocating the History of Science* (2015). She is currently working on R. Feldhay, J. Renn, M. Schemmel, and M. Valleriani, eds, *Emergence and Expansion of Pre-classical Mechanics* (2017).

ROBERT MORRISON is a professor of religion at Bowdoin College in Brunswick, Maine. He is currently working on scholarly intermediaries between the Ottoman Empire and the Veneto region of Italy. His most recent book is *The Light of the World: Astronomy in al-Andalus* (2016).

F. JAMIL RAGEP is Canada Research Chair in the History of Science in Islamic Societies, McGill University. He is currently working on projects involving the relation of Islamic astronomy to Copernicus, science education in Islam, and a database of Islamic scientific manuscripts. He recently published "Ibn al-Shāṭir and Copernicus: The Uppsala Notes Revisited," *Journal for the History of Astronomy* 47, no. 4 (2016).

SALLY P. RAGEP is a senior researcher at the Institute of Islamic Studies, McGill University, and an executive board member of the Islamic Scientific Manuscripts Initiative project, a collaborative effort to make accessible online information on all Islamic manuscripts in the exact sciences. She has recently published *Jaghmīnī's Mulakhkhaṣ: An Islamic Introduction to Ptolemaic Astronomy* (2016).

MICHAEL H. SHANK is a professor emeritus in the recently terminated Department of the History of Science at the University of Wisconsin-Madison. He co-edited, with David C. Lindberg, *The Cambridge History of Science*, vol. 2, *Medieval Science* (2013), and translated, from the French, Roshdi Rashed, *Classical Mathematics from Al-Khwarizmi to Descartes* (2015).

EDITH DUDLEY SYLLA is a professor emerita of history at North Carolina State University, Raleigh. She has recently published "Averroes and Fourteenth-Century Theories of Alteration: Minima Naturalia and the Distinction between Mathematics and Physics," in Paul J.J.M. Bakker, ed., *Averroes' Natural Philosophy and Its Reception in the Latin West* (2015).

Index

273 asl – hypothesis [not model]
n10
182 Reinbach – derives like Arabic ones
131 Regiomontanus quoting Arab
improvement of science